Thermal and Nonthermal Encapsulation Methods

T0252606

Advances in Drying Science & Technology

Series Editor: Arun S. Mujumdar

PUBLISHED TITLES

Thermal and Nonthermal Encapsulation Methods

Edited by
Magdalini K. Krokida

CRC Press
Taylor & Francis Group
Boca Raton London New York

CRC Press is an imprint of the
Taylor & Francis Group, an **informa** business

CRC Press
Taylor & Francis Group
6000 Broken Sound Parkway NW, Suite 300
Boca Raton, FL 33487-2742

First issued in paperback 2017

ISBN-13: 978-1-138-03543-0 (hbk)
ISBN-13: 978-0-367-88679-0 (pbk)

Library of Congress Cataloging-in-Publication Data

Names: Krokida, Magdalini K., editor.
Title: Thermal and nonthermal encapsulation methods / editor, Magdalini K. Krokida.
Description: Boca Raton : CRC Press, Taylor & Francis Group, 2017. | Series: Advances in drying science & technology | Includes bibliographical references and index.
Identifiers: LCCN 2017026591| ISBN 9781138035430 (hardback : acid-free paper) | ISBN 9781315267883 (ebook)
Subjects: LCSH: Food--Drying. | Freeze-drying.
Classification: LCC TP371.5 .T45 2017 | DDC 664/.0284--dc23
LC record available at https://lccn.loc.gov/2017026591

Visit the Taylor & Francis Web site at
http://www.taylorandfrancis.com

and the CRC Press Web site at
http://www.crcpress.com

Dedication

To Despoina Krokida and Maria-Despoina Stefanakis

Contents

Preface

Encapsulation is a topic of interest in a wide range of scientific and industrial areas for the protection and controlled release of various substances. Its application varies from pharmaceutics and nutraceuticals to agriculture. Since encapsulated materials can be protected from external conditions, encapsulation enhances their stability and maintains their viability. Various methods for encapsulation have been established, characterized by both advantages and disadvantages. Depending on the operating conditions and the raw materials of each process, products with varying characteristics and final properties are developed.

In this book, a comprehensive review of various techniques used for encapsulation is presented. Various thermal and nonthermal encapsulation methods (freeze drying and microwave freeze drying; spray drying, spray chilling, and spray cooling; electrospinning and electrospraying; extrusion; osmotic dehydration; air suspension coating; pan coating; vacuum drying) that are applied in various industries are presented. The fundamentals, principles, and applications of each method together with the current state of the art are presented in detail. The book comprises a methodological approach that could contribute to the design of novel products and processes. Each chapter contains figures, tables, illustrations, and examples of the application of each studied method, leading to increased understanding of these processes.

Chapter 1 describes, in general, the most commonly used encapsulation techniques, the principles and parameters that affect encapsulation efficiency, and the properties of the final products, as well as methods to evaluate the physicochemical properties, structure, and function of capsules.

The following six chapters are devoted to novel and conventional encapsulation methods. Chapter 2 reviews general principles, process parameters, types, and applications of freeze drying and microwave freeze drying. The need of drying methods for encapsulation as well as the specific principles and applications of freeze drying encapsulation are also discussed. Microwave freeze drying is also described as a promising encapsulation method. Chapter 3 describes and analyzes the principles of spray drying, spray cooling, and spray chilling techniques, as well as their various applications. In addition, spray drying, spray cooling, and spray chilling as microencapsulation methods are extensively discussed. In Chapter 4, a review of electrohydrodynamic processing for the micro- and nanoencapsulation of active substances and sensitive compounds is presented. The basic principles of electrohydrodynamic processing, the parameters affecting the processes, and their applications are described in detail. Chapter 5 describes extrusion processes as encapsulation methods for incorporating bioactive compounds into various food products. In Chapter 6, the basic principles of osmotic dehydration, process parameters, mechanisms, and applications are presented. Emphasis is given to novel approaches, such as the combination of osmotic dehydration with encapsulation procedures, which seems promising. Chapter 7 covers several physicomechanical methods, such as air-suspension coating, pan coating, and vacuum drying, that are widely used as microencapsulation techniques in many industries.

Finally, Chapter 8 describes and compares encapsulation technologies from an engineering aspect, that is, the impact of process conditions on products' properties. The chapter also provides examples of the main patents based on the presented encapsulation techniques.

The book is a methodological tool that presents conventional and novel methods of encapsulation. I hope that it will contribute to understanding in depth the basic principles and the applications of the relevant methods, and it is addressed to anyone that would like to identify a suitable method of encapsulation for a certain application.

Magdalini K. Krokida

Editor

Magdalini K. Krokida is associate professor in the School of Chemical Engineering at the National Technical University of Athens (NTUA), Athens, Greece. She has been a member of the Laboratory of Process Analysis and Design since 2002. Her field of expertise covers design and optimization of physical processes, design and development of new products (especially foods), determination of the thermophysical properties of materials, and evaluation of the economic and environmental impact of several products and processes. Professor Krokida has published considerably, including two books, seven chapters in international publications, and more than 100 articles in international peer-reviewed journals, with more than 3000 citations. She has participated in numerous research projects, both as a scientific advisor and as a researcher. She is a reviewer of ten international journals and has participated as a member of the organizing committee in several scientific congresses. Since 2010, she has been a representative of Greece at the International Scientific Committee of Drying (EFCF Working Party on Drying) and is an ambassador of the Global Harmonization Initiative (GHI) in Greece.

Contributors

Costas G. Biliaderis
Department of Food Science and
  Technology
Aristotle University of Thessaloniki
Thessaloniki, Greece

Balanč Bojana
Department of Chemical Engineering
University of Belgrade
Belgrade, Serbia

Bugarski Branko
Department of Chemical Engineering
University of Belgrade
Belgrade, Serbia

M.A. Busolo
Novel Materials and Nanotechnology
  Group
Institute of Agrochemistry and Food
  Technology
Paterna, Spain

S. Castro
Novel Materials and Nanotechnology
  Group
Institute of Agrochemistry and Food
  Technology
Paterna, Spain

E. Dermesonlouoglou
School of Chemical Engineering
National Technical University of Athens
Athens, Greece

Panagiota Eleni
School of Chemical Engineering
National Technical University of Athens
Athens, Greece

M. Giannakourou
Department of Food Technology
Technological Educational Institute of
  Athens
Athens, Greece

Drvenica Ivana
Department of Chemical Engineering
University of Belgrade
Belgrade, Serbia

Maciej Jaskulski
Faculty of Process and Environmental
  Engineering
Lodz University of Technology
Łódź, Poland

Trifković Kata
Department of Chemical Engineering
University of Belgrade
Belgrade, Serbia

Abdolreza Kharaghani
Thermal Process Engineering
Otto von Guericke University Magdeburg
Magdeburg, Germany

Magdalini K. Krokida
School of Chemical Engineering
National Technical University of Athens
Athens, Greece

J.M. Lagaron
Novel Nanomaterials and
  Nanotechnology Group
Institute of Agrochemistry and Food
  Technology
Paterna, Spain

Andriana Lazou
Department of Food Technology
Technological Educational Institute of
  Athens
Athens, Greece

Ioannis Mourtzinos
Department of Food Science and
  Technology
Aristotle University of Thessaloniki
Thessaloniki, Greece

Vasiliki P. Oikonomopoulou
School of Chemical Engineering
National Technical University of Athens
 Athens, Greece

P. Taoukis
School of Chemical Engineering
National Technical University of Athens
Athens, Greece

Evangelos Tsotsas
Thermal Process Engineering
Otto von Guericke University Magdeburg
Magdeburg, Germany

Lević Steva
Institute of Food Technology and
  Biochemistry
University of Belgrade
Belgrade, Serbia

Đorđević Verica
Department of Chemical Engineering
University of Belgrade
Belgrade, Serbia

Nedović Viktor
Institute of Food Technology and
  Biochemistry
University of Belgrade
Belgrade, Serbia

1 Principles and Applications of Encapsulation Technologies to Food Materials

Ioannis Mourtzinos and Costas G. Biliaderis

CONTENTS

1.1 INTRODUCTION

Encapsulation in the food science area is defined as the process of entrapment of food materials in solid, liquid, or gaseous forms (termed as internal phase, core, fill, payload phase, incipient, (bio)active agent, etc.), within small capsules made from secondary materials, often referred to as carrier or wall, shell, excipient, or encapsulant, with the aim to protect the entrapped constituents from harsh environments and release them at controlled rates over prolonged periods of time or at specific target sites (e.g., in the gastrointestinal track). The capsule wall structure is often made up of food grade polymers, such as proteins (gelatin, milk whey proteins, etc.) and polysaccharides (gum arabic, alginates, pectins, chitosan, cellulose and starch derivatives, etc.) as well as lipids (fats and waxes), used either alone or in mixtures and layers, to enhance the performance of the capsules for controlled release of the active constituents (Augustin & Hemar, 2009); the latter can be nutrients,

1

bioactive proteins or peptides, natural pigments, various food additives (flavoring compounds, colorants, acidulants), probiotics, etc. Physical, chemical, or physico-chemical interactions between components involved in wall formation are critical for the stability of the capsule and its ability to preserve the encapsulated substances and make them available at the target site and at a specific rate. Thus, the encapsulation process, depending on the conditions, can be reversible and the entrapped molecules or other materials could be released depending on the microenvironment of the food that a capsule is being incorporated or upon digestion of the product in the human digestive track.

The composition, chemical properties, and physical state of the wall materials used in capsule manufacturing are critical determinants of the matrix functionality. Ideally, the coating material(s) should possess most of the following characteristics (Desai & Park, 2005a; Zuidam & Shimoni, 2010): (1) ease of material handling throughout the encapsulation process (mostly related to solubility, interfacial activity, and rheological properties); (2) ability to effectively disperse or emulsify the encapsulated constituents and stabilize the composite structure; (3) lack of reactivity with the active core materials (during processing and storage); (4) ability to "seal" and protect the active constituents during product manufacturing and storage of the capsules alone or as part of a composite food matrix; (5) proper swelling, solubility, dispersibility under certain conditions for controlled release of the encapsulated constituents, reflecting the structural modification of the wall material when exposed to different environmental stresses (pH, ionic strength, temperature, hydrolytic enzymes, etc.); and (6) inexpensive raw materials and of food-grade status. As a single material does not possess all the desired properties, blends of multiple wall materials in one formulation are frequently employed. Even when a wall formulation is established, it is rather unique for a particular type of core material and effectively functional within the constraints of the specific encapsulation method used.

Encapsulation methods can be categorized in several ways, with most descriptions referring to:

- Specific infrastructure that is used during the process of encapsulation (e.g., spray drying, fluidized bed drying).
- Wall/core materials used for the encapsulation (encapsulation with cyclodextrins, micro-hydrogel particles using polysaccharides, e.g., alginate, chitosan, or proteins).
- The size of the formed capsules (microencapsulation, nanoencapsulation). Capsule particle size varies from submicrometers to several millimeters; most capsules are being classified in three range sizes, namely, nanocapsules (<1 μm), microcapsules (1–1000 μm), and macrocapsules (>1000 μm). The size range of the capsules varies with the encapsulation method employed (Zuidam & Shimoni, 2010); most of the methods (spray drying, freeze drying, fluidized bed coating, spray chilling/cooling, extrusion, microsphere preparation, emulsification, liposome entrapment, etc.), depending on the conditions used, are capable of producing microcapsules ranging from 10 to 1000 μm, and only inclusion complexation, emulsification, and nanoparticle technologies can provide smaller capsules ranging in size from 0.01 to 10 μm.

- The physicochemical phenomena/mechanisms involved in the encapsulation process (emulsification, inclusion complex formation, ionic crosslinking, interfacial polymerization, cocrystallization, etc.).
- Cost, ranging from cheap techniques such as pan coating, microgel bead formation to expensive techniques such as coacervation and liposome entrapment.

The aim of encapsulation in food product development is multidimensional. Usually it is applied for protection and effective delivery of active additives/ingredients of foods (e.g., bioactives) for one or more of the following reasons:

- Easy handling and effective incorporation of bioactives in different food matrices, for example, water-insoluble and lipid-soluble compounds could be transferred in water-based foods.
- Protection of bioactives during processing and upon storage under conditions that can affect their stability (light, heat, moisture, O_2, enzymes, other chemicals). By diminishing the exposure of sensitive molecules to adverse environmental factors, the stability and shelf life of the bioactives can be increased.
- Controlled and/or targeted release of bioactive ingredients when the intensity and time span for their activity (odor, taste, biological function) are crucial to be controlled in a composite food matrix during storage, upon consumption of the product, and passing through the varying environments of the human digestive track (pH, digestive fluids, etc.).
- Reduced evaporation of volatile ingredients such as flavorings or essential oils.
- Masking of undesirable flavors (e.g., bitterness of protein hydrolysates and bioactive peptides from different sources).
- Improved food processing characteristics of encapsulated ingredients (desired texture properties, reduced hygroscopicity, and controlled flowability).
- Transformation of bioactive ingredients into raw materials that can be easily incorporated in food formulations to modify organoleptic properties or exert a physiological function in the human body.

Recently, state-of-the-art encapsulation techniques have emerged by exploiting innovations in several scientific fields such as chemistry, physical chemistry, chemical engineering, and biochemistry. There is also an expansion of encapsulation procedures with numerous applications in the food processing area, since the growing market of bioactive ingredients and functional foods has become a major driving force behind innovation in the food industry during the past decade. Ingredients that provide health benefits, namely, bioactives, have indeed broadened the role of conventional foods from satisfying hunger, providing the necessary nutrients and sensorial perception, extending shelf life, and meeting safety requirements to the prevention of nutrition-related chronic diseases and improvement of human well-being (Gul, Singh, & Jabeen, 2016). However, the enrichment of conventional foods with isolated bioactives has limitations as some bioactives are not soluble/dispersible, stable, and active in different food matrices. Therefore, the main challenge for food manufacturers is to develop new products containing bioactives without adversely affecting the sensorial quality of the product. The bioactives must also remain stable during food processing and storage of the food, and following ingestion they have to be protected during the gastrointestinal transit till they reach the desired site in the body where they become bioavailable. Many of these limitations could be overcome by preparing bioactives in encapsulated forms (Đorđević et al., 2015).

There is no universal method for the encapsulation of bioactives due to their enormous chemodiversity, reflecting on their stability and physical properties, as well as the physicochemical complexity of the different food matrices intended to act as vehicles for encapsulated bioactives. Therefore, selection of a suitable method can vary from case to case depending on the physical and chemical characteristics of the bioactive, the carrier food, and the required release mechanism of the active component (Desai & Park, 2005a). Taking as an example ascorbic acid, it has been encapsulated with more than 10 different techniques, including spray drying, fluidized bed coating, liposome formation, etc., using several wall materials such as maltodextrins, gum arabic, rice starch, gelatin, modified starch, modified chitosan, pea protein isolate/concentrate, methacrylate copolymers, medium-chain triacylglycerol, β-cyclodextrin, alginate, ethyl cellulose, and carboxymethyl cellulose (Abbas, Da Wei, Hayat, & Xiaoming, 2012). The practical implications of food ingredient encapsulation for improved functionality together with the evolution of sophisticated techniques and scale-up processing at industrial level have made the microencapsulation a very attractive research field, both for academia and the food industry; this is attested by the exponential increase in the number of scientific reports published during the last 25 years on food-related applications of different microencapsulation technologies (Figure 1.1).

The scientific approaches undertaken do not only focus on capsule formation but also on the structure and characteristics of the final capsule as a vehicle to protect

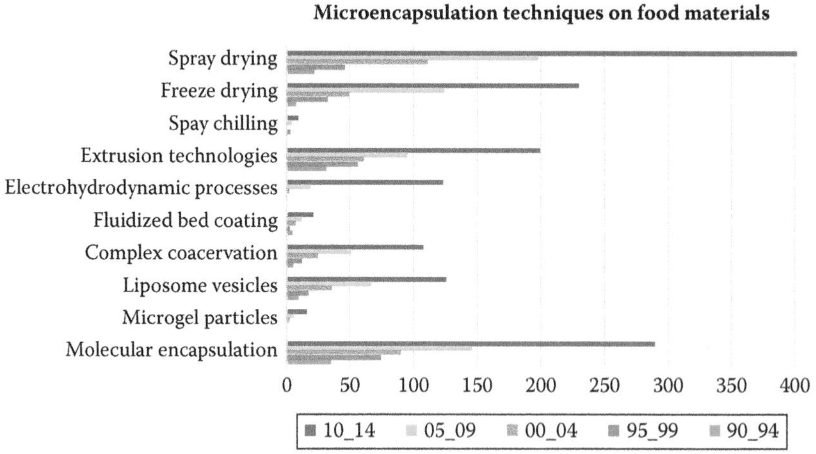

FIGURE 1.1 Evolution of the number of scientific articles on applications of encapsulation techniques on food systems (ingredients) during the last 25 years. (Data complied from Web of Science.)

and effectively deliver the active compound(s). There are several methods used to characterize the physical properties and the quality of the final capsule as well as the enhancement in performance of the encapsulated materials. These are summarized in Table 1.1.

The selection of wall materials remains a crucial decision in every encapsulation process as it has a large impact on the performance of the final capsule. It depends basically on the nature of the encapsulated core material, the objective of microencapsulation process, and the desired barrier properties. In general, wall materials used can be classified into four categories: water-soluble polymers, water-insoluble polymers, water-soluble nonpolymeric materials, and water-insoluble nonpolymeric materials (Kuang, Oliveira, & Crean, 2010).

Usually, food grade hydrophilic wall materials are carbohydrates and proteins, whereas common hydrophobic materials are lipids. Modified carbohydrates (e.g., cellulose and starch derivatives) are also used in food encapsulation, as native carbohydrates do not usually possess interfacial properties, the exception being gum arabic, a natural multifunctional hydrocolloid emulsifier (a mixture of a lower molecular mass polysaccharide and a higher molecular mass hydroxyproline-rich glycoprotein component). On the other hand, proteins, being amphiphilic macromolecules, possess the physicochemical and functional characteristics necessary to encapsulate both hydrophobic and hydrophilic molecules, whereas lipids are used for the encapsulation of hydrophobic substances. Maillard reaction products of amphiphilic nature, obtained by the heat treatment of proteins with carbohydrates at low moisture contents (covalently linked conjugates by Schiff base formation between reducing end groups of the carbohydrate component and the amine groups of proteins), have been proposed as wall matrices for encapsulation of oxidation-sensitive ingredients, for example, fish oils (Augustin, Sanguansri, Margetts, & Young, 2001). For encapsulation of hydrophilic bioactives, wall materials with good film-forming properties are employed. For the encapsulation of hydrophobic bioactives, encapsulants with both emulsifying and good film-forming properties are preferred. Such a double requirement may be either accomplished by an amphiphilic biopolymer having also good encapsulating properties (e.g., gum arabic, proteins, modified starches, or other modified polysaccharides) or a natural (lecithin) or synthetic surfactant, used as an emulsifying agent, together with a bulk agent (low-molecular weight carbohydrate) (Encina, Vergara, Giménez, Oyarzún-Ampuero, & Robert, 2016). The large availability of food-grade biomolecules and their modified derivatives that differ in physicochemical characteristics and functional properties by tailoring sizes, charges, chemical groups, film-forming abilities, and permeability properties provides alternative approaches to produce novel carriers able to encapsulate any type of active compounds (Santiago & Castro, 2016).

The principles of major techniques and their applicability in encapsulation of food ingredients are shortly presented below. The most commonly applied technologies are spray drying, freeze drying, spray chilling/spray cooling, extrusion technologies, coacervation, electro-hydrodynamic processes, biopolymeric microgel particles, molecular inclusion, and liposome formation, as schematically outlined in Figure 1.2. More detailed information on these processes and relevant food applications can be found in the references cited herein, in several comprehensive

TABLE 1.1
Methods of Characterization of Microcapsules

Encapsulation and/or Material Property	Method of Assessment	Reference (Examples)
Encapsulation efficiency	The amount of encapsulated bioactive with respect to the initial amount of bioactive used	Risch and Reineccius (1988); Santos, Bozza, Thomazini, & Favaro-Trindade (2015)
Particle size measurements	Dynamic light scattering, refractive index, or optical microscopy to determine the average particle size of the obtained powder or microcapsule	Comunian et al. (2013)
Zeta potential	Electrophoretic mobility measurements—Laser Doppler velocimetry measurements	Desai & Park (2005b)
Morphological characterization of the microcapsules	Optical, confocal, or scanning electronic (SEM) microscopy	Comunian & Favaro-Trindade (2016)
Moisture content	Measurement of the weight loss percentage of the powders, after drying at 105°C until constant weight	AOAC (1997)
Solubility	Powder sample added into the blender The formed dispersion centrifuged and the supernatant transferred to Petri dishes and dried	Cano-Chauca, Stringheta, Ramos, & Cal-Vidal (2005)
Structure modification during encapsulation	Fourier transform infrared spectroscopy	Hu et al. (2016); Pasukamonset, Kwon, & Adisakwattana (2016)
Behavior of the material in relation to weight loss	Thermogravimetric analysis	Fernandes et al. (2016); Silva, Azevedo, Cunha, Hubinger, & Meireles (2016)
Bulk density	Pycnometric method using toluene	Bhandari, Dumoulin, Richard, Noleau, & Lebert (1992)
Wettability	The time necessary for 1 g of powder to disappear from the surface of water (100 mL, 20°C)	Fuchs et al. (2006)
Moisture adsorption isotherm in order to establish water sorption properties under different storage conditions	Using salt solutions saturated with water activity varying from 0.12 to 0.98 at 25°C	Fernandes et al. (2016)
Crystallinity of the powder	X-ray powder diffraction (XRPD)	Cano-Chauca et al. (2005); Fernandes et al. (2017)
Physical state of the wall material and core	Differential scanning calorimetry (DSC) to measure phase transitions of the wall and core materials	Awad, Helgason, Weiss, Decker, & McClements (2009)
Changes in protein conformation	UV-vis absorption measurements, DSC	Amara, Eghbal, Degraeve, & Gharsallaoui (2016)
Formation of inclusion complex	^{1}H NMR	Anselmi et al. (2008); Zhao, Chao, Du, & Huang (2011)

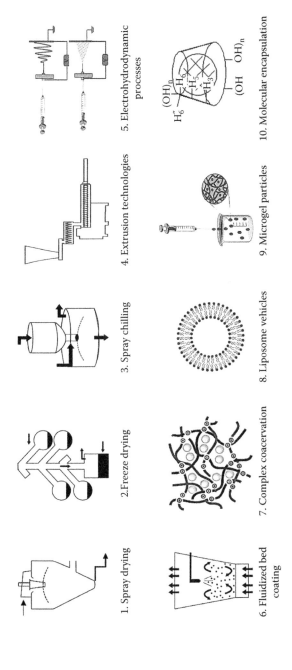

FIGURE 1.2   Most commonly applied microencapsulation technologies to food and pharmaceutical materials.

reviews (Appelqvist, Golding, Vreeker, & Zuidam, 2007; Barbosa-Cánovas, Ortega-Rivas, Juliano, & Yan, 2005; Comunian & Favaro-Trindade, 2016; Desai & Park, 2005a; Đorđević et al., 2015; Drosou, Krokida, & Biliaderis, 2017; Fang & Bhandari, 2012; Farjami & Madadlou, 2017; Fu, Sarkar, Bhunia, & Yao, 2016; Ghorani & Tucker, 2015; Gouin, 2004; Kuang, Oliveira, & Crean, 2010; Mao, Roos, Biliaderis, & Miao, 2017; McClements, 2005; Oxley, 2012; Santiago & Castro, 2016; Shewan & Stokes, 2013; Zuidam & Shimoni, 2010) as well as in the following chapters of this book.

1.2 SPRAY DRYING

Spray drying is the most extensively applied (90% of produced microencapsulates) encapsulation technique in the food industry because it is flexible, continuous, and rather inexpensive (10–30 times cheaper compared to freeze drying). The method involves the dissolution of the bioactive in an aqueous dispersion of the wall material. The dispersion (infeed) is subsequently atomized with a high pressure nozzle or a centrifugal wheel and sprayed over a stream of hot air. As a result, the droplets are dehydrated in the heated chamber of the spray drier and converted into capsules with diameters from 10 to 200 μm, and then collected at the bottom of the drier.

The characteristics of the wall material are crucial to the final encapsulation efficiency and stability of encapsulated ingredients. In most instances, hydrophilic carbohydrates and proteins are used as wall materials, for example, gum arabic, chemically modified starches (e.g., octyl-substituted derivatives for encapsulation of flavor oils), and starch hydrolysis products (maltodextrins), milk and soy proteins. The use of other natural hydrocolloids (gums) is often limited due to solubility and viscosity constraints, and in order to further improve the encapsulation efficiency, small molecular weight carbohydrates (e.g., lactose, trehalose, or maltodextrins) can be included as part of the wall formulation to enhance the quality and stability of the encapsulated constituents (Zhao & Tang, 2016). The aqueous dispersion of the carbohydrates, used as wall materials, undergoes a state transformation, so-called glass transition, when the droplets are rapidly evaporated in the spray dryer, leading to an amorphous (glassy) state; ultimately, the glass transition temperature of the capsule wall material is largely affected by the molecular size of the constituents used in the formulation and the residual moisture content of the capsule. With higher molecular weight hydrocolloids used as wall materials (higher Tg), there is greater physical stability toward humidification and structural collapse of the carrier matrix. Even if the wall materials are largely hydrophilic, the core material could be either hydrophilic or hydrophobic. Hydrophobic molecules are usually emulsified in oil-in-water emulsions and subsequently spray dried (Aghbashlo, Mobli, Madadlou, & Rafiee, 2013). The use of modified wall materials, for example, mixtures of gums and maltodextrins, can greatly alter the surface morphology, hygroscopicity, water dispersibility, and solubility of the capsule as well as the protection/retention of the encapsulated compounds (Krishnan, Kshirsagar, & Singhal, 2005; Loksuwan, 2007). The problems of stickiness during spray drying of sugar-rich foods such as fruit juices, honey, and some starch derivatives (glucose syrups/maltodextrin at higher dextrose equivalent values) are well recognized and have been related to the low Tg of these low molecular weight materials (Bhandari & Howes, 1999). The most common approach to dry such products has been to add high molecular weight additives (i.e., raising the Tg of the composite matrix) and/or operate the dryer at relatively lower temperatures to avoid collapse phenomena driven by the glass-rubber transition.

Among the limitations of the spray drying method is the degradation of thermally sensitive bioactives as during spray drying, the compounds are exposed to high temperatures (Gharsallaoui, Roudaut, Chambin, Voilley, & Saurel, 2007). Another disadvantage is the nonuniform size and shape of the dried particles as well as particle aggregation. Thus, the microcapsules may have teeth or concavities on the outer surface (Buffo, Probst, Zehentbauer, Luo, & Reineccius, 2002). For improving particle wetting and dispersibility–solubility, spray-dried powders can be subjected to a secondary agglomeration or granulation step by fluidized bed coating technologies (Barbosa-Cánovas, Ortega-Rivas, Juliano, & Yan, 2005; Ortega-Rivas, 2005). Agglomeration is itself a process to instantize food powders by improving their wettability and is performed by applying heat or rewetting the powder particles or combining both to form larger size aggregates with porous structure.

Spray drying has been used for the encapsulation of probiotics as the process improves probiotic survival and viability (Bustamante, Oomah, Rubilar, & Shene, 2017; Bustamante, Villarroel, Rubilar, & Shene, 2015). The selection of encapsulating media is also crucial as it has a direct impact on protection against oxygen, acidic pH, and salt concentration. Several encapsulating media (maltodextrins, proteins, reconstituted skim milk, polysaccharides, etc.) have been used to protect probiotics from thermal destruction and dehydration damage during spray drying, to maintain their viability upon storage, and to become active under the gastrointestinal conditions of the host.

Spay drying techniques have been applied to convert liquid flavorings (synthetic flavorings, essential oils, and natural oleoresins) to dry powders, providing convenience and preparations with reduced volatility when incorporated in food formulations; these dry preparations can be used as food flavorings in different finished products (Fernandes et al., 2013). Most constituents of essential oils are fairly sensitive to adverse environmental conditions such as light, heat, and moisture. Thus, the process of spray drying for these ingredients involves an emulsification step of the lipophilic flavoring compounds before spray drying (Osorio-Tobón, Silva, & Meireles, 2016). Emulsions with weaker physical stability (dependent on oil droplet size, droplet surface charge, interfacial properties, and viscosity) result in encapsulated spray-dried powders with higher surface oil content and lower volatile retention (Drosou, Krokida, & Biliaderis, 2017), for example, the higher surface oil on particles produced from emulsions with large droplet size has been attributed to droplet breakdown during the atomization step (Jafari, Assadpoor, He, & Bhandari, 2008). Overall, the quality of the obtained product and the powder encapsulation efficiency depend on the operating conditions such as inlet and outlet temperature of the spray drier, feed temperature and flow rate, as well as on the emulsion properties (nature of the oil phase and droplet size, the type of wall materials used, weight ratio of wall to core, total solids content, and viscosity of the atomizing fluid) (Langrish & Premarajah, 2013). Encapsulation of flavoring oils by double emulsion, which can be subsequently spray dried, has been explored in the case of orange oil for better protection and delayed release properties of the volatiles (Edris & Bergnstahl, 2001).

Spray drying was also proposed as a suitable method for the microencapsulation of polyphenols (Murali, Kar, Mohapatra, & Kalia, 2014). It is known that the stability of polyphenols can be affected by many factors, including light, enzymes such as polyphenol oxidase and peroxidase, ascorbic acid, pH, moisture, and oxygen. Microencapsulation of polyphenols may be a good alternative for maintaining the stability and antioxidant potency of these bioactive compounds when extracted from plant materials (Rigon & Zapata Noreña, 2016).

Encapsulated polyphenols were recently tested in ready-to-eat products. Polyphenols of pomegranate peels were microencapsulated in maltodextrins by spray drying and used to fortify an ice cream product. Microencapsulation indeed seemed to improve the stability of phenolics upon storage and their *in vitro* bioactivity when included in the ice cream mixture. Moreover, the formulated product with encapsulated polyphenols showed high acceptance by consumers, implying an effective masking of off-flavor notes (Çam, İçyer, & Erdoğan, 2014). Even propolis, a sticky and resinous bee product of low water solubility and with undesirable taste and aroma, was transformed into a more attractive material for fortification of food products by encapsulation in different maltodextrin matrices, with or without added gums (Busch et al., 2017).

Water-free spray drying (using ethanol as organic solvent) has been proposed as an alternative technique for the encapsulation of hydrophobic molecules. The benefits of this approach arise from the avoidance of emulsion preparation. Moreover, if nitrogen is also used as a drying medium, the oxidation during spray drying can be largely minimized. This technique has already been used in the pharmaceutical field (Duan, Vogt, Li, Hayes, & Mansour, 2013) and proposed as replacement for applications relevant to the food industry (Encina et al., 2016).

1.3   FREEZE DRYING

Freeze drying (also known as lyophilization) is a drying process for the long-term preservation of heat-sensitive food materials based on the phenomenon of sublimation, that is, the removal of solvent (water) from the solid state (ice) to the vapor phase under reduced pressure and low temperature (Ray, Raychaudhuri, & Chakraborty, 2016). Freeze drying and spray drying together are the most commonly used industrial technologies for encapsulation of many heat-sensitive food ingredients and drugs. Encapsulation by freeze drying is achieved as the core materials homogenize in the matrix solution and then co-lyophilize; the homogenized sample is first frozen at temperatures between −90°C and −40°C and then dried by sublimation at low temperatures (−80°C to −30°C). Following drying, the amorphous cake can be broken into smaller size particles by grinding or milling devices.

A successful freeze drying process preserves most of the initial raw material properties such as molecular structure and conformation (important in the case of enzymes for activity), appearance, taste, color, flavor, texture, and biological function (Ceballos, Giraldo, & Orrego, 2012). The efficiency of protection and controlled release mainly depend on the composition and structure of the wall material(s) used for encapsulation; to stabilize freeze-sensitive agents, it may be desirable to incorporate in the solution, along with the other wall materials, small amounts of a cryoprotectant (e.g., sorbitol, disaccharides such as trehalose, maltodextrins). The most commonly used wall materials are gum arabic, maltodextrins, modified starches or other polysaccharides, and whey proteins. The major disadvantages of freeze drying are the high energy input and long processing time (~1 day). That leads to a 10–30 times higher cost than spray drying, the second most popular drying technique. In addition, following processing a rather open barrier structure is generated between the active agent and its surroundings; such a highly porous wall offers poor protection (e.g., toward oxidation) when a prolonged release of an active compound is required. Nevertheless, the freeze drying process is a relatively simple technique and is particularly suited for the encapsulation of aromatic compounds, provided an appropriate carrier matrix is chosen.

Even though freeze drying is operationally a well-established technique, crucial parameters with a direct impact on the quality of the resultant capsules, such as the core material concentration (weight ratio with the encapsulant), the nature and composition of the wall materials, and operational variables (chamber pressure and freezing rate and temperature during sublimation in relation to the Tg) should be optimized with response surface methodology (RSM) to maximize the quality/performance of the end product (Ramírez, Giraldo, & Orrego, 2015). RSM is indeed a very practical and frequently used optimization tool for modeling experimental data derived from such processing protocols by analyzing several variables at a time and using special experimental designs to reduce the number of required tests (Youssefi, Emam-Djomeh, & Mousavi, 2009) with the aim to identify optimum processing conditions.

Freeze drying could be used for the encapsulation of water-soluble compounds (Fang & Bhandari, 2010). Moreover, lipophilic ingredients could be encapsulated via freeze drying after emulsification (o/w emulsions). A common merit of the freeze drying process is that an amorphous state of the encapsulation matrix is easily obtained, even when easy-to-crystallize materials, such as sucrose and trehalose, are used as matrix formers (Nakayama et al., 2015).

Recent applications of freeze drying include the stabilization of plant anthocyanins, making them attractive food ingredients, as they are natural colorants with high antioxidant activity (Celli, Dibazar, Ghanem, & Brooks, 2016; Murali et al., 2014). Freeze drying has also been proven a convenient processing route for the preparation of cyclodextrin inclusion complexes in dried forms. Inclusion complex formation followed by freeze drying was recently applied to encapsulate carvacrol, a natural antimicrobial, offering new opportunities for the development of alternative food preservatives (Santos, Kamimura, Hill, & Gomes, 2015). Moreover, water soluble extracts of natural colorants derived from saffron and beetroot or even isolated colorants, such as curcumin, have been encapsulated in different agents by means of freeze drying (Cano-Higuita, Malacrida, & Telis, 2015; Chranioti, Nikoloudaki, & Tzia, 2015). The microencapsulation does improve their stability and facilitates their use in food formulations. A recent interesting application was the development of powders containing oil at high contents (~70 wt%, in either liquid or solid form), using double emulsions stabilized by quinoa-modified starch granules (modification with octenyl succinate groups to improve their emulsifying properties) and freeze drying the emulsions (Marefati, Sjöö, Timgren, Dejmek, & Rayner, 2015). The Pickering double emulsions formed had high encapsulation efficiency and as freeze-dried powders can be quite versatile ingredients in formulating food and pharmaceutical products.

Different preservation techniques when applied to microbial cells are known to cause structural and physiological cell damage by several mechanisms, ultimately leading to decrease of cell viability. Freeze drying, alone or combined with another encapsulation method, is often applied for protection and long-term storage (viability) of microbial cultures, including probiotics. In a recent study, concentrated cultures of *Lactobacillus reuteri* have been first encapsulated in microbeads (ionotropic gelation using mixed solutions of xanthan gum and gellan gum, co-gelled with $CaCl_2$) and subsequently freeze dried (Juarez Tomas, De Gregorio, Leccese Terraf, & Nader-Macias, 2015). In some of the formulated freeze-dried microbial cultures, lactose and skim milk have been included as cryoprotectants. The survival of the encapsulated *L. reuteri* during lyophilization and storage at 4°C significantly improved in the presence of the cryoprotectants.

1.4   SPRAY CHILLING

Spray chilling, also known as spray cooling, spray congealing, or prilling, is an encapsulation technique that usually involves the solidification of an atomized liquid spray into particles due to the cocurrent or countercurrent flow of a cool air stream; this is the least expensive encapsulation process, making particles from a few microns to several millimeters. Spray chilling often involves the addition of the component of interest to a molten fat carrier, where it can be either dissolved in the lipid phase (hydrophobic compounds) or simply be dispersed (suspended) as dry separate particles or incorporated in emulsion droplets. This is the reason that spray chilling/cooling is typically referred to as "matrix" encapsulation (i.e., discrete particles of the active material are buried randomly in the solidified fat matrix) in contrast to "true" encapsulation processes where the microcapsules possess the typical core/shell structure. A variety of atomization methods are available for feeding the mixture of liquefied fat with the active ingredients to be encapsulated, including pressure nozzles, vibrating nozzles, and spinning (rotating) disc atomizers (Oxley, 2012). When the nebulized liquid is put into contact with the cold environment, at temperatures below the melting point of the matrix material, the vehicle solidifies, leading to solid lipid microparticles (insoluble in water due to the lipid coating). Microspheres are the most common particle morphology noted with this technique, with the active ingredient(s) being dispersed more or less homogeneously throughout the encapsulating matrix. The difference from spray drying is that the cooled air is used for particle solidification and not for solvent evaporation, that is, in spray chilling there is no mass transfer (no evaporation from the atomized droplets). With this technique, it is feasible to encapsulate water-soluble vitamins, minerals, enzymes, acidulants, and other food ingredients.

Common carriers are lipid ingredients such as fatty acids, wax, palm oil, beeswax, and cocoa butter (Fang & Bhandari, 2012). The main advantages of the technique are the avoidance of solvent use and the low cost. Moreover, it is a continuous process, scalable, and only relatively mild conditions are used (Okuro, de Matos, & Favaro-Trindade, 2013a). Spray chilling has been applied for encapsulation of ascorbic acid using in the feed dispersions fully hydrogenated palm oil and vegetable glycerol monostearate as carriers (de Matos, Comunian, Thomazini, & Favaro-Trindade, 2017). The technique was also employed for encapsulation of functional ingredients like probiotic microorganisms and prebiotics (Okuro, Thomazini, Balieiro, Liberal, & Fávaro-Trindade, 2013b). According to Gouin (2004), the release kinetics of encapsulated ingredients from sprayed chilled particles, when they are brought into contact with a foodstuff, may be governed by different mechanisms, for example, slow diffusion of water through the shell imperfections affecting particle integrity, osmotic forces, mechanical disruption of the particles, etc. Moreover, due to the nature of the process, a significant amount of the encapsulated ingredient is located at or near the surface of the particle, thus being in direct access to the food environment surrounding the particle. Modulation of the release kinetics can be further effected by modifying the crystalline structure of the shell material (different types of polymorphs and crystal sizes can be generated by proper selection of cooling conditions and by formulation of the fat composition with additives favoring one crystalline form over another).

Researchers compared spray chilling (stearic acid + hydrogenated vegetable fat as carrier) with the spray drying technique (gum arabic as wall material) by preparing encapsulated forms of ascorbic acid using the two methods and tested their efficiency in retaining ascorbic acid during the baking of a biscuit product; although both methods showed high encapsulation efficiencies (>97%) and similar particle morphologies (spherical), the spray-dried microparticles (of smaller diameter, ~9 μ vs. 31 μ for the spray chilled microcapsules) were more effective in protecting vitamin C during baking (Alvim, Stein, Koury, Dantas, & Cruz, 2016). By comparison with biscuit preparations fortified with nonencapsulated ascorbic acid, it was concluded that both microencapsulation methods are effective in the protection of the active ingredient during thermal processing of the fortified product. Consoli, Grimaldi, Sartori, Menegalli, and Hubinger (2016) also proposed the technique of spray chilling as a means to encapsulate phenolic compounds that can be transformed to solid lipid microparticles, using gallic acid as a model phenolic compound. From a product application viewpoint, the use of spray-chilled particles carrying bioactives might be quite challenging in certain food formulations, due to the relatively large particle size that can have detrimental effects on microstructure, texture, or the physical stability of the end product.

1.5 EXTRUSION TECHNOLOGIES

Extrusion microencapsulation technologies are distinguished to melt injection, co-extrusion, and extrusion. In melt injection, a melt composed of a carbohydrate-based mix (sugars, maltodextrins, starches, and other gums) with the actives is pressed through one or more filters and then quenched to a glassy state (a physical state of relatively low molecular mobility) by a cold, dehydrating solvent (e.g., isopropanol) that can be also combined with liquid nitrogen spraying. The coating material hardens on contact with the dehydrating solvent, thereby encapsulating the active ingredients. Encapsulates made by melt injection are water soluble, with particle sizes from 200 to 2000 µm, and the payload is rather low, ranging between 8% and 20%; with modified carbohydrate wall materials (e.g., hydrophobically modified starches), the payload can further increase (up to 40%) without loss of stability for the flavor incipient (Mutka & Nelson, 1988). When the process involves an extruder with one or more rotating screws in a continuous process (thermomechanical mixing within the extruder's barrel) the method is called melt extrusion (Zuidam & Shimoni, 2010); several additives (plasticizers and other constituents) are often included into the feed in order to modify the physical properties of the molten material during the process as well as the end-product performance. Transport of the feed material through the barrel takes place by the rotating screws that promote homogenization of the ingredients and compress the extrudate, as the molten material moves toward the exit of the barrel, that is, at the die. In most commercial extruders, the barrel is "divided" into sections to allow for segmental control of temperature in the different processing stages of the extrusion process. The extrusion conditions (temperature, water content, feed composition, viscosity of the molten material, etc.) and the die configuration (orifice geometry) determine the shape of the final product (e.g., rope, thread, and sheet). A chopper/cutter device can be attached to the die to reduce the size (0.5–1.5 mm) of the exiting extrudate. Alternatively, grinders or mills might be employed as postpreparation equipment to further decrease the particle size of the microcapsules.

Extrusion microencapsulation has been successfully applied to entrap unstable (oxidation-prone) volatiles (flavor compounds) in glassy carbohydrate matrices; apparently, the diffusion rate of O_2 through hydrophilic glassy biopolymer matrices (Biliaderis, Lazaridou, & Arvanitoyannis, 1999) is very low, thereby offering remarkable protection to extruded flavor oils, compared to other encapsulation technologies like spray drying or fluidized bed coating, processes that also promote the formation of glassy carbohydrate shells. In the case of extrusion, the microstructure is more compact and hence exhibits improved gas barrier properties. Most studied flavors are citrus oils, and orange terpenes are used as a model flavor system in order to evaluate the efficiency of the encapsulation process (Tackenberg, Krauss, Schuchmann, & Kleinebudde, 2015). Flavor compounds are effectively entrapped in several carbohydrate amorphous matrices such as starch, modified starches, maltodextrins, corn syrup, sucrose, and gums (Castro et al., 2016). Emulsifiers are also used in the feed mix to increase the compatibility between the matrix and the incipient. Other additives that are often used in the encapsulation process include antioxidants (ascorbic acid, citric acid, erythorbic acid, and tocopherols) and plasticizers. Plasticizers are low molecular weight compounds, such as water or polyols (e.g., glycerol), that can act as processing aids during melt extrusion and have an impact on other physical properties of the extrudates (lowering the glass transition temperature of the material, improving solubility, and modifying the extrudate surface morphology) (Vieira, da Silva, dos Santos, & Beppu, 2011). Extrusion technologies have also been tested for encapsulation of enzymes and proteins; in this case, the process involves milder conditions (moderate thermal treatment and applied shear forces) and the actives can even be incorporated at a later stage of the extrusion process (last section of the barrel).

Extrusion technologies have many advantages for the encapsulation of microbes. A low temperature process is often adopted, which is relatively gentle (van Lengerich, 2001), leading to the entrapment of the microbial cells in a plasticized composite matrix of starch/flour and fat. It does not also involve the use of deleterious solvents and can be done in both aerobic and anaerobic environments, the latter being particularly beneficial when anaerobic microorganisms are being incorporated into food products. The modification to accomplish this is rather simple, that is, the entire extrusion device is placed in a sterile cabinet where oxygen is substituted by nitrogen.

1.6   ELECTROHYDRODYNAMIC PROCESSES

Electrohydrodynamic processes are nonthermal methods to prepare fibers and/or capsules in the submicron range, presenting a large surface-to-volume ratio, through the action of an externally applied strong electric field between two electrodes, imposed on a polymer solution or melt, that is, the electrostatic forces produce electrically charged jets from viscoelastic polymer solutions that by evaporation of the solvent (drying) lead to ultrathin structures deposited on a grounded collector (Li & Xia, 2004). These processes are referred to as electrospinning when ultrathin continuous fibers are obtained, whereas when size-reduced capsules are generated, the process is called electrospraying. The difference between electrospinning and electrospraying is based on the degree of molecular cohesion in the raw material, a property that is most readily controlled by variation in concentration and viscosity of the polymeric solution (Drosou, Krokida, & Biliaderis, 2017; Ghorani & Tucker, 2015). Encapsulation of active compounds into fibers or powders is achieved by either direct incorporation of the bioactives into the polymer solution (co-solubilization of core and wall materials) or through a coaxial electrohydrodynamic process where the polymer and the core material are introduced into the processing zone of the equipment from separate solutions, particularly when component molecular immiscibility is encountered (Angeles, Cheng, & Velankar, 2008).

The electrohydrodynamic processes have been attracting lately much attention for the production of polymer fibers or particles as they can produce structured materials with diameters in the range from several micrometers down to tens of nanometers (Drosou, Krokida, & Biliaderis, 2017). Crucial parameters that determine the fiber and particle dimensions as well as morphologies, physical state, and product performance are related to the polymer solution properties (i.e., concentration, viscosity, molecular weight, and conformation of the macromolecular chains, surface tension, solvent volatility, and electrical conductivity), processing conditions (applied voltage, flow rate, tip-to-collector distance, needle diameter), or other regulating the electrohydrodynamic process parameters like humidity and temperature of the surroundings (Drosou, Krokida, & Biliaderis, 2017; Wongsasulak, Kit, McClements, Yoovidhya, & Weiss, 2007).

Common materials used in electrospinning are proteins such as whey proteins, soy protein, egg albumin, collagen, and casein (López-Rubio & Lagaron, 2012; Nieuwland et al., 2013). Carbohydrates such as guar-gum, pullulan, chitosan, inulin, alginate, and dextrans have also been used either alone or in combination with proteins for the electrospinning of outer wall materials to produce nanofibers (Alborzi, Lim, & Kakuda, 2013; Fabra, López-Rubio, & Lagaron, 2014). Compared to fiber products prepared from synthetic polymer solutions, there is growing interest for bionanofibers because of their biodegradability and biocompatibility. However, electrospinning of biopolymer solutions has proven to be challenging because of the limiting solubility of biopolymers in most organic solvents, the polyelectrolyte nature in aqueous media, poor molecular flexibilities, formation of three-dimensional networks via hydrogen bonding, and most importantly insufficient chain entanglements to facilitate the electrospinning process. Therefore, researchers have focused on the use of composite blends of biopolymers with other polymers and/or other processing aids (plasticizers, surfactants) that are compatible for the formation of electrospun fibers with enhanced material properties such as higher tensile strength and intact micro/nanostructures.

Recently, highly valued antioxidants such as curcumin and gallic acid have been encapsulated by electrohydrodynamic techniques (Dhandayuthapani et al., 2012; Neo et al., 2013). The results indicated that the loaded antioxidants maintained their phenolic structures and antioxidant activity after electrospinning, with adjustable release profiles, thereby further expanding their potential for food and drug applications. Vitamins also have been encapsulated with these techniques, for example, vitamin B was stabilized by encapsulation in electrospun fibers (Alborzi et al., 2013; Madziva, Kailasapathy, & Phillips, 2005; Wongsasulak, Pathumban, & Yoovidhya, 2014). Electrospun nanofibers exhibit properties such as high porosity, large surface area per unit mass, high gas permeability, and small interfibrous pore size, with most of these properties being appropriate when considering these materials as carriers for delivery of bioactive compounds. Electrospinning is also a useful technique for the encapsulation of probiotics using hydrocolloids as wall materials. Electrosprayed hydrocolloid-based encapsulation structures demonstrated the ability to prolong the survival of bifidobacteria under different temperature and relative humidity conditions (López-Rubio, Sanchez, Wilkanowicz, Sanz, & Lagaron, 2012). Extensive lists of applications of electrohydrodynamic techniques for the encapsulation of bioactive compounds, using various polymeric wall materials, are presented in a recent review by Drosou, Krokida, and Biliaderis (2017). Overall, the nanostructured fiber and particle morphologies produced by electrohydrodynamic techniques offer tunable release kinetics for encapsulated actives, pertinent to diverse applications in the biomedical and food production sectors.

1.7 FLUIDIZED BED COATING

Fluidized bed coating is an encapsulation technique for the production of coated particles by spraying an encapsulating agent onto a fluidized powder bed, that is, suspended powder particles by an air stream are coated with a shell material applied in an atomized liquid form. The solid particles are suspended by high velocity air in a humidity- and temperature-controlled chamber where the coating liquid formulation, with droplet size of one order of magnitude smaller than the fluidized solid particles, is applied to achieve uniform and complete particle coating. Important processing variables in fluid coating like solid particle circulating rate, nozzle atomization pressure, coating liquid viscosity, spray rate, humidity of the inlet air, and the chamber temperature affect the agglomeration and film forming around the particles, and thus determine coating efficiency (Ray et al., 2016); in general, 10–50% of coating is applied, and a spherical particle morphology and dense structures are preferred for effective coating. Different fluidized bed coating apparatus are available, allowing the application of the atomized coating liquid in different ways, by changing the position of the nozzle (Desai & Park, 2005a): (1) top spray (conventional method), (2) bottom spray (known as the Wurster system), and (3) tangential spray. Continuous fluidized bed systems have also been developed for a large material throughput. A typical fluidized bed apparatus can process particles with size ranging from 100 µ to a few millimeters. The most important parameters in a fluidized bed operation that can be optimized, dependent on each specific application, are inlet air temperature, air velocity, coating liquid temperature and spray rate, atomization pressure, relative humidity, and temperature of inlet air (Desai & Park, 2005b).

Coating materials for fluidized bed coating must be thermally stable, good film formers and their aqueous solutions must have viscosities that facilitate pumping and atomization; the most commonly used materials for particle coating are natural gums and cellulose derivatives, proteins, starches (pregelatinized, chemically modified), and maltodextrins. Recently, sweet whey powder, which is a cheap waste product from cheese manufacturing plants, consisting mainly of lactose, has been used as coating material in order to establish an acidic resistant formulation for dietary probiotics with a pH-controlled release of the core enclosed bacteria (Schell & Beermann, 2014).

Fluidized bed coating has been extensively used to encapsulate micronutrients (e.g., vitamins, minerals, and various premixes), antimicrobials (e.g., sorbates, propionates, lactic acid), processing aids (e.g., acids, leavening agents), and salts. The technique has also been applied for the encapsulation of herbal extracts; rosemary was used as model plant source. The results showed that granules with improved physicochemical properties can be produced, extending the application range of such plant extracts from foods to drugs and cosmetics (Benelli & Oliveira, 2016). In the same direction of fulfilling the demand of consumers for natural food ingredients, colorants such as carotenoids have been recently encapsulated with a fluidized bed coating process and tested in real food matrices (Coronel-Aguilera & San Martín-González, 2015).

As alternative coating liquids, molten lipids (hydrogenated fats, waxes, or emulsifiers) can be employed in fluidized bed operations; however, in this case heating of the pumped liquid, the nozzle, and the atomizing air is required for preventing premature solidification of the lipid material before it reaches the powder particles (Desai & Park, 2005a); once in the chamber, control of lipid solidification can be effected by the inlet air temperature that is below (by ~10–20°C) the melting point of the coating lipid phase, otherwise sticky particles and extensive agglomeration may occur (Zuidam & Shimoni, 2010). Other patented innovative processes have been developed for particle coating with fats and waxes using supercritical fluids (CO_2) as solvent for the ingredients of the coating liquid (Klose, 1992; Pacifico, Wu, & Fraley, 2001; Wu, Roe, Gimino, Seriburi, Martin, & Knapp, 2002). In all these processes, the coating formulation is concentrated, since no solvent is employed, as in aqueous coating solutions, and therefore minimum energy input is needed and much shorter processing times are sufficient to accomplish the coating stage.

1.8 COMPLEX COACERVATION

Complex coacervation is an encapsulation technique in which most frequently two combined polymers are used as encapsulating agents; often, one polymer is a protein and the other a polysaccharide. The two polymers are oppositely charged and electrostatic attractions are encountered among them that lead to the formation of complex structures, that is, a phase separation process between oppositely charged macropolyions occurs. Other weak interactions may also contribute to coacervate formation such as hydrogen bonding, van der Waals forces, and hydrophobic interactions (Ach et al., 2015). The self-assembled structural entities are readily controlled and triggered by changes in the environmental conditions (pH, ionic strength, temperature, shear conditions) as well as the total polymer concentration, molecular size, and conformation of the polymeric chains, charge density, and the molar ratio of the interacting macromolecular species (Xiao, Liu, Zhu, Zhou, & Niu, 2014).

Biopolymer coacervation begins with the dissolution of the two polymers in an aqueous medium. The next step may involve the addition and emulsification of a lipophilic material in the aqueous solution; apparently, coacervate complexes are very effective in stabilizing emulsions since the complexes adsorb at the interfaces and thereby increase the thickness of the interfacial layers. By adjusting the pH, the electrostatic forces exerted between the interacting species eventually lead to a two-phase system, the lower coacervate phase rich in biopolymers and the lighter upper phase that is depleted in both biopolymers (serum phase). Typically, the coacervate phase is a highly viscous or gel-like material that still contains large amounts of water; apparently, the coacervates are highly dynamic structures, evolving with time and being amenable to rearrangements in response to modifications of the environmental conditions (e.g., pH changes in relation to the pI and pK values of the protein and polysaccharides used, ionic strength, and temperature). The gel-like material that is allowed to separate and sediment can be further "hardened" by modifying the ionic environment and might be even dried (spray drying, fluidized bed drying, or freeze drying) to generate microcapsules. As different operating conditions, such as pH, nature, and weight ratio of the interacting polymers (wall material) and the encapsulated constituents, the total wall material concentration in the aqueous solution and temperature, affect the efficiency of encapsulation, the process could be optimized using response–surface methodology techniques (Gouin, 2004; Sánchez, García, Calvo, Bernalte, & González-Gómez, 2016).

Biopolymer-based coacervate assemblies can be employed to protect and deliver a range of sensitive functional ingredients with bioactivity as well as to mask any off-taste flavor of the bioactive compounds. A wide range of materials can be used, including several polysaccharides (e.g., modified starches, cereal cell wall polysaccharides, low-methoxyl pectins, chitosan, gum arabic, and other gums) and proteins (e.g., whey proteins, caseins, and soy proteins). Some examples of mixed biopolymer systems used for coacervate production are gum arabic/gelatin, gum arabic/whey proteins, gum arabic/chitosan, chitosan/whey proteins, chitosan/pectins, chitosan/alginates, etc. Some of the biopolymers utilized to fabricate coacervates have themselves nutritional and functional properties (e.g., pectins) in addition to acting as wall materials.

Structured delivery systems using biopolymer coacervates can include emulsion droplets to incorporate bioactive agents that are polar, nonpolar, and amphiphilic within the same system (Moschakis, Murray, & Biliaderis, 2010). This method is primarily used for encapsulating lipophilic compounds, such as essential oils, vegetable oils, fish oils, palm oil, and lipophilic extracts from shrimp waste containing astaxanthin (Gomez-Estaca, Comunian, Montero, Ferro-Furtado, & Favaro-Trindade, 2016; Jun-xia, Hai-yan, & Jian, 2011; Rutz et al., 2017; Shen et al., 2016). Nevertheless, the process has also been applied for encapsulation of hydrophilic constituents, for example, vitamin C (Comunian et al., 2013). This method is quite useful for thermosensitive materials, because of the absence of thermal treatment. A recent study reports on the encapsulation of the antimicrobial peptide nisin to use it as natural antimicrobial in foods (Calderón-Oliver, Pedroza-Islas, Escalona-Buendía, Pedraza-Chaverri, & Ponce-Alquicira, 2017).

Coacervates produced from whey protein isolate and gum arabic were employed to entrap live probiotic bacteria cultures and protect them from harsh environments often encountered during food processing (e.g., heating, high salt concentrations, and low pH), storage, and upon their transit through the gastrointestinal tract (low pH of gastric juice and bile salts in the upper gut) (Bosnea, Moschakis, & Biliaderis, 2014); notably, the encapsulated probiotic cells exhibited enhanced viability upon heat treatment (63–65°C), at low pH (exposure for 3 h at pH 2.0), as well as at different NaCl concentrations (at pH 4.0). The findings suggested that complex coacervation has the potential to deliver alive probiotic cells in low pH food systems, such as fermented dairy products, salty processed foods, acidic juices, as well as mild heat-treated food products.

The microencapsulation of bioactive ingredients by utilizing complex coacervates has several advantages. The conditions adopted for coacervate production are fairly mild, suitable for the encapsulation of heat-sensitive ingredients. The coacervate complexes can also dissolve (disintegrate) under certain environmental conditions to ensure the controlled release of the carriers (bacteria/bioactives) at the specific target point, for example, probiotic bacteria in the human lower gut so that they can colonize the intestinal tract and provide health benefits to the host. Typically, coacervates are liquid-in-nature macromolecular assemblies and therefore they can be incorporated in many liquid-like food systems (dairy products, fruit juices) without adversely affecting their mechanical properties and sensorial attributes, for example, no graininess is perceived during mastication (Bosnea, Moschakis, & Biliaderis, 2014, 2017). The wall material in complex coacervate structures can be even cross-linked chemically (glutaraldehyde) or enzymatically (transglutaminase) to modify the structure, improve hardening and stability of the coacervate, and to fine-tune the release kinetics of the entrapped bioactives (Gouin, 2004; Zuidam & Shimoni, 2010). Moreover, coacervation coupled with other methods of encapsulation (e.g., ionotropic gelation) can be utilized as a potential novel technique for enhancing protection, target release of the active ingredients, and improving compatibility of the materials used with other constituents in a composite food structure (Bosnea, Moschakis, & Biliaderis, 2017).

1.9   ENTRAPMENT IN LIPOSOME VESICLES

Liposomes are self-assembled spherical vesicles (25 nm to 1 µm in diameter) with one or more phospholipid bilayers separating the inner aqueous environment (liquid core) from the outer aqueous medium. Liposome entrapment as encapsulation technique, although its application in food systems has not yet been fully exploited by the food industry, as in the case of pharmaceuticals and cosmetics, is an evolving field for the development of delivery systems to improve nutrient/bioactive functionality in food products due to its biocompatibility, biodegradability, and absence of toxicity (Gouin, 2004; Singh, Thompson, Liu, & Corredig, 2012; Zuidam & Shimoni, 2010). Through liposome entrapment a wide variety of bioactive compounds could be encapsulated due to the amphiphilicity of the phospholipid encapsulating material (da Silva Malheiros, Daroit, & Brandelli, 2010). Hydrophobic materials may be incorporated in the lipid membrane (maximum loading does not exceed ~25% of the phospholipids, without causing destabilization), whereas hydrophilic molecules can be entrapped in the aqueous phase inside the liposome structure; liposomes can thus carry simultaneously both hydrophilic and hydrophobic compounds.

The basic components of liposomes are phospholipids (from egg yolk, soy, and milk) that upon dispersion in an aqueous medium have the tendency to aggregate and form bilayers to minimize contact between the hydrophobic fatty acid chains and the water phase. When energy is provided into the system (mechanical, ultrasound, heat, high pressure homogenization, and microfluidization), the edges of the bilayer sheets close forming tiny vesicles; different liposome structures can be formed, based on the type of energy and phospholipids (saturation and length of fatty acid chains, size and charge of head group) used, affecting their physical characteristics, including physical and chemical stability, protection of the encapsulated material, and its rate of release (Kim & Baianu, 1991; New, 1990). Based on the size and lamellarity (number of concentric bilayers contained within each vesicle), liposomes can be found in four categories: small unilamellar vesicles (SUVs), large unilamellar vesicles (LUVs), multilamellar vesicles (MLVs), and multivesicular vesicles (MVVs) (Sharma & Sharma, 1997), the LUVs being the most appropriate liposomes for food-related applications because of their high encapsulation efficiency, good stability, and simple methods for production. Most of the liposome preparations are kept in relatively dilute aqueous dispersions, although dry liposome microencapsulates can be also obtained by freeze drying.

Conventional methods of liposomes preparation are thin-film hydration, reverse phase evaporation vesicles, and membrane extrusion. Nowadays, "green" techniques containing less organic solvent usage in the liposomal microencapsulation process have been introduced, including microfluidics, rapid expansion of supercritical solutions, supercritical reverse phase evaporation, and several dense gas processes (Tsai & Rizvi, 2016).

Applications of liposome entrapment technique in the food industry include encapsulation of acids, alcalis, amino acids, proteins, enzymes, vitamins, minerals, antimicrobials, and flavor compounds (Gouin, 2004; Kirby, 1991; Mozafari et al., 2008). The entrapment of natural antioxidants in liposome vesicles has been suggested as an effective means of preventing oxidation in oil-in-water emulsions, where both α-tocopherol (in the membrane) and ascorbic acid (entrapped in the aqueous regions of the liposome structure) can act in a synergistic mode (Kirby, 1991). The concurrent encapsulation of vitamin E and vitamin C has also been reported, with the liposome structures protecting both vitamins from heat degradation (Marsanasco, Márquez, Wagner, Alonso, & Chiaramoni, 2011); the authors suggested the potential application of these vitamin carrier systems in juices and other fortified beverages subjected to thermal treatments (e.g., pasteurization). Liposomes were also used to protect thermo-sensitive molecules of bioactive oils, such as essential oils, from degradation and oxidation (Rodríguez, Martín, Ruiz, & Clares, 2016). Another example of a novel approach is the production of phenolipids, meaning the incorporation of phenolics into phospholipids. Phenolipids could extend the application of encapsulated phenolics in the foods (Ramadan, 2012). Dairy and meat processing applications have been reported for entrapment of proteolytic and lipolytic enzymes in cheese production (improve the ripening process and flavor profiles) or in meat-tenderization (Kheadr, Vuillemard, & El-Deeb, 2003; Laloy, Vuillemard, Dufour, & Simard, 1998; Lee, Jin, Hwang, & Lee, 2000).

Recent investigations also focused on the incorporation of encapsulated forms of bioactives in real food matrices. The behavior of tea catechins, a well-studied group of antioxidants, encapsulated within soy lecithin liposomes was studied in a low-fat hard cheese system (Rashidinejad, Birch, Sun-Waterhouse, & Everett, 2014). The masking effect of nanoliposomes on the odor of fish oil was also examined using yogurt as a food matrix and the results revealed that encapsulation in liposomes have a positive effect on the sensory profile of fish oil (Ghorbanzade, Jafari, Akhavan, & Hadavi, 2017). Liposomes modified with polysaccharides, either by the development of a cross-linked alginate gel core (Smith, Jaime-Fonseca, Grover, & Bakalis, 2010) or by coating with an ionic polysaccharide (Filipović-Grcić, Škalko-Basnet, & Jalsenjak, 2001; Liu & Park, 2010) may provide alternative possibilities for designing delivery systems with improved retention and better controlled release properties of the bioactives, as evidenced with simulated gastric digestion protocols.

There is relatively little information concerning the fate of liposomes and their encapsulated core material following processing and storage of complex matrices such as foods. The liposome membranes are semipermeable, allowing the transfer of small molecules, including the entrapped compounds. For permeability of hydrophilic molecules, the following general sequence has been suggested: water > small nonelectrolytes > anions > cations > large nonelectrolytes > large polyelectrolytes > proteins (Frezard, 1999). A number of mechanisms have also been identified as the underlying processes for the release of bioactive molecules entrapped in liposome structures as discussed in several review articles (Pothakamury & Barbosa-Cánovas, 1995; Singh, Thompson, Liu, & Corredig, 2012). These include (1) diffusion of the entrapped hydrophilic molecules—affected by the solubility in the phospholipid bilayer and the membrane permeability; (2) osmotically controlled release—with osmotic gradient across the membrane, liposomes swell (or may even burst) and membrane permeability increases; (3) physical destabilization of the bilayer by fusion (colloidal instability) or by the action of hydrolytic enzymes (phospholipases); (4) ionic destabilization or pH-driven release due to charge changes on the liposomal structures; (5) temperature release—with raising temperature membrane permeability increases and if liposomes are coated with another layer of impermeable material that melts, upon melting it would permit other release mechanisms (e.g., diffusion) to proceed.

1.10   HYDROGEL-BASED MICROGEL PARTICLES

A hydrogel refers to a network structure formed by polymers that physically traps large amounts of water; as a result, it represents a unique system with mechanical and chemical properties between those of solids and liquids, having a high capacity of loading other molecules or cellular entities within its structure. Biopolymeric microgels, made from proteins and/or polysaccharides, have been gaining increased attention for encapsulation and controlled release applications in the pharmaceutical and food industries (Farjami & Madadlou, 2017; Gouin, 2004; Shewan & Stokes, 2013; Zuidam & Shimoni, 2010). The microgels consist of physically, chemically, or enzymatically cross-linked biopolymers that form a stable three-dimensional structure, due to the presence of covalent linkages or strong noncovalent interactions (ionic, H-bonding, hydrophobic interactions, etc.), which has high water absorption capacity and the ability to swell-deswell in response to externally applied stimuli, that is, alterations in ionic strength, pH, temperature, and solvent quality. Apparently, the porous aqueous microenvironment created by the microgels is quite useful and beneficial for the encapsulation of bioactive substances, usually of larger size, including peptides, proteins, enzymes, drugs, and whole cells. Depending on the biopolymers used and the cross-linking mechanism involved, it is feasible to design engineered particles with controllable and environment-responsive properties via modulation of the inter-polymer and solvent-polymeric chain interactions/associations. For food-related applications, polymers that are commonly used as materials for microgel production are polymeric carbohydrates (alginates, cellulose and starch derivatives, carrageenan, agarose, and chitosan) and proteins (whey proteins, gelatin, gluten, soy protein, corn zein, and casein); depending on production method, colloidal hydrogel systems can be in the nanoscale range, with diameters <0.5 μm, and microgels that have diameters between 0.5 and 5 μm. The techniques used to fabricate microgel particles can be categorized into two general groups, those based on molecular associations (either through interchain associations of a single type of biopolymer or via associative complexation in mixed biopolymer systems, for example, phase separation due to complex coacervation) and those referred to as mechanical methods, which combine first a processing step for "breaking" a bulk solution/dispersion into a large number of small-sized droplets, using mechanical devices, and subsequently a physicochemical method to induce cross-linking and microgel formation. Among the latter, microspheres can be made by several routes (Farjami & Madadlou, 2017; Gouin, 2004; Shewan & Stokes, 2013): (a) extrusion or drop-based methods (most common method, at least on small batch scale), carried out with relatively simple devices and involving mild conditions (no heating); (b) atomization methods, where small size droplets can be either dehydrated by spray drying and then rapidly rehydrated and cross-linked by addition in an aqueous solution carrying a gelling co-solute (e.g., calcium for alginate microspheres production) or spray chilled by solidification of the liquid droplets by rapid cooling at temperatures below the gelling point of the biopolymer used as matrix material (for gelatin 35–40°C, agar 32–45°C, carrageenan 40–70°C, pectin 42–68°C); (c) shearing methods, by application of shear forces to a gelling biopolymer system while it is undergoing the sol-gel transition, either thermally or chemically (i.e., typically leads to anisotropic microgel particles, and often it is referred to as "fluid gels"); and (d) emulsion-based processes, involving water-in-oil emulsions as templates to produce biopolymer particles, where the aqueous droplets can be cross-linked via enzymatic treatment, thermal denaturation, or ionotropic gelation.

The extrusion-based methods are widely used to produce hydrogel beads; the biopolymer solution carrying the active is forced through a syringe needle, a spray nozzle, a vibrating nozzle, an atomizing disk, a jet cutter, or an electrospray technique (Champagne, Blahuta, Brion, & Gagnon, 2000; Krasaekoopt, Bhandari, & Deeth, 2003; Mi, Wong, Shyu, & Chang, 1999; Prusse, Fox, Kirchhoff, Bruske, Breford, & Vorlop, 1998; Whelehan & Marison, 2011; Zuidam & Shimoni, 2010) to form droplets that subsequently gel in the hardening solution as a result of a cross-linking reaction (e.g., by enzymes, mineral ions, glutaraldehyde), for example, chitosan droplets are often hardened in a tripolyphosphate solution, alginate droplets hardened in calcium chloride solution, and κ-carrageenan in potassium chloride solution. Alternatively, a temperature change (heating or cooling by injection of the polymer droplets into a hot or a cold liquid) may be applied for induction of gelation in the case of thermally setting biopolymers (e.g., gelatin—cold set gelation, whey proteins—heat set gelation) or complexation with another polymer (Farjami & Madadlou, 2017). Droplet size (based on needle or nozzle orifice diameter), feed solution viscoelasticity, flow rate, as well as the properties of the gelling environment determine the microgel particle size; in general, particle diameters between 0.1 and 5 mm can be made. Among the above listed extrusion methods, the vibration-jet technique is regarded as the most suitable to obtain uniform, monodisperse, and small beads in up-scaled production (Heinzen, Berger, & Marison, 2004). Furthermore, the extrusion process can proceed via a concentric nozzle (co-extrusion), yielding core-shell type of encapsulates; the two streams of liquids are immiscible (lipophilic core carrying the bioactive and hydrophilic wall having the gelling polymer).

The emulsion-based techniques also include several variants. One may involve the addition of $CaCl_2$ to an emulsion of water (alginate solution) droplets containing the dissolved or suspended active ingredient in vegetable oil (continuous phase), that is, microbeads are formed by the Ca-induced gelation of the alginate droplets. Alternatively, both alginate and calcium (in an insoluble form, e.g., $CaCO_3$) are initially present in the water droplets of the water-in-oil emulsion. Upon addition of an oil-soluble acid (e.g., acetic acid), the pH in the aqueous alginate droplets decreases, solubilizing Ca^{2+} ions and thereby inducing gelation of the alginate droplets. Capillary microfluidic devices provide a flexible method for fabricating monodisperse microgel particles, through emulsification of a gelling solution within the oil phase, followed by physical or chemical cross-linking of the aqueous droplets (Luo & Chen, 2012). Actually, the dispersed aqueous phase may consist of one stream containing the biopolymer solution (gelling solution carrying the active) or multiple streams of the biopolymer and its cross-linking agent going through the capillary droplet generator. On the other hand, the stream of the continuous phase (oil with a surfactant) breaks up the gelling solution into droplets and carries them as separate entities till they solidify in the microfluidic device, for example, gelation of droplets containing pectin and/or alginate and $CaCO_3$ (as cross-linking agent in inactive form) is induced by the diffusion of acetic acid from the oil continuous phase into the polysaccharide droplets where the pH decreases and the liberated (solubilized) Ca^{2+} bridges the polysaccharide chains (ionotropic gelation) (Marquis, Davy, Cathala, Fang, & Renard, 2015). Similarly, membrane emulsification involves the use of membranes through which the aqueous solution of the gelling biopolymer permeates under a certain level of applied pressure; as a result, formation of uniform-sized aqueous droplets takes place, which are released/dispersed into the oil (continuous) phase on the other side of the membrane. The biopolymer droplets are then hardened by physical or chemical cross-linking, producing microgels of monodisperse size distribution (Oh, Drumright, Siegwart, & Matyjaszewski, 2008; Oh, Lee, & Park, 2009). This technique was utilized to produce chitosan microgels (Wang, Ma, & Su, 2005). The chitosan/acetic acid aqueous solution was forced by pressure (N_2 gas) to permeate through a porous glass membrane into a liquid paraffin and petroleum ether mixture containing an emulsifier, to form a W/O emulsion. Then, the uniform size droplets were solidified using glutaraldehyde as cross-linking agent. The above described emulsion methods generally produce smaller size microspheres (10 µm to 1 mm), compared to the extrusion processes (0.2–5 mm), and they are easier to scale up (Zuidam & Shimoni, 2010); however, when an emulsion method is employed, there is often a need to wash out the surface oil (residual) from the beads. For a particular biopolymer system, modification of the microgel structure and thereby its mechanical properties and permeability can be manipulated by polymer and cross-linking agent concentrations in their respective solutions, as well as by further posthardening treatments with additional coating layers using other interacting biopolymers (e.g., ionic interactions), and/or with microgel network formation to yield larger hydrogel assemblies.

The use of microgels in food products does offer several opportunities to modify the mechanical properties (rheology), replace fat, contribute to satiety control, and act as delivery vehicles for larger size components (e.g., bioactive peptides and proteins, probiotics) aiming at targeted release (e.g., in the small intestine or colon). However, developing feasible processes at an industrial-scale level by which the designed microgels could be effective carriers of bioactives in foods, particularly for entrapment of small molecules (vitamins, flavor compounds, phytonutrients), and be compatible with other ingredients in the formulated products, without influencing adversely the mouthfeel perception (texture, taste) of the food, still remains a challenge.

1.11   MOLECULAR ENCAPSULATION

For the sake of completeness, the method of molecular encapsulation or inclusion complexation is discussed. Molecular encapsulation is a process during which a single molecule is incorporated as guest in another host molecule that contains a cavity in its structure. Most common host molecules with applications in foods are cyclodextrins (CDs), which are cyclic nonreducing oligosaccharides with a torus-like structure; the most common native CDs are composed of 6, 7, and 8 glucosyl units interconnected via α-($1 \rightarrow 4$) glycosidic linkages, and called α-, β-, and γ-CD, respectively. Although β-CD has limited aqueous solubility, derivatives like 2-hydroxypropylated-β-CD, with improved water solubility, have been synthesized (Kurkov & Loftsson, 2013). Cyclodextrins have a hydrophobic interior and a hydrophilic exterior. One of the most important characteristics of cyclodextrins is the formation of inclusion complexes with various ligands. The hydrophobic cavity of the cyclodextrin varies in size (5–8 Å) depending on the number of glucosyl units in the molecule (Zuidam & Shimoni, 2010). An active nonpolar molecule (guest), with the right size and co-solubilized or dispersed with the cyclodextrin (host) in an aqueous solution, can be reversibly entrapped into the cavity to form an inclusion complex. Loading of the guest molecules can even be achieved by mixing aqueous pastes (of 20–30% water–content) of CDs with the guest compound (e.g., flavor). The driving forces of host-guest binding are mainly due to the release of enthalpy-rich water molecules from the cavity and noncovalent interactions including van der Waals interactions, hydrogen bonding, and hydrophobic interactions. Generally, the higher the hydrophobicity and the smaller the guest molecule is, the greater the affinity for the CDs (Del Valle, 2004).

Other examples of molecular inclusion complexes include those of amylose (the linear starch polymer) with lipids (monoacyl lipids, for example, lysophospholipids, free fatty acids) or other small hydrophobic ligands (Biliaderis, 2009). These complexes are known to exist in the form of helical structures (left-handed single helices), of the so-called V-conformation, which have six residues per turn (V_6) for complexes with aliphatic alcohols and monoacyl lipids; all complexing ligands induce chain helicity and are believed to occupy the central cavity of the helix. When the ligand is bulkier than a hydrocarbon chain, helices of seven or eight glucose residues per turn are also feasible. A detailed description of the supramolecular structure of amylose-lipid complexes in thermally processed foods is a difficult task, because helices can exist in various states of aggregation, depending on the thermomechanical history of the product. Application of calorimetry and x-ray diffraction to studies on starch–lipid interactions revealed at least two structurally distinct forms, named as polymorphs I and II (Biliaderis, 1991; Biliaderis & Galloway, 1989). Form I (of low *Tm*) was described as an aggregated state where ordered (helical) chain segments have insufficient packing, thus exhibiting an amorphous pattern by x-ray diffraction. In contrast, form I1 (of high *Tm*) appeared as a polycrystalline aggregate structure with well-developed long-range order, showing the typical reflection of the V-pattern by x-rays. Understanding the supramolecular structure, stability, and transformations between the various forms of amylose-lipid complexes is of great fundamental and technological importance, considering the multifunctional role of lipids in starch-based products. For example, incorporation of monoglycerides in the dough is known to retard starch retrogradation and bread firming. Similarly, monoglycerides added to dried potato granules prevent stickiness. Moreover, improvements in structural integrity of cereal kernels during cooking (e.g., parboiled rice), as well as decreased swelling, solubilization, and thickening power of starch, have all been ascribed to starch–lipid interactions. From a nutritional view point, there is an interest in the complexes of amylose with lipids as they protect unsaturated fatty acids and other sensitive to O_2 ligands from oxidation. They also exhibit a slower rate of digestion with the pancreatic α-amylase, particularly the form II structure, constituting another type of the so-called "resistant starch" fraction in thermally processed food products (Biliaderis, 1992). Shimoni et al. (2007) used this property of amylose to encapsulate a wide range of bioactive agents, by employing a dual feed jet homogenization, and produced particles ranging from 400 nm to a few tens of microns. When the complexes are digested in the human digestive track by amylolytic enzymes, the bioactive ligands are expected to be released.

Polyphenols from fruits and plants can be encapsulated in cyclodextrins as the benzene ring of polyphenols is an ideal guest for cyclodextrins. Several examples include the encapsulation of gallic acid (Aytac, Kusku, Durgun, & Uyar, 2016), hydroxycinnamic acids, such as caffeic acid, p-coumaric acid, ferulic acid (Kfoury et al., 2016), as well as ellagic acid, which is an important antioxidant polyphenol of many fruits like pomegranates, raspberries, strawberries, and blackberries (Kurkov & Loftsson, 2013). Olive leaf extracts (rich in oleuropein) were also encapsulated in β-CD (Mourtzinos, Salta, Yannakopoulou, Chiou, & Karathanos, 2007). Inclusion complex formation with polyphenols was found to increase their antioxidant activity, which may be attributed to an increase in their solubility (Fang & Bhandari, 2010).

Moreover, several food flavors have been encapsulated with cylodextrins (anethole, thymol and geraniol, citral, citronellal, menthol, linalool—two enantiomers of a naturally occurring terpene alcohol—and a number of other essential oils including anise, sage, cinnamon, jasmine, bergamot, orange, lemon, lime, onion, garlic, mustard, and marjoram oils), in order to decrease their volatility during thermal processing of the food product and to improve their chemical stability when exposed to air, light, moisture, and heat (Marques, 2010). Encapsulated whole essential oils (trans-cinnamaldehyde, eugenol, cinnamon bark, and clove bud extracts) with cyclodextrins exhibited beneficial properties such as increased water solubility and antimicrobial activity when incorporated into foods (Comunian & Favaro-Trindade, 2016; Fu, Sarkar, Bhunia, & Yao, 2016). Other lipophilic ingredients that do not contain a benzene ring have also been encapsulated with cyclodextrins; most of them are well-known bioactives, such as omega-3 (Vestland, Jacobsen, Sande, Myrset, & Klaveness, 2015) and phytosterols (Meng, Pan, & Liu, 2012), and by complexation they can be transformed into a powdered form, easy to handle from a product formulation viewpoint. Finally, vitamin D has been encapsulated using this technique, by solubilizing the active compound in ethanol (Soares, Murhadi, Kurpad, Chan She Ping-Delfos, & Piers, 2012). Although encapsulation of flavor volatiles and other bioactives in β-cyclodextrin is quite effective (good loading capacity and protection of the aroma compounds), the use of this encapsulant is still limited for food applications, possibly due to the regulatory constraints in some countries; in EU, β-cyclodextrin is allowed in limited number of products at specified concentrations, <1 g/kg, whereas a GRAS status is given for α-, β-, and γ-cyclodextrins by FDA in the United States, and in Japan cyclodextrins are regarded as natural products (Yuliani, Bhandary, Rutgers, & D' Arcy, 2004; Zuidam & Shimoni, 2010).

1.12   CONCLUDING REMARKS

There is a growing interest in supplementing foods with natural food additives, bioactives and nutraceuticals, such as vitamins, colorants, flavor compounds, ω-3 fatty acids, probiotics, antioxidants, and many other phytochemicals. Many of these ingredients need to be introduced into food products in forms that are protected from environmental stresses during processing and storage (e.g., O_2, low pH, and heating), when becoming part of the food matrix (possible interactions with other food ingredients) as well as to resist the gastric environment (low pH, hydrolytic enzymes), in order to be effective and exert their functionality at an acceptable level where and when it is needed (e.g., specific target release in the upper or the lower part of the intestinal track). This necessitates the use of microencapsulated food ingredients and bioactives to provide solutions for existing problems in functional food product manufacturing and control of their release for enhancing the health-promoting potential of the formulated product. A dry powder form, involving spray or freeze drying, the two most commonly used drying techniques, is often a preferred delivery system for the actives, because of ease of handling and lower storage and transport costs. However, even with these long-used processing technologies, there is a multitude of factors that should be optimized to transform a liquid dispersion of a bioactive into a stable solid form; the wall material composition and its effectiveness to efficiently entrap and protect the core material, to remain nonreactive with the entrapped compounds, and without affecting the sensorial characteristics of the product during storage and consumption, are major requirements in this context. Quite often, with a single material in the wall formulation all the desired properties are not met and thus a mixture of ingredients must be used. Moreover, optimization of all processing parameters, such as type and concentration of the wall materials, by taking into account the nature of the core material, ratio of core/wall, and conditions of the drying process, is carried out in order to have an efficient encapsulation. For certain applications, there is even a need to develop more structurally complex delivery systems (multilayered structures amenable to swell and/or slow release of the entrapped components) that can be fabricated with simple or sophisticated equipment, and the challenge becomes to utilize food grade materials, preferably natural and of low cost. A number of new techniques have been developed and/or others further evolved, including different extrusion and co-extrusion technologies, electrohydrodynamic processing, and coacervation and hydrogel microsphere formation using biopolymers. There is more room to improve our understanding of the solution behavior and state transformations of biopolymers as exposed to different processing environments (heating, cooling, pH-ionic strength, drying, high shear conditions, pressure, etc.); the conformational responses undergone by proteins and polysaccharides as a result of different processing protocols, alone or in mixtures, does offer many possibilities to structure core-shell microcapsules and develop new delivery systems for food bioactives. From a food engineer point of view, there is also a challenge to move the processing from the lab-bench to full-scale production of the final fortified products. The economic aspects of the encapsulation system must be considered to ensure it is cost-effective for the food industry. Even if encapsulation requires extra production steps and the usage of energy and new materials, the cost of encapsulated active ingredients could be lower as they could be used in lower dosages. A characteristic example is the encapsulation of food flavors that allows the use of minor amounts of them when used for flavoring a food matrix, compared to the non-encapsulated forms that are volatile and much higher concentrations are required. For most of the encapsulation systems developed, there is a need to establish the links between microstructure and functionality. Moreover, issues like the encapsulation efficiency, the stability of the bioactive components in the carrier matrix during processing and when incorporated into the food system, the interactions between the core and the wall components, as well as the mechanisms of release are important and will assist to produce tailor-made microcapsules for the fortification of food products. Finally, as we move toward the development of more efficient delivery systems and taking into account the increasing consumer interests/demand for "personalized nutrition" strategies, there should be more effort put into assessing the real health impact of the microencapsulated bioactives using properly designed *in vitro* and *in vivo* studies, and how efficacy relates with the food structure and the release properties of the encapsulate.

REFERENCES

Abbas, S., Da Wei, C., Hayat, K., & Xiaoming, Z. (2012). Ascorbic acid: Microencapsulation techniques and trends—A review. *Food Reviews International, 28*(4), 343–374.

Ach, D., Briançon, S., Dugas, V., Pelletier, J., Broze, G., & Chevalier, Y. (2015). Influence of main whey protein components on the mechanism of complex coacervation with Acacia gum. *Colloids and Surfaces A: Physicochemical and Engineering Aspects, 481,* 367–374.

Aghbashlo, M., Mobli, H., Madadlou, A., & Rafiee, S. (2013). Influence of wall material and inlet drying air temperature on the microencapsulation of fish oil by spray drying. *Food and Bioprocess Technology, 6*(6), 1561–1569.

Alborzi, S., Lim, L.-T., & Kakuda, Y. (2013). Encapsulation of folic acid and its stability in sodium alginate-pectin-poly(ethylene oxide) electrospun fibres. *Journal of Microencapsulation, 30*(1), 64–71.

Alvim, I. D., Stein, M. A., Koury, I. P., Dantas, F. B. H., & Cruz, C. L. de C. V. (2016). Comparison between the spray drying and spray chilling microparticles contain ascorbic acid in a baked product application. *LWT—Food Science and Technology, 65,* 689–694.

Amara, C. B., Eghbal, N., Degraeve, P., & Gharsallaoui, A. (2016). Using complex coacervation for lysozyme encapsulation by spray-drying. *Journal of Food Engineering, 183,* 50–57.

Angeles, M., Cheng, H.-L., & Velankar, S. S. (2008). Emulsion electrospinning: Composite fibers from drop breakup during electrospinning. *Polymers for Advanced Technologies*, *19*, 728–733.

Anselmi, C., Centini, M., Maggiore, M., Gaggelli, N., Andreassi, M., Buonocore, A., & Facino, R. M. (2008). Non-covalent inclusion of ferulic acid with α-cyclodextrin improves photo-stability and delivery: NMR and modeling studies. *Journal of Pharmaceutical and Biomedical Analysis*, *46*(4), 645–652.

AOAC (1997). *Official Methods of Analysis of AOAC International*, 3rd rev, AOAC, Gaithersburg, MD.

Appelqvist, I. A. M., Golding, M., Vreeker, R., & Zuidam, N. J. (2007). Emulsions as delivery systems in foods. In J. M. Lakis (Ed.), *Encapsulation and controlled release technologies in food systems* (pp. 41–81). Blackwell Publishing, Ames, IA.

Augustin, M. A., & Hemar, Y. (2009). Nano- and micro-structured assemblies for encapsulation of food ingredients. *Chemical Society Reviews*, *38*, 902–912.

Augustin, M. A., Sanguansri, L., Margetts, C., & Young, B. (2001). Microencapsulation of food ingredients. *Food Australia*, *53*, 220–223.

Awad, T. S., Helgason, T., Weiss, J., Decker, E. A., & McClements, D. J. (2009). Effect of omega-3 fatty acids on crystallization, polymorphic transformation and stability of tripalmitin solid lipid nanoparticle suspensions. *Crystal Growth & Design*, *9*(8), 3405–3411.

Aytac, Z., Kusku, S. I., Durgun, E., & Uyar, T. (2016). Encapsulation of gallic acid/cyclodextrin inclusion complex in electrospun polylactic acid nanofibers: Release behavior and antioxidant activity of gallic acid. *Materials Science and Engineering: C*, *63*, 231–239.

Barbosa-Cánovas, G. V., Ortega-Rivas, E., Juliano, P., & Yan, H. (2005). *Food powders: Physical properties, processing, and functionality*. Kluwer Academic/Plenum Publishers, New York.

Benelli, L., & Oliveira, W. P. (2016). System dynamics and product quality during fluidized bed agglomeration of phytochemical compositions. *Powder Technology*, *300*, 2–13.

Bhandari, B. R., Dumoulin, E. D., Richard, H. M. J., Noleau, I., & Lebert, A. M. (1992). Flavor encapsulation by spray drying: Application to citral and linalyl acetate. *Journal of Food Science*, *57*(1), 217–221.

Bhandari, B. R., & Howes, T. (1999). Implication of glass transition for the drying and stability of dried foods. *Journal of Food Engineering*, *40*, 71–79.

Biliaderis, C. G. (1991). The structure and interactions of starch with food constituents. *Canadian Journal of Physiology and Pharmacology*, *69*, 60–78.

Biliaderis, C. G. (1992). Structures and phase transitions of starch in food systems. *Food Technology*, *46*(6), 98–100, 102, 104, 106, 108, 109, 145.

Biliaderis, C. G. (2009). Structural transitions and related physical properties of starch. In R. L. Whistler & J. N. BeMiller (Eds.), *Starch: Chemistry and technology*, 3rd edition (pp. 293–372). Academic Press, Burlington, MA.

Biliaderis, C. G., & Galloway, G. (1989). Crystallization behavior of amylose-V complexes: Structure-functional property relationships. *Carbohydrate Research*, *189*, 31–48.

Biliaderis, C. G., Lazaridou, A., & Arvanitoyannis, I. (1999). Glass transition and physical properties of polyol-plasticized pullulan-starch blends at low moisture. *Carbohydrate Polymers*, *40*, 29–47.

Bosnea, L., Moschakis, T., & Biliaderis, C. G. (2014). Complex coacervation as a novel microencapsulation technique to improve viability of probiotics under different stresses. *Food and Bioprocess Technology*, *7*, 2767–2781.

Bosnea, L., Moschakis, T., & Biliaderis, C. G. (2017). Microencapsulated cells of *Lactobacillus paracasei* subsp. paracasei in complex coacervates and their function in a yogurt matrix. *Food & Function*, *8*(2), 554–562.

Buffo, R. A., Probst, K., Zehentbauer, G., Luo, Z., & Reineccius, G. A. (2002). Effects of agglomeration on the properties of spray-dried encapsulated flavours. *Flavour and Fragrance Journal*, *17*(4), 292–299.

Busch, V. M., Pereyra-Gonzalez, A., Šegatin, N., Santagapita, P. R., Poklar Ulrih, N., & Buera, M. P. (2017). Propolis encapsulation by spray drying: Characterization and stability. *LWT—Food Science and Technology*, *75*, 227–235.

Bustamante, M., Oomah, B. D., Rubilar, M., & Shene, C. (2017). Effective *Lactobacillus plantarum* and *Bifidobacterium infantis* encapsulation with chia seed (*Salvia hispanica* L.) and flaxseed (*Linum usitatissimum* L.) mucilage and soluble protein by spray drying. *Food Chemistry*, *216*, 97–105.

Bustamante, M., Villarroel, M., Rubilar, M., & Shene, C. (2015). *Lactobacillus acidophilus* La-05 encapsulated by spray drying: Effect of mucilage and protein from flaxseed (*Linum usitatissimum* L.). *LWT—Food Science and Technology*, *62*(2), 1162–1168.

Calderón-Oliver, M., Pedroza-Islas, R., Escalona-Buendía, H. B., Pedraza-Chaverri, J., & Ponce-Alquicira, E. (2017). Comparative study of the microencapsulation by complex coacervation of nisin in combination with an avocado antioxidant extract. *Food Hydrocolloids*, *62*, 49–57.

Çam, M., İçyer, N. C., & Erdoğan, F. (2014). Pomegranate peel phenolics: Microencapsulation, storage stability and potential ingredient for functional food development. *LWT—Food Science and Technology*, *55*(1), 117–123.

Cano-Chauca, M., Stringheta, P. C., Ramos, A. M., & Cal-Vidal, J. (2005). Effect of the carriers on the microstructure of mango powder obtained by spray drying and its functional characterization. *Innovative Food Science and Emerging Technologies*, *6*(4), 420–428.

Cano-Higuita, D. M., Malacrida, C. R., & Telis, V. R. N. (2015). Stability of curcumin microencapsulated by spray and freeze drying in binary and ternary matrices of maltodextrin, gum arabic and modified starch. *Journal of Food Processing and Preservation*, *39*(6), 2049–2060.

Castro, N., Durrieu, V., Raynaud, C., Rouilly, A., Rigal, L., & Quellet, C. (2016). Melt extrusion encapsulation of flavors: A review. *Polymer Reviews*, *56*(1), 137–186.

Ceballos, A. M., Giraldo, G. I., & Orrego, C. E. (2012). Effect of freezing rate on quality parameters of freeze dried soursop fruit pulp. *Journal of Food Engineering*, *111*(2), 360–365.

Celli, G. B., Dibazar, R., Ghanem, A., & Brooks, M. S.-L. (2016). Degradation kinetics of anthocyanins in freeze-dried microencapsulates from lowbush blueberries (*Vaccinium angustifolium* Aiton) and prediction of shelf-life. *Drying Technology*, *34*(10), 1175–1184.

Champagne, C. P., Blahuta, N., Brion, F., & Gagnon, C. (2000). A vortex-bowl disk atomizer system for the production of alginate beads in a 1500-liter fermentor. *Biotechnology and Bioengineering*, *68*, 681–688.

Chranioti, C., Nikoloudaki, A., & Tzia, C. (2015). Saffron and beetroot extracts encapsulated in maltodextrin, gum arabic, modified starch and chitosan: Incorporation in a chewing gum system. *Carbohydrate Polymers*, *127*, 252–263.

Comunian, T. A., & Favaro-Trindade, C. S. (2016). Microencapsulation using biopolymers as an alternative to produce food enhanced with phytosterols and omega-3 fatty acids: A review. *Food Hydrocolloids*, *61*, 442–457.

Comunian, T. A., Thomazini, M., Alves, A. J. G., de Matos Jr, F. E., de Carvalho Balieiro, J. C., & Favaro-Trindade, C. S. (2013). Microencapsulation of ascorbic acid by complex coacervation: Protection and controlled release. *Food Research International*, *52*(1), 373–379.

Consoli, L., Grimaldi, R., Sartori, T., Menegalli, F. C., & Hubinger, M. D. (2016). Gallic acid microparticles produced by spray chilling technique: Production and characterization. *LWT—Food Science and Technology*, *65*, 79–87.

Coronel-Aguilera, C. P., & San Martín-González, M. F. (2015). Encapsulation of spray dried β-carotene emulsion by fluidized bed coating technology. *LWT—Food Science and Technology*, *62*(1, Part 1), 187–193.

da Silva Malheiros, P., Daroit, D. J., & Brandelli, A. (2010). Food applications of liposome-encapsulated antimicrobial peptides. *Trends in Food Science & Technology*, *21*(6), 284–292.

de Matos Jr, F. E., Comunian, T. A., Thomazini, M., & Favaro-Trindade, C. S. (2017). Effect of feed preparation on the properties and stability of ascorbic acid microparticles produced by spray chilling. *LWT—Food Science and Technology*, *75*, 251–260.

Del Valle, E. M. M. (2004). Cyclodextrins and their uses: A review. *Process Biochemistry*, *39*(9), 1033–1046.

Desai, K. G. H., & Park, H. J. (2005a). Recent developments in microencapsulation of food ingredients. *Drying Technology*, *23*(7), 1361–1394.

Desai, K. G. H., & Park, H. J. (2005b). Encapsulation of vitamin C in tripolyphosphate cross-linked chitosan microspheres by spray drying. *Journal of Microencapsulation*, *22*(2), 179–192.

Dhandayuthapani, B., Anila, M., Ravindran Girija, A., Yutaka, N., Venugopal, K., Yasuhiko, Y., & Sakthikumar, D. (2012). Hybrid fluorescent curcumin loaded zein electrospun nanofibrous scaffold for biomedical applications. *Biomedical Materials*, *7*(4), 045001.

Đorđević, V., Balanč, B., Belščak-Cvitanović, A., Lević, S., Trifković, K., Kalušević, A., & Nedović, V. (2015). Trends in encapsulation technologies for delivery of food bioactive compounds. *Food Engineering Reviews*, *7*(4), 452–490.

Drosou, C. G., Krokida, M. K., & Biliaderis, C. G. (2017). Encapsulation of bioactive compounds through electrospinning/electrospraying and spray drying: A comparative assessment of food-related applications. *Drying Technology*, *35*(2), 139–162.

Duan, J., Vogt, F. G., Li, X., Hayes, D., & Mansour, H. M. (2013). Design, characterization, and aerosolization of organic solution advanced spray-dried moxifloxacin and ofloxacin dipalmitoylphosphatidylcholine (DPPC) microparticulate/nanoparticulate powders for pulmonary inhalation aerosol delivery. *International Journal of Nanomedicine*, *8*, 3489–3505.

Edris, A., & Bergnstahl, B. (2001). Encapsulation of orange oil in a spray dried double emulsion. *Nahrung/Food*, *45*, 133–137.

Encina, C., Vergara, C., Giménez, B., Oyarzún-Ampuero, F., & Robert, P. (2016). Conventional spray-drying and future trends for the microencapsulation of fish oil. *Trends in Food Science & Technology*, *56*, 46–60.

Fabra, M. J., López-Rubio, A., & Lagaron, J. M. (2014). On the use of different hydrocolloids as electrospun adhesive interlayers to enhance the barrier properties of polyhydroxyalkanoates of interest in fully renewable food packaging concepts. *Food Hydrocolloids*, *39*, 77–84.

Fang, Z., & Bhandari, B. (2010). Encapsulation of polyphenols—A review. *Trends in Food Science & Technology*, *21*(10), 510–523.

Fang, Z., & Bhandari, B. (2012). Encapsulation techniques for food ingredient systems. In B. Bhandarj & Y.H. Roos (Eds.), *Food Materials Science and Engineering* (pp. 320–348). Wiley-Blackwell, Chichester, West Sussex, UK.

Farjami, T., & Madadlou, A. (2017). Fabrication of biopolymer microgel and microgel-based hydrogels. *Food Hydrocolloids*, *62*, 262–272.

Fernandes, R. V. de B., Borges, S. V., Botrel, D. A., Silva, E. K., da Costa, J. M. G., & Queiroz, F. (2013). Microencapsulation of rosemary essential oil: Characterization of particles. *Drying Technology*, *31*(11), 1245–1254.

Fernandes, R. V. de B., Botrel, D. A., Silva, E. K., Borges, S. V., de Oliveira, C. R., Yoshida, M. I., & de Paula, R. C. M. (2016). Cashew gum and inulin: New alternative for ginger essential oil microencapsulation. *Carbohydrate Polymers*, *153*, 133–142.

Fernandes, R. V. de B., Silva, E. K., Borges, S. V., de Oliveira, C. R., Yoshida, M. I., da Silva, Y. F., & Botrel, D. A. (2017). Proposing novel encapsulating matrices for spray-dried ginger essential oil from the whey protein isolate-inulin/maltodextrin blends. *Food and Bioprocess Technology*, *10*, 115–130.

Filipović-Grcić, J., Škalko-Basnet, N., & Jalsenjak, I. (2001). Mucoadhesive chitosan-coated liposomes: Characteristics and stability. *Journal of Microencapsulation*, *18*, 3–12.

Frezard, F. (1999). Liposomes: From biophysics to the design of peptide vaccines. *Brazilian Journal of Medical and Biological Research*, *32*, 181–189.

Fu, Y., Sarkar, P., Bhunia, A. K., & Yao, Y. (2016). Delivery systems of antimicrobial compounds to food. *Trends in Food Science & Technology*, *57(Part A)*, 165–177.

Fuchs, M., Turchiuli, C., Bohin, M., Cuvelier, M. E., Ordonnaud, C., Peyrat-Maillard, M. N., & Dumoulin, E. (2006). Encapsulation of oil in powder using spray drying and fluidised bed agglomeration. *Journal of Food Engineering*, *75*(1), 27–35.

Gharsallaoui, A., Roudaut, G., Chambin, O., Voilley, A., & Saurel, R. (2007). Applications of spray-drying in microencapsulation of food ingredients: An overview. *Food Research International*, *40*(9), 1107–1121.

Ghorani, B., & Tucker, N. (2015). Fundamentals of electrospinning as a novel delivery vehicle for bioactive compounds in food nanotechnology. *Food Hydrocolloids*, *51*, 227–240.

Ghorbanzade, T., Jafari, S. M., Akhavan, S., & Hadavi, R. (2017). Nano-encapsulation of fish oil in nano-liposomes and its application in fortification of yogurt. *Food Chemistry*, *216*, 146–152.

Gomez-Estaca, J., Comunian, T. A., Montero, P., Ferro-Furtado, R., & Favaro-Trindade, C. S. (2016). Encapsulation of an astaxanthin-containing lipid extract from shrimp waste by complex coacervation using a novel gelatin–cashew gum complex. *Food Hydrocolloids*, *61*, 155–162.

Gouin, S. (2004). Microencapsulation: Industrial appraisal of existing technologies and trends. *Trends in Food Science & Technology*, *15,* 330–347.

Gul, K., Singh, A. K., & Jabeen, R. (2016). Nutraceuticals and functional foods: The foods for the future world. *Critical Reviews in Food Science and Nutrition*, *56*(16), 2617–2627.

Heinzen, C., Berger, A., & Marison, I. (2004). Use of vibration technology for jet break-up for encapsulation of cells and liquids in monodisperse microcapsules. In V. Nedovic & R. Willaert (Eds.), *Fundamentals of Cell Immobilization Biotechnology*, Vol. 8A (pp. 257–275). Kluwer Academic Publisher, Dordrecht, the Netherlands.

Hu, L., Zhang, J., Hu, Q., Gao, N., Wang, S., Sun, Y., & Yang, X. (2016). Microencapsulation of brucea javanica oil: Characterization, stability and optimization of spray drying conditions. *Journal of Drug Delivery Science and Technology*, *36*, 46–54.

Jafari, S. M., Assadpoor, E., He, Y., & Bandari, B. (2008). Encapsulation efficiency of food flavours and oils during spraying drying. *Drying Technology*, *26*, 816–835.

Juarez Tomas, M. S., De Gregorio, P. R., Leccese Terraf, M. C., & Nader-Macias, M. E. (2015). Encapsulation and subsequent freeze-drying of *Lactobacillus reuteri* CRL 1324 for its potential inclusion in vaginal probiotic formulations. *European Journal of Pharmaceutical Sciences*, *79*, 87–95.

Jun-xia, X., Hai-yan, Y., & Jian, Y. (2011). Microencapsulation of sweet orange oil by complex coacervation with soybean protein isolate/gum arabic. *Food Chemistry*, *125*(4), 1267–1272.

Kfoury, M., Lounès-Hadj Sahraoui, A., Bourdon, N., Laruelle, F., Fontaine, J., Auezova, L., & Fourmentin, S. (2016). Solubility, photostability and antifungal activity of phenylpropanoids encapsulated in cyclodextrins. *Food Chemistry*, *196*, 518–525.

Kheadr, E. E., Vuillemard, J. C., & El-Deeb, S. A. (2003). Impact of liposome-encapsulated enzyme cocktails in cheddar cheese ripening. *Food Research International*, *36*, 241–252.

Kim, H. Y., & Baianu, I. C. (1991). Novel liposome microencapsulation techniques for food applications. *Trends in Food Science & Technology*, *2*, 55–61.

Kirby, C. (1991). Microencapsulation and controlled delivery of food ingredients. *Food Science and Technology Today*, *5*, 74–78.

Klose, R. E. (1992). Encapsulated bioactive substances. PCT WO 92/21249.

Krasaekoopt, W., Bhandari, B., & Deeth, H. (2003). Evaluation of encapsulation techniques of probiotics for yoghurt. *International Dairy Journal, 13*, 3–13.

Krishnan, S., Kshirsagar, A. C., & Singhal, R. S. (2005). The use of gum Arabic and modified starch in the microencapsulation of a food flavoring agent. *Carbohydrate Polymers, 62*, 309–315.

Kuang, S. S., Oliveira, J. C., & Crean, A. M. (2010). Microencapsulation as a tool for incorporating bioactive ingredients into food. *Critical Reviews in Food Science and Nutrition, 50*(10), 951–968.

Kurkov, S. V., & Loftsson, T. (2013). Cyclodextrins. *International Journal of Pharmaceutics, 453*(1), 167–180.

Laloy, E., Vuillemard, J. C., Dufour, P., & Simard, R. (1998). Release of enzymes from liposomes during cheese ripening. *Journal of Controlled Release, 54*, 213–222.

Langrish, T. A. G., & Premarajah, R. (2013). Antioxidant capacity of spray-dried plant extracts: Experiments and simulations. *Advanced Powder Technology, 24*, 771–779.

Lee, D. H., Jin, B. H., Hwang, Y. I., & Lee, S. C. (2000). Encapsulation of bromelain in liposome. *Journal of Food Science and Nutrition, 5*, 81–85.

Li, D., & Xia, Y. (2004). Electrospinning of nanofibers: Reinventing the wheel? *Advance Materials, 16*, 1151–1170.

Liu, N., & Park, H. J. (2010). Factors effect on the loading efficiency of vitamin C loaded chitosan-coated nanoliposomes. *Colloids and Surfaces B: Biointerfaces, 76*, 16–19.

Loksuwan, J. (2007). Characteristics of micro-encapsulated β-carotene formed by spray drying with modified tapioca starch, native tapioca starch and maltodextrin. *Food Hydrocolloids, 21*, 928–935.

López-Rubio, A., & Lagaron, J. M. (2012). Whey protein capsules obtained through electrospraying for the encapsulation of bioactives. *Innovative Food Science & Emerging Technologies, 13*, 200–206.

López-Rubio, A., Sanchez, E., Wilkanowicz, S., Sanz, Y., & Lagaron, J. M. (2012). Electrospinning as a useful technique for the encapsulation of living bifidobacteria in food hydrocolloids. *Food Hydrocolloids, 28*(1), 159–167.

Luo, R.-C., & Chen, C.-H. (2012). Structured microgels through microfluidic assembly and their biomedical applications. *Soft, 1*, 1–23.

Madziva, H., Kailasapathy, K., & Phillips, M. (2005). Alginate–pectin microcapsules as a potential for folic acid delivery in foods. *Journal of Microencapsulation, 22*(4), 343–351.

Mao, L., Roos, Y. H., Biliaderis, C. G., & Miao, S. (2017). Food emulsions as delivery systems of flavor compounds: A review. *Critical Reviews in Food Science and Nutrition 57*, 3173–3187.

Marefati, A., Sjöö, M., Timgren, A., Dejmek, P., & Rayner, M. (2015). Fabrication of encapsulated oil powders from starch granule stabilized W/O/W Pickering emulsions by freeze-drying. *Food Hydrocolloids, 51*, 261–271.

Marques, H. M. C. (2010). A review on cyclodextrin encapsulation of essential oils and volatiles. *Flavour and Fragrance Journal, 25*(5), 313–326.

Marquis, M., Davy, J., Cathala, B., Fang, A., & Renard, D. (2015). Microfluidics assisted generation of innovative polysaccharide hydrogel microparticles. *Carbohydrate Polymers, 116*, 189–199.

Marsanasco, M., Márquez, A. L., Wagner, J. R., Alonso, S. V., & Chiaramoni, N. S. (2011). Liposomes as vehicles for vitamins E and C: An alternative to fortify orange juice and offer vitamin C protection after heat treatment. *Food Research International, 44*, 3039–3046.

McClements, D. J. (2005). *Food emulsions: Principles, practices and techniques.* CRC Press, Boca Raton, FL.

Meng, X., Pan, Q., & Liu, Y. (2012). Preparation and properties of phytosterols with hydroxy-propyl β-cyclodextrin inclusion complexes. *European Food Research and Technology*, *235*(6), 1039–1047.

Mi, F.-L., Wong, T.-B., Shyu, S.-S., & Chang, S. F. (1999). Chitosan microspheres: Modification of polymeric chem-physical properties of spray-dried microspheres to control the release of antibiotic drug. *Journal of Applied Polymer Science*, *71*, 747–759.

Moschakis, T., Murray, B. S., & Biliaderis, C. G. (2010). Modifications in stability and structure of whey protein-coated o/w emulsions by interacting chitosan and gum arabic mixed dispersions. *Food Hydrocolloids*, *24*, 8–17.

Mourtzinos, I., Salta, F., Yannakopoulou, K., Chiou, A., & Karathanos, V. T. (2007). Encapsulation of olive leaf extract in β-cyclodextrin. *Journal of Agricultural and Food Chemistry*, *55*(20), 8088–8094.

Mozafari, M. R., Khosravi-Darani, K., Borazan, G. G., Cui, J., Pardakhty, A., & Yurdugul, S. (2008). Encapsulation of food ingredients using nanoliposome technology. *International Journal of Food Properties*, *11*(4), 833–844.

Murali, S., Kar, A., Mohapatra, D., & Kalia, P. (2014). Encapsulation of black carrot juice using spray and freeze drying. *Food Science and Technology International*, *21*(8), 604–612.

Mutka, J. R., & Nelson, D. B. (1988). Preparation of encapsulated flavors with high flavor level. *Food Technology*, *42*, 154–157.

Nakayama, S., Kimura, Y., Miki, S., Oshitani, J., Kobayashi, T., Adachi, S., & Imamura, K. (2015). Influence of sugar surfactant structure on the encapsulation of oil droplets in an amorphous sugar matrix during freeze-drying. *Food Research International*, *70*, 143–149.

Neo, Y. P., Ray, S., Jin, J., Gizdavic-Nikolaidis, M., Nieuwoudt, M. K., Liu, D., & Quek, S. Y. (2013). Encapsulation of food grade antioxidant in natural biopolymer by electro-spinning technique: A physicochemical study based on zein–gallic acid system. *Food Chemistry*, *136*(2), 1013–1021.

New, R. (1990). *Liposomes—A practical approach*. IRL Press, Oxford, U.K.

Nieuwland, M., Geerdink, P., Brier, P., van den Eijnden, P., Henket, J. T. M. M., Langelaan, M. L. P., & Martin, A. H. (2013). Food-grade electrospinning of proteins. *Innovative Food Science & Emerging Technologies*, *20*, 269–275.

Oh, J. K., Drumright, R., Siegwart, D. J., & Matyjaszewski, K. (2008). The development of microgels/nanogels for drug delivery applications. *Progress in Polymer Science*, *33*, 448–477.

Oh, J. K., Lee, D.I., & Park, J. M. (2009). Biopolymer-based microgels/nanogels for drug delivery. *Progress in Polymer Science*, *34*, 1261–1282.

Okuro, P. K., de Matos, F. E., & Favaro-Trindade, C. S. (2013a). Technological challenges for spray chilling encapsulation of functional food ingredients. *Food Technology and Biotechnology*, *51*(2), 171–182.

Okuro, P. K., Thomazini, M., Balieiro, J. C. C., Liberal, R. D. C. O., & Fávaro-Trindade, C. S. (2013b). Co-encapsulation of *Lactobacillus acidophilus* with inulin or polydextrose in solid lipid microparticles provides protection and improves stability. *Food Research International*, *53*(1), 96–103.

Ortega-Rivas, E. (2005). Handling and processing of food powders and particulates. In C. Onwulata (Ed.), *Encapsulated and powdered foods* (pp. 75–144). CRC Press, Boca Raton, FL.

Osorio-Tobón, J. F., Silva, E. K., & Meireles, M. A. A. (2016). 3—Nanoencapsulation of flavors and aromas by emerging technologies. In A. Grumezescu, (Ed.), *Encapsulations* (pp. 89–126), Academic Press, London, UK.

Oxley, J. D. (2012). Spray cooling and spray chilling for food ingredient and nutraceutical encapsulation. In N. Garti & D. J. McClements (Eds.), *Encapsulation technologies and delivery systems for food ingredients and nutraceuticals* (pp. 110–130). WoodHead Publishing Ltd., Oxford, U.K.

Pacifico, C. J., Wu, W. H., & Fraley, M. (2001). U.S. Patent 6,251,478 B1.

Pasukamonset, P., Kwon, O., & Adisakwattana, S. (2016). Alginate-based encapsulation of polyphenols from *Clitoria ternatea* petal flower extract enhances stability and biological activity under simulated gastrointestinal conditions. *Food Hydrocolloids, 61*, 772–779.

Pothakamury, U. R., & Barbosa-Cánovas, G. V. (1995). Fundamental aspects of controlled release in foods. *Trends in Food Science & Technology, 6*, 397–406.

Prusse, U., Fox, B., Kirchhoff, M., Bruske, F., Breford, J., & Vorlop, K. D. (1998). New process (jet cutting method) for the production of spherical beads from highly viscous polymer solutions. *Chemical Engineering Technology, 21*, 29–33.

Ramadan, M. F. (2012). Antioxidant characteristics of phenolipids (quercetin-enriched lecithin) in lipid matrices. *Industrial Crops and Products, 36*(1), 363–369.

Ramírez, M. J., Giraldo, G. I., & Orrego, C. E. (2015). Modeling and stability of polyphenol in spray-dried and freeze-dried fruit encapsulates. *Powder Technology, 277*, 89–96.

Rashidinejad, A., Birch, E. J., Sun-Waterhouse, D., & Everett, D. W. (2014). Delivery of green tea catechin and epigallocatechin gallate in liposomes incorporated into low-fat hard cheese. *Food Chemistry, 156*, 176–183.

Ray, S., Raychaudhuri, U., & Chakraborty, R. (2016). An overview of encapsulation of active compounds used in food products by drying technology. *Food Bioscience, 13*, 76–83. doi: 10.1016/j.fbio.2015.12.009.

Rigon, R. T., & Zapata Noreña, C. P. (2016). Microencapsulation by spray-drying of bioactive compounds extracted from blackberry (*Rubus fruticosus*). *Journal of Food Science and Technology, 53*(3), 1515–1524.

Risch, S. J., & Reineccius, G. A. (1988). Flavor encapsulation. In *ACS Symposium Series*, vol. 370, American Chemical Society, Washington, DC.

Rodríguez, J., Martín, M. J., Ruiz, M. A., & Clares, B. (2016). Current encapsulation strategies for bioactive oils: From alimentary to pharmaceutical perspectives. *Food Research International, 83*, 41–59.

Rutz, J. K., Borges, C. D., Zambiazi, R. C., Crizel-Cardozo, M. M., Kuck, L. S., & Noreña, C. P. Z. (2017). Microencapsulation of palm oil by complex coacervation for application in food systems. *Food Chemistry, 220*, 59–66.

Sánchez, F. M., García, F., Calvo, P., Bernalte, M. J., & González-Gómez, D. (2016). Optimization of broccoli microencapsulation process by complex coacervation using response surface methodology. *Innovative Food Science & Emerging Technologies, 34*, 243–249.

Santiago, L. G., & Castro, G. R. (2016). Novel technologies for the encapsulation of bioactive food compounds. *Current Opinion in Food Science, 7*, 78–85.

Santos, E. H., Kamimura, J. A., Hill, L. E., & Gomes, C. L. (2015a). Characterization of carvacrol beta-cyclodextrin inclusion complexes as delivery systems for antibacterial and antioxidant applications. *LWT—Food Science and Technology, 60*(1), 583–592.

Santos, M. G., Bozza, F. T., Thomazini, M., & Favaro-Trindade, C. S. (2015b). Microencapsulation of xylitol by double emulsion followed by complex coacervation. *Food Chemistry, 171*, 32–39.

Schell, D., & Beermann, C. (2014). Fluidized bed microencapsulation of *Lactobacillus reuteri* with sweet whey and shellac for improved acid resistance and in-vitro gastro-intestinal survival. *Food Research International, 62*, 308–314.

Sharma, A., & Sharma, U. S. (1997). Liposomes in drug delivery: Progress and limitations. *International Journal of Pharmaceutics, 154*(2), 123–140.

Shen, L., Chen, J., Bai, Y., Ma, Z., Huang, J., & Feng, W. (2016). Physical properties and stabilization of microcapsules containing thyme oil by complex coacervation. *Journal of Food Science, 81*(9), N2258–N2262.

Shewan, H. M., & Stokes, J. R. (2013). Review of techniques to manufacture micro-hydrogel particles for the food industry and their applications. *Journal of Food Engineering, 119*, 781–792.

Shimoni, E., Lesmes, U., & Ungar, Y. (2007). Non-covalent complexes of bioactive agents with starch for oral delivery. Patents WO2007122624 and US 2006794110.

Silva, E. K., Azevedo, V. M., Cunha, R. L., Hubinger, M. D., & Meireles, M. A. A. (2016). Ultrasound-assisted encapsulation of annatto seed oil: Whey protein isolate versus modified starch. *Food Hydrocolloids, 56*, 71–83.

Singh, H., Thompson, A., Liu, W., & Corredig, M. (2012). Liposomes as food ingredients and nutraceutical delivery systems. In N. Garti & D. J. McClements (Eds.), *Encapsulation technologies and delivery systems for food ingredients and nutraceuticals* (pp. 287–318). WoodHead Publishing Ltd., Oxford, U.K.

Smith, A. M., Jaime-Fonseca, M., Grover, L. M., & Bakalis, S. (2010). Alginate-loaded liposomes can protect encapsulated alkaline phosphatase functionality when exposed to gastric pH. *Journal of Agricultural and Food Chemistry, 58*, 4719–4724.

Soares, M. J., Murhadi, L. L., Kurpad, A. V., Chan She Ping-Delfos, W. L., & Piers, L. S. (2012). Mechanistic roles for calcium and vitamin D in the regulation of body weight. *Obesity Reviews, 13*(7), 592–605.

Tackenberg, M. W., Krauss, R., Schuchmann, H. P., & Kleinebudde, P. (2015). Encapsulation of orange terpenes investigating a plasticisation extrusion process. *Journal of Microencapsulation, 32*(4), 408–417.

Tsai, W.-C., & Rizvi, S. S. H. (2016). Liposomal microencapsulation using the conventional methods and novel supercritical fluid processes. *Trends in Food Science & Technology, 55*, 61–71.

van Lengerich, B. H. (2001). Encapsulation of sensitive components into a matrix to obtain discrete shelf-stable particles. PCT WO 2001025414 A1.

Vestland, T. L., Jacobsen, Ø., Sande, S. A., Myrset, A. H., & Klaveness, J. (2015). Compactible powders of omega-3 and β-cyclodextrin. *Food Chemistry, 185*, 151–158.

Vieira, M. G. A., da Silva, M. A., dos Santos, L. O., & Beppu, M. M. (2011). Natural-based plasticizers and biopolymer films: A review. *European Polymer Journal, 47*(3), 254–263.

Wang, I. Y., Ma, G. H., & Su, Z. G. (2005). Preparation of uniform sized chitosan microspheres by membrane emulsification technique and application as a carrier of protein drug. *Journal of Controlled Release, 106*, 62–75.

Whelehan, M., & Marison, I. W. (2011). Microencapsulation using vibrating technology. *Journal of Microencapsulation, 28*, 669–688.

Wongsasulak, S., Kit, K. M., McClements, D. J., Yoovidhya, T., & Weiss, J. (2007). The effect of solution properties on the morphology of ultrafine electrospun egg albumen–PEO composite fibers. *Polymer, 48*(2), 448–457.

Wongsasulak, S., Pathumban, S., & Yoovidhya, T. (2014). Effect of entrapped α-tocopherol on mucoadhesivity and evaluation of the release, degradation, and swelling characteristics of zein–chitosan composite electrospun fibers. *Journal of Food Engineering, 120*, 110–117.

Wu, W. H., Roe, W. S., Gimino, V. G., Seriburi, V., Martin, D. E., & Knapp, S. E. (2002). Low melt encapsulation. PCT QO 00/74499.

Xiao, Z., Liu, W., Zhu, G., Zhou, R., & Niu, Y. (2014). A review of the preparation and application of flavour and essential oils microcapsules based on complex coacervation technology. *Journal of the Science of Food and Agriculture, 94*(8), 1482–1494.

Youssefi, S., Emam-Djomeh, Z., & Mousavi, S. M. (2009). Comparison of artificial neural network (ANN) and response surface methodology (RSM) in the prediction of quality parameters of spray-dried pomegranate juice. *Drying Technology, 27*(7–8), 910–917.

Yuliani, S., Bhandary, B., Rutgers, R., & D' Arcy, B. (2004). Application of microencapsulated flavor to extrusion product. *Food Reviews International, 20*, 163–185.

Zhao, W., Chao, J., Du, R., & Huang, S. (2011). Spectroscopic studies on the inclusion behavior between caffenic acid and γ-cyclodextrin. *Journal of Inclusion Phenomena and Macrocyclic Chemistry, 71*(1), 25–34.

Zhao, X.-H., & Tang, C.-H. (2016). Spray-drying microencapsulation of CoQ10 in olive oil for enhanced water dispersion, stability and bioaccessibility: Influence of type of emulsifiers and/or wall materials. *Food Hydrocolloids, 61*, 20–30.

Zuidam, N. J., & Shimoni, E. (2010). Overview of microencapsulates for use in food products or processes and methods to make them. In N. J. Zuidam & V. Nedovic (Eds.), *Encapsulation technologies for active food ingredients and food processing* (pp. 3–29). Springer, New York.

2 Freeze Drying and Microwave Freeze Drying as Encapsulation Methods

*Vasiliki P. Oikonomopoulou
and Magdalini K. Krokida*

CONTENTS

2.1 INTRODUCTION

Drying is a physical process used for the removal of water from various materials. Drying processes are used for the reduction of weight and volume of the products, helping with storage and transportation, and reducing the relative costs (Maroulis and Saravacos 2003, Marques et al. 2006). In food industry, drying was firstly applied for the dehydration of vegetables and nowadays its use has been extended to a wide variety of products (Barbosa-Canovas and Vega-Mercado 1996). Drying contributes to the extension of products' shelf-life. The reduction of water activity (below 0.5) limits the development of pathogens and spoilage microorganisms and reduces the enzymatic activity and chemical reactions inside the materials. Apart from being used as a dehydration and preservation method, drying can also be applied for the protection of bioactive compounds by their encapsulation in a stable matrix.

Encapsulation is used to protect and control the release of sensitive materials, such as flavors, probiotics, vitamins, antioxidants, colorants, enzymes, etc. (Chen et al. 2013). It is applied in various industrial sectors, such as food, nutraceutical, and pharmaceutical industries, for protecting the encapsulated compounds from adverse environmental conditions and controlling their release during transportation, storage, and consumption (Barbosa-Canovas et al. 2005, Fang and Bhandari 2012). The encapsulated materials can, however, face various problems, such as permanent aggregation, chemical instability, and leakage of the encapsulated compounds. In that context, drying can overcome these problems and contribute to the preservation of the stability of encapsulated products, as well as the controlled release of the bioactive substances (Chacon et al. 1999, Anandharamakrishnan 2014). Various drying methods are widely used as encapsulation methods, such as freeze drying, spray drying, fluidized bed drying, etc. This chapter is dedicated to the description and analysis of the freeze drying method that has a wide industrial application as encapsulation method and is used for the production of products with high value and functionality, as well as the newly proposed microwave freeze drying method.

2.2 FREEZE DRYING PROCESS

2.2.1 INTRODUCTION

Freeze drying was firstly used, in the 1940s, for the production of dry plasma and afterward for drying antibiotics and biological materials. It was introduced in the food industry in the early 1950s to produce high-quality products, due to the low temperatures and pressures applied. Nowadays, it is applied to various high-value products, such as foods (coffee or tea extracts, fruits, vegetables, meat, fish, etc.), biomaterials, pharmaceuticals, and other thermolabile materials (Barbosa-Canovas and Vega-Mercado 1996, Marques and Freire 2005, Barbosa-Canovas et al. 2005, Nastaj and Witkiewicz 2011).

The basic principle of the freeze drying process is based on the sublimation of ice of a frozen product. The low temperatures applied, the absence of liquid water, as well as the absence of oxygen, due to the low vacuum, decrease destabilization and degradation due to oxidation or chemical interactions and preserve the physical, chemical, and biological properties of the raw materials (Barbosa-Canovas and Vega-Mercado 1996, Ratti 2013). Freeze-dried final products contain extremely low levels of moisture (1–4%). Concerning food products, freeze drying is ideal for the elongation of their shelf-life, making them able to be stored for long periods (Barbosa-Canovas and Vega-Mercado 1996, Marques and Freire 2005, Barbosa-Canovas et al. 2005, Nastaj and Witkiewicz 2011). Their low moisture and water activity limits the deterioration reactions, such as nonenzymatic browning, protein denaturation, and enzymatic reactions, which cause degradation of the product. Low temperatures also help in the preservation of nutrients, color, aroma, taste, and biological activity (Krokida et al. 1998, Duan et al. 2010a, Oikonomopoulou et al. 2011).

Freeze drying leads to the maintenance of initial shape and dimensions of the products, when compared with conventional methods, such as hot air drying (Nastaj and Witkiewicz 2011). The low mobility of ice limits shrinkage phenomena and leads to the formation of a cake structure with many pores, which is referred to as "sponge." The high porosity offers the ability of quick and almost complete rehydration, resulting in products that recover their initial structure and appear similar to the raw materials (Barbosa-Canovas and Vega-Mercado 1996, Barbosa-Canovas et al. 2005, Bonazzi and Dumoulin 2011). As a result, freeze-dried products are considered superior to those dried using conventional techniques (Ratti 2001).

2.2.2 PRINCIPLES OF FREEZE DRYING

The freeze drying process is carried out in three stages (Krokida and Maroulis 1997, Krokida et al. 1998, Anandharamakrishnan 2014):

1. The product is first frozen (freezing).
2. Ice is removed by sublimation (primary drying) due to the low temperatures and pressures applied, directly from the solid to the gas phase. The material is simultaneously cooled, since the latent heat of sublimation is required for the sublimation of ice.
3. Finally, the unfrozen bound water is removed by desorption (secondary drying).

2.2.2.1 Freezing Process

The first step of freeze drying process is the freezing of the product. Prior to drying, the material is frozen at low temperatures and must be kept below its triple point. The water is converted into ice, and some soluble solid components and lipids are crystallized. Freezing temperature is reduced until eutectic solutions are also crystallized. Freezing separates aqueous solutions in two parts, the ice and the freeze concentrate

(Roos 1995), which solidifies between the ice crystals (Ratti 2013). The unfrozen water containing amorphous solutes is in a solid state and must be maintained below a critical temperature to avoid melting or structure collapse during drying (Fang and Bhandari 2012). Freezing rate cannot be easily calculated, since the properties and the ratio of frozen and unfrozen regions vary with temperature (Geankoplis 1993). Many models have been developed, and Plank's model (Plank 1913) is the most widely used, despite its low accuracy (Barbosa-Canovas et al. 2005). Plank's model is given by the equation

$$t = \frac{\rho \Delta H_f}{T - T_a} \left(\frac{Pd}{h} + \frac{Rd^2}{\lambda} \right) \tag{2.1}$$

where
   t is the freezing time (s)
   T is the freezing temperature (K)
   T_a is the temperature of the environment (K)
   d is the material thickness (m)
   ΔH_f is the latent heat of fusion (kJ/kg)
   ρ is the density of the unfrozen region (kg/m^3)
   h is the heat transfer coefficient (J/(s·m^2·K))
   λ is the thermal conductivity (J/(s·m·K))
   P, R are constants depending on the geometry of the material

The application of this model depends on the latent heat and the thermal conductivity values while it does not consider the variation in temperature and the time for heat removal (Heldman 1992, López-Leiva and Hallström 2003, Barbosa-Canovas et al. 2005).

Freezing can take place outside or inside the freeze drying chamber (Ratti 2013) and can be carried out in two modes; either the material is in direct contact with the cooling device or the material is divided from the cooling device with an obstacle and freezing is carried out in indirect contact (Heldman 1992, Barbosa-Canovas et al. 2005). The freezing process affects the properties and the quality of the products, as well as the primary drying rate.

2.2.2.2 Drying

During drying, ice sublimation is carried out below the triple point and the applied pressure maintains the water pressure at this condition (Nastaj and Witkiewicz 2011). This condition is necessary for succeeding sublimation and not evaporation of the water (Roos 1995).

Sublimation is based on the vapor pressure difference between the ice in the inner part of the material and the drying chamber. Heat is supplied to convert ice into vapor (Ratti 2013). For small quantities of material, the required heat is given either by the material or the environment inside the chamber, whereas for large amounts, the required heat is given with heating. At the beginning, the drying rate is high, due to the limited resistance in heat and mass flow. As drying progresses the rate decreases. A dry layer is formed in the material that increases during drying and restricts heat and mass flow (Barbosa-Canovas and Vega-Mercado 1996, Barbosa-Canovas et al. 2005). The drying rate can be increased using an inert gas, such as nitrogen, inside the drying chamber (Zouboulis et al. 2009). Drying is carried out in two stages, the primary and secondary drying stages.

2.2.2.2.1 Primary Drying (Sublimation)

The primary drying stage is characterized by the sublimation of ice. The pressure reduces, and when the required heat of sublimation is supplied, the ice starts to sublime from the surface of the material and the sublimation progresses in the inner part. Due to the low pressure, the vapor is removed from the surface without causing melting of ice. During freeze drying, the energy is given to the material mainly by radiation and conduction from the heated plate, where convection is negligible due to increased vacuum (Ratti 2013). Mass transfer occurs mainly by diffusion of vapor from the sublimation front to the drying chamber. Heat and mass transfer occur simultaneously (Geankoplis 1993). The sublimation occurs due to the difference between the water vapor pressure at the sublimation front and the water vapor pressure of the environment inside the dryer (Barbosa-Canovas and Vega-Mercado 1996, Barbosa-Canovas et al. 2005).

The sublimation front starts at the material surface and ice is removed from the surface or close to it. The front continues to the center of the material, creating pores or gaps. The water vapors that are sublimed pass through the pores to the dryer. The size of these pores depends on the size of ice crystals formed during freezing, as well as the drying conditions, with the assumption that no collapse of the product happens (Michailidis et al. 2008). As mentioned, at the beginning of drying, the rate is high. However, the formation of a dry layer at the surface of the material, with increasing thickness, reduces the drying rate (Barbosa-Canovas et al. 2005, Michailidis et al. 2008, Ratti 2013). The increase of the thickness of this dry layer during drying causes the decrease of mass transfer, due to the decrease of water diffusion from the sublimation front to the material surface (Barbosa-Canovas and Vega-Mercado 1996). The development of the dry layer can be avoided using rotating chambers.

The increase of temperature causes the increase of drying rate. The temperature should be higher than the ice temperature, in order to offer the required heat, but not too high to cause ice melting, thermal degradation, and product deterioration. The vapor pressure in the ice front should also be higher than in the dryer, in order to cause vapor flow from the inner part to the material surface (Geankoplis 1993).

At the end of primary drying, 85–95% of the water has sublimed. The properties and structure of the final dried products depend on the temperature and pressure applied during drying. The formation of porous structure depends on the prevention of collapse phenomena during drying. To avoid collapse, the temperature of the product should be lower than its collapse temperature, which is related to the glass transition temperature of the product (Ratti 2013). If the temperature is higher than the glass transition temperature of the material, the mobility inside the material increases, leading to reduction of viscosity. At this condition, the material cannot support its own weight, and the cell walls collapse. The temperature at which structure collapse occurs decreases with the decrement of molecular weight and the increment of moisture content (Anandharamakrishnan 2014).

2.2.2.2.2 Secondary Drying (Desorption)

At the end of primary drying, all the ice has sublimed and the moisture is attributed to the unfrozen bound water. The removal of the unfrozen bound water takes place by desorption and at the end of this stage, the water content of the final product is extremely low (1–4%). This stage may start before the completion of primary drying, but it is predominant at the end of sublimation (Ratti 2013) and lasts one third of the whole drying time (Anandharamakrishnan 2014). Desorption is accomplished with the increase of temperature in a controlled way, which still remains below the glass transition temperature of the material, and the decrease of the vapor pressure (Barbosa-Canovas and Vega-Mercado 1996, Michailidis et al. 2008). The pressure is the driving force for the water vapor desorption (Morais et al. 2016).

The water vapor that is removed is then transferred to a condenser and condensates as ice. The condenser is used to avoid the return of vapors to the material and their transfer to the vacuum pump, causing corrosion of the pump and decrease of the drying rate due to the reduction of the vacuum (Geankoplis 1993, Barbosa-Canovas et al. 2005). At the end of drying, the ice in the condenser is defrosted with the use of hot water, hot air, steam, or heated jackets and removed (Barbosa-Canovas et al. 2005).

2.2.3 FREEZE DRYING PROCESS PARAMETERS

The quality of freeze-dried products is influenced by two main parameters: the freezing temperature and the applied pressure during the process.

2.2.3.1 Freezing Temperature

Freezing affects the quality, texture, and properties of the dried products. The freezing temperature controls the size of the ice crystals that are formed during freezing, which subsequently affects the structure, the color, and the aroma of the final dried products (Land 1991).

Freezing is usually conducted with 0.5 to 3.0 cm/h freezing rate (Barbosa-Canovas et al. 2005). At low freezing temperatures, ice nuclei are formed quickly and the ice development is slow, resulting in the formation of many ice crystals with small diameters. The sublimation of these ice crystals leads to dried products with a large amount of small pores. Under these conditions, the mechanical stresses are low, and the collapse of structure is avoided; however, it can result in greater aggregation (Fang and Bhandari 2012). On the other hand, at higher freezing temperatures, slow freezing is favored and a lower amount of larger crystals is created. The sublimation of these ice crystals leads to dried products with larger pores but causes damage of the cell walls.

The selection of the freezing temperature depends on the properties of raw materials and the desired characteristics of the final products (Barbosa-Canovas et al. 2005). Specifically, for foods and biomaterials, freezing temperatures are in the range of −50°C to −80°C, to avoid damage of cell walls (Nastaj and Witkiewicz 2011). Sometimes, an annealing process (maintenance at a specific subfreezing temperature for a specific period) can be applied to control the size of ice crystals. Annealing is used in order to enhance crystallization and promote ice crystal size, in order to create pores and gaps to the material facilitating water vapor transfer during drying (Morgan et al. 2006, Fang and Bhandari 2012, Ratti 2013). Another parameter that affects the growth of ice crystals is the degree of supercooling (difference between the equilibrium freezing point and the temperature of ice nucleation) (Ratti 2013). In recent years, other techniques, such as mechanical agitation, ultrasonic control, pressure shift freezing, and application of electric and magnetic fields have been applied in order to control the size of ice crystals (Rhim et al. 2011).

2.2.3.2 Freeze Drying Pressure

The collapse of structure during drying depends on the viscosity of the solids in the material. The viscoelastic properties of the materials are influenced by the pressure and temperature during drying, and, as a result, the ice crystals that sublime create pores or gaps with variable characteristics. If the temperature is higher than the glass transition temperature of the material, the amorphous parts become soft and the molecular movement is increased. Then, the melting of ice is more intense, leading to plasticization of the matrix and decrease in the material's viscosity. The material cannot support its own weight and collapses. The collapse is greater when the temperature difference between the drying chamber and the glass transition of the material is higher. On the other hand, at temperatures below the glass transition, the mobility is low, and the high viscosity of the concentrated amorphous solution prevents shrinkage. As a result, during freeze drying the temperature is maintained below or close to the glass transition temperature of the material, leading to limited shrinkage and highly porous structure (Barbosa-Canovas and Vega-Mercado 1996, Krokida and Maroulis 1997, Krokida et al. 1998, Oikonomopoulou et al. 2011).

2.2.4 Types of Freeze Dryers

Freeze dryers can be classified into the following types (Barbosa-Canovas et al. 2005, Nireesha et al. 2013).

2.2.4.1 Batch Freeze Dryers

In batch dryers, many vessels containing the product are placed in a dryer. The frozen material is placed onto flat trays inside the chamber. Heat is provided by radiation and conduction and both drying stages (primary and secondary) take place, leading to low moisture levels. Batch drying is commonly used in pharmaceutical industry.

2.2.4.2 Continuous Freeze Dryers

Continuous freeze dryers are of lower cost than batch dryers. They are classified into

1. Static (tray) dryers: The product is placed on static trays that move through the dryer.
2. Dynamic dryers (trayless): The product is moved through the dryer using belt conveyors, circular or vibrating plates, etc.

2.2.4.3 Improvements on Freeze Dryers

The increased cost of the freeze drying process compared with other drying methods has resulted in the development of various approaches used in connection with freeze drying, such as spray freeze drying, adsorption freeze drying, fluid-bed freeze drying, microwave freeze drying, etc. (Ratti 2001, Claussen et al. 2007, Fang and Bhandari 2012, Ratti 2013).

For example, spray freeze drying is a novel process for producing high-quality products. It is carried out in three steps: Firstly, the atomization of a liquid takes place to form droplets, then the droplets are frozen, and finally the ice is sublimed. In this process, the heat and mass transfer are improved and the dried product is characterized by advanced quality. The particles produced have controlled size distribution and high specific surface area, leading to enhanced characteristics and reconstitution properties (Mumenthaler and Leuenberger 1991, Fang and Bhandari 2012). Adsorption freeze drying uses an absorbent for condensation, resulting in reduction of operational costs, however, the quality is lower than conventional freeze-dried products (Ratti 2013).

In addition, microwave freeze drying can be applied to various products, such as foods and pharmaceuticals, to decrease the cost, energy consumption, and the drying time. This process is described extendedly in Section 2.3.

2.2.5 APPLICATION OF FREEZE DRYING

Freeze drying can be applied in various industrial sectors. It is used in food industry, for the dehydration of coffee and for the production of instant coffee, for the dehydration of some aromatic herbs, as well as the dehydration of fruits and vegetables used in the preparation of instant foods such as soups, cereals, snacks, ice creams, and confectionery creams (Hammami and René 1997). Freeze-dried foods are widely consumed by astronauts and hikers due to their low volume and weight. Apart from the food industry, freeze drying is also used for the production of dry plasma and pharmaceutical supplements, as well as for the preservation of wooden materials and other materials of organic origin (textiles) that are placed in the water. In some cases, it is used for the recovery of books and files that have been destroyed due to floods, as well as for the production of porous materials that are used in various industrial applications, such as catalysts or adsorbents (Zouboulis et al. 2009). Finally, nowadays, freeze drying is successfully used as a gentle encapsulation technique, as described in the following sections, for the preservation of thermolabile materials like proteins, enzymes, microorganisms, flavors, etc., that have wide use.

2.3 MICROWAVE FREEZE DRYING PROCESS

2.3.1 PRINCIPLES

Freeze drying leads to products with high quality, however, it is an expensive and energy consuming method and requires a long time (Ratti 2001). As mentioned, many methods have been used to overcome these drawbacks. Microwave freeze drying is one of them that has been widely studied since the mid-twentieth century (Nastaj and Witkiewicz 2011). Microwave freeze drying is a microwave-assisted drying method operating below the freezing point of water (Gaukel et al. 2017). It is considered a fourth-generation drying method (Vega-Mercado et al. 2001) and can be applied to various products, such as foods (seafood, foamed milk, skimmed milk, meat, solid soups, fruits, vegetables) and pharmaceuticals, to achieve a quick dehydration with low cost and low energy consumption compared to the conventional freeze drying method (Duan et al. 2010a). In addition, the quality of the products produced is high. During drying, the drying rate depends on the low heat transfer rate to the frozen regions. Microwaves can enhance heat and mass transfer (Sunderland 1980, Nastaj and Witkiewicz 2011), reducing the drying time by up to 75% (Duan et al. 2010a).

The microwave freeze drying principle is based on the penetration of microwaves into the ice and the volumetric heating of the material without the need for a thermal gradient. As a result, the problem of heat conduction at the dried layer is being overcome (Ang et al. 1977). Frozen regions are heated and the energy required for sublimation is conducted to them (Sunderland 1980). Microwaves can be effectively applied under the vacuum, reducing freeze drying time (Zhang et al. 2007, Duan et al. 2010b). Microwave heating uses electromagnetic waves of high frequencies (300 MHz to 300 GHz). The molecular dipoles of samples absorb the microwaves and are quickly heated (Barbosa-Canovas and Vega-Mercado 1996, Barbosa-Canovas et al. 2005, Duan et al. 2010a). In the conventional freeze drying process, drying starts from the surface and the sublimation front is progressing to the inner part of the material leading to increased drying periods. In microwave freeze drying, heat is generated throughout the whole volume of the product and sublimation occurs from the whole volume, increasing the drying rate. Typical microwave freeze drying periods are between 1 to 4 h compared to conventional freeze drying methods that can take from 20 to 60 h.

Microwave freeze drying leads to products with advanced quality that are superior to freeze-dried products. The short duration of the secondary stage leads to retention of volatiles (Duan et al. 2010a). The products can easily dissolve in aqueous solutions and have good texture. The cost of the equipment is higher than the freeze drying method but not too high, so it can be used instead of conventional technologies (Nastaj and Witkiewicz 2011).

2.3.2 Microwave Freeze Drying Process Parameters

The supplied energy and the efficiency of the microwave freeze drying process is based on various parameters, such as the characteristics and dielectric properties of the products, the microwave intensity, and the drying pressure (Toledo 1991, Zhang et al. 2006, Nastaj and Witkiewicz 2011).

2.3.2.1 Microwave Power

The drying rate increases with the increase of microwave power (Wray and Ramaswamy 2015). Wang et al. (2009) showed that drying time decreased with the increase of electric field and a very low microwave power led to long drying periods.

2.3.2.2 Pressure

Increased pressure during microwave freeze drying leads to slower drying rate (Wray and Ramaswamy 2015). Low pressures are mainly used to avoid ice melting, permitting use of higher microwave powers to reduce drying time. However, very low pressures are not useful. In addition, higher pressures increase the probability of plasma discharge (Ang et al. 1977).

2.3.2.3   Sample Characteristics

The drying rate increases with the decrease of material thickness (Wray and Ramaswamy 2015). Thick materials have higher resistance to heat and mass flow, decreasing drying rate (Wang et al. 2009). Small samples are preferred in microwave freeze drying since they are easily dried, using higher microwave powers (Ang et al. 1977). Microwaves cannot penetrate thick materials and the sublimation heat is supplied by conduction, which decreases the drying rate (Wang et al. 2009). In addition, microwaves have different depths of penetration regarding the dielectric properties of the samples. Based on the dielectric properties, a changed microwave intensity can lead to advanced product quality and reduced drying time (Michailidis and Krokida 2014).

2.3.3   Types of Microwave Freeze Dryers

Microwave freeze dryers can be classified into two types (Duan et al. 2010a). In the first, freeze drying is conducted simultaneously with the microwave process. The heat is supplied by a microwave field and the freeze drying rate is enhanced. The cost of the equipment is not too high compared with the conventional method (Duan et al. 2010a, Jiang et al. 2011). In the second type, freeze drying and microwave application are carried out in two separate steps. Freeze drying is used to reach moisture levels between 20% and 40%, and then the remaining water is removed with microwave drying (Barbosa-Canovas and Vega-Mercado 1996, Barbosa-Canovas et al. 2005). This method is conducted in two dryers. The moisture content of the material that is moved to the second dryer (Chou and Chua 2001) affects the quality of the product (Duan et al. 2010a).

2.3.4   Advantages and Disadvantages of Microwave Freeze Drying Process

Microwave freeze drying has several advantages in comparison with the freeze drying method. Microwave freeze drying is faster, the cost is low, the energy consumed is decreased, and it leads to increased production volumes (Owusu-Ansah 1991). Microwaves heat the product volumetrically throughout its whole volume, enhancing the process. In addition, microwave freeze drying has the capability of selective heating as well as quick dissipation of energy through the material (Feng and Tang 1998, Duan et al. 2010a).

On the other hand, the distribution of microwaves is not uniform, leading to varying temperature inside the material, causing ice melting, overheating, and quality deterioration (Zhang et al. 2006). In addition, plasma can occur in the chamber under high vacuum, which leads to energy losses and degradation of the material (Lombraña et al. 2001). To avoid plasma discharge, chamber pressure and microwave power should be optimized. Finally, the temperature of the materials cannot be easily measured during drying since microwaves affect the signal of the thermocouples (Duan et al. 2010a,b).

2.4 DRYING PROCESSES AS ENCAPSULATION METHODS

2.4.1 INTRODUCTION

In recent years, there has been an increasing demand for the development of functional and health promoting products that contain bioactive ingredients and nutrients, such as vitamins, minerals, antioxidants, flavors, colorants, enzymes, etc. These ingredients are, however, not stable and can degrade during processing, long-term storage, and consumption, losing their properties and functionality (Fang and Bhandari 2012). As a result, encapsulation methods have been developed to protect these substances from chemical interactions with other components, adverse environmental conditions, and also to enhance their controlled release.

Encapsulation is the process of coating or entrapping an active substance that is in a solid, liquid, or gas form (core material), within a more stable material (wall material) (Chen et al. 2013). The encapsulation process is used in various industrial sectors, such as food, nutraceutical, and pharmaceutical industries, offering controlled release of the encapsulated compounds, protection from the environmental conditions (temperature, moisture, oxidation, and light), masking of flavors, etc. (Barbosa-Canovas et al. 2005, Fang and Bhandari 2012). Encapsulated products are characterized by various parameters, such as shape, particle size, particle size distribution, wall thickness, and morphology. Particles of various forms, such as spheres with uniform thickness, particles containing a core with irregular shape, etc., can be produced depending on the core and wall materials and the method used. The characteristics of the produced particles influence the encapsulation efficiency, stability, and release rate of the bioactive substances (Arshady 1993, Poshadri and Kuna 2010).

Various encapsulation techniques have been developed, including extrusion, electrohydrodynamic processes, spray drying, spray chilling and spray cooling, freeze drying, coacervation, air suspension coating, co-crystallization, etc. (Barbosa-Canovas et al. 2005).

However, the encapsulated materials can face various problems: they might aggregate, be chemically unstable, and leak the encapsulated substance. To overcome these difficulties, moisture should be removed and products should be dried (Anandharamakrishnan 2014). In addition, in the case of food products (soup mixes, beverages, etc.), the flavoring agents must be added in a dry form (Barbosa-Canovas et al. 2005). The use of drying processes as encapsulation methods has been of growing interest in the last years. Several drying techniques, such as freeze drying, spray drying, fluidized bed coating, etc., are used to incorporate bioactive substances in various products, enhancing their stability and controlled release and protecting them from degradation (Ray et al. 2016). Drying methods as encapsulation techniques can be applied either alone or in connection with other encapsulation methods.

Drying techniques are simple, economic, and offer high stability due to the low moisture content of the final particles. In addition, they offer good dissolution in aqueous solutions (Fang and Bhandari 2012, Anandharamakrishnan 2014). Prior to drying, emulsions, dispersions, or suspensions with the active substances are created. Their properties (type of wall material, ratio of wall/core, viscosity, droplet size, type and concentration of emulsifiers, etc.) affect the properties of the final products. The drying process and conditions affect the matrix characteristics and the stability of the encapsulated substance (Chen et al. 2013). The wall materials selected must be non-toxic and cost effective, must have advanced emulsion properties, and be compatible with the core substance (Fang and Bhandari 2012, Ezhilarasi et al. 2013b).

Among the drying methods, freeze drying is a widely used drying method for encapsulation that leads to high-quality products (Schwegman 2009, Chen et al. 2013).

2.4.2 Freeze Drying as Encapsulation Method

2.4.2.1 Principles of Freeze Drying Encapsulation

Freeze drying is a simple and mild technique for the encapsulation of various ingredients, including aromatic materials (water-soluble extracts and natural aromas), lipids, vitamins, antioxidants, probiotics, as well as drugs and nutraceuticals (Desai and Jin Park 2005, Fang and Bhandari 2012). The basic steps for freeze drying encapsulation are (Fang and Bhandari 2010, Zuidam and Shimoni 2010, Ray et al. 2016) the following:

- Core materials are dissolved, homogenized, dispersed, or emulsified in wall materials. The emulsification step can be performed using several devices, such as mixers, pressurized or ultrasonic homogenizers, etc. (Malacrida et al. 2015).
- The mixture is frozen.
- The frozen mixture is freeze dried under vacuum to produce a porous structure.
- The produced material is ground, if necessary.

Freeze drying is suitable for producing microcapsules (>20 µm), as well as nano-capsules (below 400 nm) and lead to increased encapsulation efficiency (about 70%) (Zuidam and Shimoni 2010, Ezhilarasi et al. 2013b, Anandharamakrishnan 2014). This method leads to products with advanced quality since freeze drying encapsulation is carried out under vacuum and low temperatures, as well as with increased stability of the encapsulated substance (Fang and Bhandari 2012, Murali et al. 2014). Due to the low temperatures applied, this method is suitable for thermolabile materials (Chen et al. 2013). Freeze drying encapsulation can provide protection to the core material from oxidation, limit its interaction with other components, and control its release rate (Pothakamury and Barbosa-Cánovas 1995, Mahfoudhi and Hamdi 2015). In addition, freeze drying preserves the shape and structure of the particles since they are formed during freezing (Garza Sáenz 2013). Freeze drying can also be used for the separation of nanoparticles formed by other nanoencapsulation processes, as well as for the production of powdered encapsulated materials (Ezhilarasi et al. 2013b).

The properties of freeze-dried encapsulated materials depend on various parameters. The encapsulation efficiency, the stability during storage, and the controlled release of the encapsulated compounds are influenced by the characteristics of wall and core material, their compatibility, and the emulsification technique (Miao et al. 2008, Mehrnoush et al. 2011, Santagapita et al. 2011, Anandharamakrishnan 2014, Ray et al. 2016). The selection of the core and wall materials depends on several factors, such as the preservation of their properties after the end of the process, the stability during storage, and the desirable release rate (Fang and Bhandari 2012).

Various materials can be used as matrices, such as proteins, emulsifying starches, Arabic gum, gelatin, corn syrup, maltodextrins, disaccharides, pullulan, etc. (Onwulata et al. 1995, Kim et al. 2000, Hogan et al. 2001, Mahfoudhi and Hamdi 2015). For food applications, the wall materials must be food-grade, biodegradable, resistant to the gastrointestinal tract, have emulsifying properties, and protect the core from the external environment (Ramírez et al. 2015).

The quality and size of the produced particles are influenced by the freezing rate (Ezhilarasi et al. 2013b), as well as the pressure and temperature during freeze drying (Fang and Bhandari 2012). The formation of ice crystals during freezing affects emulsion properties. During freezing, ice is separated from other solids and the concentrated droplets can start to aggregate during drying. Cryoprotectants (monosaccharides, disaccharides, polysaccharides, etc.) can protect the system from this condition. Cryoprotectants prevent the stresses that occur during freezing and drying that can lead to physical and chemical damage of the product. The concentration of cryoprotectant depends on the composition of the system and the freezing rate. Cryoprotectants can also be used to stabilize sensitive bioactive compounds, such as probiotics, enzymes, or liposomes. Other materials, such as buffers (phosphate, citrate, etc.) and salts (sodium chloride, potassium chloride), are added as protectants (Zuidam et al. 2003, Morgan et al. 2006, Fang and Bhandari 2012, Ezhilarasi et al. 2013b, Anandharamakrishnan 2014, Morais et al. 2016, Ray et al. 2016). The solubility, type, and concentration of cryoprotectants influence the final properties of the particles.

The shelf-life of freeze-dried products depends on storage temperature. Freeze-dried products are characterized by increased stability when the storage is performed at temperatures below their glass transition temperature (Roy and Gupta 2004).

2.4.2.2 Uses of Freeze Drying Encapsulation

Freeze drying is used for the encapsulation of various flavors, probiotics, enzymes, oils, antioxidants, and other bioactive compounds providing protection from degradation and adverse environmental conditions, such as light, moisture, and oxygen. Table 2.1 presents examples of the application of freeze drying for encapsulation.

In details, freeze drying can be used for the encapsulation of *enzymes* for enhancing their stability, controlling their release during storage and consumption, and improving their catalytic properties. Cryoprotectants can be used to protect enzymes from deactivation (Fang and Bhandari 2012). The efficiency of the enzymes can be affected by the drying conditions, the raw materials characteristics, the final moisture content, and the storage conditions (Liu et al. 2011).

TABLE 2.1

Studies on Freeze Drying Process for Encapsulation

Wall Material	Core Material	Effect	References
Maltodextrin DE20	Pomace extracts	Increased stability during storage and protection of the substance	Delgado-Vargas et al. (2000), Francis and Markakis (1989)
Maltodextrin, Arabic gum	Grape juice powder	Enhancement of encapsulation of grape juice, increased stability during storage	Gurak et al. (2013)
Almond gum or Arabic gum	β-Carotene	Almond gum led to higher protection of β-carotene, powders could color homogenously the food system	Mahfoudhi and Hamdi (2015)
Pullulan	Anthocyanin extract	Color degradation was delayed, good antiradical activity during storage	Gradinaru et al. (2003)
Maltodextrin DE5-8, DE18.5	Polyphenol-rich raspberry extract	Increased stability, protection to oxidation during storage, preservation of antioxidant activity	Laine et al. (2008)
Maltodextrin DE10, DE20, DE30	Anthocyanins from black rice bran	High encapsulation efficiency and total anthocyanin content	Laokuldilok and Kanha (2016)
Cress seed gum (CSG), Arabic gum (AG), maltodextrin (MD) DE20, DE7	Saffron petal's extract	AG-MDE20 mixture presented the highest anthocyanin content	Jafari et al. (2016), Mahdavee Khazaei et al. (2014)
Gum ghatti, Arabic gum, and soy protein isolate (SPI)	α-Linolenic acid rich oil	0.32 ratio gum/SPI, 2 ratio wall/core, 10,000 rpm homogenization speed led to maximum encapsulation efficiency	Naik et al. (2014)
Maltodextrin	Extracts of olive kernel and leaves	82.39–92.12% and 87.98–91.06% encapsulation efficiency for olive kernel and leaves, respectively	Chanioti et al. (2016)
Whey protein isolate (WPI), maltodextrin, their combination (1:1 ratio)	*Garcinia cowa* fruit extract	Particle size of 15–100 μm, WPI exhibited better encapsulation efficiency	Ezhilarasi et al. (2013a)
Maltodextrin DE10	Red wine polyphenols	No loss of total polyphenols, free-flowing powder, stability during storage	Sanchez et al. (2013)

(*Continued*)

TABLE 2.1 (*Continued*)
Studies on Freeze Drying Process for Encapsulation

Wall Material	Core Material	Effect	References
Modified starch, gelatin	Turmeric oleoresin	30% modified starch, 1% gelatin, mechanical homogenization led to best encapsulation efficiency	Malacrida et al. (2015)
Whey protein isolate (WPI), soluble corn fiber (SCF)	Fish oil co-encapsulated with phytosterol esters and limonene	Good redispersion properties	Chen et al. (2013)
Chickpea or lentil protein isolate and maltodextrin DE9, DE18	Flaxseed oil	Increase in oil concentration resulted in increase in diameter and decrease in encapsulation efficiency, protection against oxidation during storage	Karaca et al. (2013)
Sodium caseinate, maltodextrin, carboxymethyl cellulose, lecithin	Walnut oil or extra-virgin olive oil	1:1.5 ratio oil/wall material and maltodextrin, carboxymethyl cellulose, and lecithin resulted in highest encapsulation yield	Calvo et al. (2011, 2012)
Gelatin/sucrose	Oregano essential oil	High antioxidant and antimicrobial activity	Beirão da Costa et al. (2012)
Lactose, maltodextrin	Fish oil	Protection to oxidation	Heinzelmann and Franke (1999)
Chitosan, modified starch, maltodextrin DE21	Fennel essential oil	Chitosan was more efficient with good encapsulation ability	Chranioti and Tzia (2014)
Maltodextrin (MD) and Arabic gum (AG)	Polyphenols	100% AG and 10–20% encapsulant or 80–100% MD, and freezing rates higher than 0.65°C/min led to higher gallic acid content	Ramírez et al. (2015)
Skim milk	Sodium alginate with or without inulin	Inulin enhanced the viability and functionality of *L. plantarum*	Wang et al. (2016)
Pea protein isolate-alginate hydrogel	*Lactobacillus casei* ATCC 393	High yield of 85.69%	Xu et al. (2016)
Chitosan, xanthan, β-cyclodextrin	Gallic acid	No loss of antioxidant activity	da Rosa et al. (2013)
β-Cyclodextrin	Blueberry juice	2–20 μm size of microparticles, powders showed good retention values of anthocyanins, high radical scavenging activity	Wilkowska et al. (2016)

(Continued)

TABLE 2.1 (*Continued*)
Studies on Freeze Drying Process for Encapsulation

Wall Material	Core Material	Effect	References
Maltodextrin, tween80	Brown seaweed pigments	Yellowish green powder pigments, high encapsulation efficiency	Indrawatia et al. (2015)
Conventional liposomes, polyethylene glycol (PEG)-coated proliposomes (PLP), tween80	Vitamin E	164 nm mean diameter and 84% encapsulation efficiency for PEG-coated lyophilized PLP, better stability for vitamin E in PLP	Zhao et al. (2011)
Zein, zein/chitosan complex, tween20	Tocopherol	Spherical nanocapsules, 200–800 nm particle size, 77–87% encapsulation efficiency	Luo et al. (2011)
Solid lipid nanoparticles, dioctyl sodium sulfosuccinate, poloxamer 188, glyceryl monostearate	Curcuminoids	Spherical particles with 450 nm mean diameter and up to 70% incorporation efficacy, prolonged release of curcuminoids (up to 12 h), stability during storage	Tiyaboonchai et al. (2007)
Poly(ε-caprolactone) (PCL), polyvinyl alcohol (PVA)	Miglyol 829 oil	PCL could be freeze dried without a cryoprotectant for 5% concentration of PVA, annealing enhanced sublimation	Abdelwahed et al. (2006)
PCL, gelatin, pluronic F68	Capsicum oleoresin	Less than 200 nm mean diameter of nanocapsules	Nakagawa et al. (2011)
Chitosan– tripolyphosphate	(+) Catechin and (−) epigallocatechin gallate	Less than 200 nm mean diameter of nanocapsules, increased protection	Dube et al. (2010)
PCL, β-cyclodextrin, pluronic F68, ethyl acetate	Fish oil	High encapsulation efficiency (99%), low leakage, less than 200 nm mean diameter, prevention of oxidation, odor masking	Choi et al. (2010)
Maltodextrin, sucrose	Blends of corn syrup solids and sugars in the aroma solution	Preservation of approximately 75% of the aroma volatiles	Kopelman et al. (1977)
Gelatin, sucrose, Arabic gum (AG)	Limonene	AG resulted in high encapsulation efficiency, high gelatin resulted in structure collapse	Kaushik and Roos (2007)

(*Continued*)

TABLE 2.1 (*Continued*)
Studies on Freeze Drying Process for Encapsulation

Wall Material	Core Material	Effect	References
Arabic gum, α-cyclodextrin	Pear aroma	High stability	Tobitsuka et al. (2006)
Native potato starch (NP), β-cyclodextrin (β-CD), maltodextrin (MD), acid-treated potato starch (ATPS), succinated potato starch (SPS)	D-Limonene	SPS showed highest encapsulation yield (93.35%)	Lee et al. (2009)
Maltodextrin, citric acid ester, chitosan	Flavor oil	Small amount of chitosan resulted in improved retention levels, 270–300 nm average particle size	Kaasgaard and Keller (2010)
Arabic gum, maltodextrin, calcium chloride	Serine protease	Good protection, 92% encapsulation yield	Mehrnoush et al. (2011)
Colloidal suspension of chitosan and xanthan gum blend	Firefly luciferase (enzyme)	Addition of nanoclay lowered the enzyme release rate	Liu et al. (2011)
Poly(lactic-co-glycolic acid)	Curcumin	Spherical shape with 264 nm mean diameter and 77% efficiency	Shaikh et al. (2009)
Chitosan-g-poly(N-vinylcaprolactam)	Curcumin	Nanocapsules were spherical with a size range of 180–220 nm	Sanoj Rejinold et al. (2011)

In addition, freeze drying is widely used to increase the stability of biological and sensitive materials (Goderska 2012) and has been used for various *probiotics*, such as *Bifidobacterium longum* (Amine et al. 2014), *Lactobacillus reuteri* (Khoramnia et al. 2011), *B. animalis* (Dianawati and Shah 2011), and *L. plantarum* (Montel Mendoza et al. 2014, Wang et al. 2016). The principle for encapsulation of probiotics using freeze drying is based firstly on freezing the probiotic and the wall material and then subjecting the mixture to freeze drying. Fast freezing is recommended for freeze drying of probiotics (Fang and Bhandari 2012). However, drying may cause damage to the cell wall and structural changes, leading to destabilization of the dried probiotic. For that reason, several agents, such as cryoprotectants, amino acids, gums, skim milk, etc., and mixtures of them, can be used to protect the probiotic during processing and storage (Carvalho et al. 2004, Saarela et al. 2006, Miao et al. 2008, Goderska 2012). For example, the addition of glucose improved the survival of *L. rhamnosus GG* and the addition of sugars or sugar alcohols was necessary for optimal survival of *L. plantarum*, and

L. rhamnosus during storage (Goderska 2012, Ying et al. 2012). *Prebiotics* can be also encapsulated together with probiotics to enhance their functionality.

Encapsulation of *flavors* is also widely applied and helps in their retention, stability, and functionality during processing and storage, in their protection from interactions with other compounds and environmental conditions (light, oxidation, etc.), in the increase of their shelf-life, as well as in their controlled release (Reineccius 1991, Madene et al. 2006, Chranioti and Tzia 2014). The retention of flavors is achieved by the low temperatures and pressures applied. The nature and composition of the materials determine their stability (Kopelman et al. 1977, Shahidi and Han 1993, Desai and Jin Park 2005). The volatile stability decreases with the increase of molecular weight of carbohydrates and the decrease of soluble solids (Shahidi and Han 1993).

The principle for flavor freeze drying encapsulation is based firstly on the formation of an emulsion or a dispersion of the bioactive substance and the wall material. The emulsion is then frozen and freeze dried and the bioactive substance is enclosed to the matrix (Kaushik and Roos 2008, Fang and Bhandari 2012). During freezing and drying, regions where volatiles and wall material are mostly concentrated are formed. High freezing temperatures result in more concentrated regions, restricting volatile losses during drying (Flink and Karel 1970, Garza Sáenz 2013). Wang et al. (2007) studied the aroma encapsulation efficiency of banana powders produced by belt drying, freeze drying, and vacuum drying and showed that freeze-dried banana powders had the best aroma, followed by vacuum dried.

The freezing of an emulsion can cause many physicochemical processes (fat crystallization, ice formation, etc.) (Walstra 2003). Protective agents can be added to limit coalescence during drying (Marefati et al. 2015). Emulsions with small particle sizes and increased stability enhance the stability of the encapsulated materials (Liu et al. 2001, Soottitantawat et al. 2003).

Freeze drying encapsulation has also been used for *phytochemicals*. The low temperatures applied and ice form of water lead to final products with advanced quality limiting their degradation (Chanioti et al. 2016). The principle of phytochemicals encapsulation is based on emulsification and freeze drying (Naik et al. 2014). This method is also used for the encapsulation of several *oils*, such as fish, flaxseed, walnut, olive oil, etc. Oils are firstly dissolved in water, frozen, and then freeze dried (Bakry et al. 2016).

2.4.2.3 Advantages and Disadvantages of Freeze Drying Encapsulation

The major advantages of freeze drying encapsulation are (Ezhilarasi et al. 2013b, Ray et al. 2016)

- Use of low temperatures that allow for the dehydration and encapsulation of sensitive ingredients
- Advanced quality of the final products
- Preservation of initial shape, texture, and color
- Preservation of biological activity of the dried products
- Particle size in the micro- and nanoscale leading to high encapsulation efficiency and increased stability
- Controlled release of the encapsulated material

The disadvantages are

* Increased cost due to the use of high energy, the use of vacuum, and high processing time (higher than 20 h). Freeze drying costs 30–50 times higher than spray drying for the encapsulation of carotenoids (Desobry et al. 1997) and 4–7 times for the encapsulation of probiotics (Chávez and Ledeboer 2007, Fang and Bhandari 2012). The cost depends on the raw materials, the final products, etc. (Sunderland 1980). The high energy costs are due to the freezing step, the sublimation heat, and the condensation of water vapor.
* The porous structure leads to limited protection of the encapsulated material when elongated release is required (Barbosa-Canovas et al. 2005, Gharsallaoui et al. 2007, Zuidam and Shimoni 2010, Fang and Bhandari 2012).

2.4.3 MICROWAVE FREEZE DRYING AS ENCAPSULATION METHOD

In order to overcome the main drawbacks of freeze drying encapsulation and decrease the cost and time, several modified freeze drying methods, such as microwave-assisted freeze drying, have been developed (Fang and Bhandari 2012). Studies on microwave encapsulation combined with freeze drying have not yet been published, to our knowledge; however, there is a strong potential for application due to the various advantages it offers.

Microwaves can be applied for producing microcapsules when the core and the wall materials have different dielectric properties. Microwaves heat the core material with the higher dielectric constant, blending the wall material and creating capsules (Abbasi and Rahimi 2008).

Microwave-assisted encapsulation has been used for the encapsulation of bioactive compounds, leading to products with high quality, low water activities, and increased stability. Anthocyanins were encapsulated in Arabic gum and maltodextrin leading to powders with advanced quality (Mohd Nawi et al. 2015). In addition, the use of microwave vacuum drying can lead to the production of flavors with increased stability compared to the conventional spray drying method (Owusu-Ansah 1991). Crude protein powders from *Ginkgo biloba* L. were produced optimizing microwave freeze drying parameters (Fan et al. 2012).

2.5 CONCLUSIONS

Various methods are applied for the encapsulation of active substances. The use of drying methods for encapsulation is of growing interest in the last years, due to the stability and advanced properties they offer. Freeze drying is a commonly used method for the encapsulation of various heat sensitive ingredients, including aromas, lipids, antioxidants, probiotics, etc., as well as drugs and nutraceuticals and can be applied in food, nutraceutical, and pharmaceutical industries. The selection of the wall material, as well as freezing and drying conditions, depends on the desired properties of the final product, the storage conditions, and applications. In addition, in order to reduce the operation cost and increase the process rate, microwave freeze drying encapsulation could be applied.

REFERENCES

Abbasi, S., and S. Rahimi. 2008. Microwave-assisted encapsulation of citric acid using hydrocolloids. *International Journal of Food Science & Technology* 43(7):1226–1232. doi: 10.1111/j.1365-2621.2007.01595.x.

Abdelwahed, W., G. Degobert, and H. Fessi. 2006. A pilot study of freeze drying of poly(epsilon-caprolactone) nanocapsules stabilized by poly(vinyl alcohol): Formulation and process optimization. *International Journal of Pharmaceutics* 309(1–2):178–188. doi: http://dx.doi.org/10.1016/j.ijpharm.2005.10.003.

Amine, K.M., C.P. Champagne, S. Salmieri, M. Britten, D. St-Gelais, P. Fustier, and M. Lacroix. 2014. Effect of palmitoylated alginate microencapsulation on viability of *Bifidobacterium longum* during freeze-drying. *LWT—Food Science and Technology* 56(1):111–117. doi: http://dx.doi.org/10.1016/j.lwt.2013.11.003.

Anandharamakrishnan, C. 2014. Drying techniques for nanoencapsulation. In *Techniques for Nanoencapsulation of Food Ingredients*, pp. 51–60. New York: Springer-Verlag.

Ang, T.K., J.D. Ford, and D.C.T. Pei. 1977. Microwave freeze-drying of food: A theoretical investigation. *International Journal of Heat and Mass Transfer* 20(5):517–526. doi: http://dx.doi.org/10.1016/0017-9310(77)90098-9.

Arshady, R. 1993. Microcapsules for food. *Journal of Microencapsulation* 10(4):413–435. doi: 10.3109/02652049309015320.

Bakry, A.M., S. Abbas, B. Ali, H. Majeed, M.Y. Abouelwafa, A. Mousa, and L. Liang. 2016. Microencapsulation of oils: A comprehensive review of benefits, techniques, and applications. *Comprehensive Reviews in Food Science and Food Safety* 15(1):143–182. doi: 10.1111/1541-4337.12179.

Barbosa-Canovas, G.V., and H. Vega-Mercado. 1996. *Dehydration of Foods*. New York: International Thomson Publishing (ITP).

Barbosa-Canovas, G.V., Ortega-Rivas, E., P. Juliano, and H. Yan. 2005. *Food Powders, Physical Properties, Processing, and Functionality*, edited by G.V. Barbosa-Canovas, New York: Kluwer Academic-Plenum Publishers.

Beirão da Costa, S., C. Duarte, A.I. Bourbon, A.C. Pinheiro, A.T. Serra, M. Moldão Martins, M.-I. Nunes Januário, A.A. Vicente, I. Delgadillo, C. Duarte, and M.-L. Beirão da Costa. 2012. Effect of the matrix system in the delivery and in vitro bioactivity of microencapsulated Oregano essential oil. *Journal of Food Engineering* 110(2):190–199. doi: http://dx.doi.org/10.1016/j.jfoodeng.2011.05.043.

Bonazzi, C., and E. Dumoulin. 2011. Quality changes in food materials as influenced by drying processes. In *Modern Drying Technology, Volume 3: Product Quality and Formulation*, edited by E. Tsotsas and A.S. Mujumdar, pp. 1–20.

Calvo, P., Á.-L. Castaño, M.-T. Hernández, and D. González-Gómez. 2011. Effects of microcapsule constitution on the quality of microencapsulated walnut oil. *European Journal of Lipid Science and Technology* 113(10):1273–1280. doi: 10.1002/ejlt.201100039.

Calvo, P., Á.-L. Castaño, M. Lozano, and D. González-Gómez. 2012. Influence of the microencapsulation on the quality parameters and shelf-life of extra-virgin olive oil encapsulated in the presence of BHT and different capsule wall components. *Food Research International* 45(1):256–261. doi: http://dx.doi.org/10.1016/j.foodres.2011.10.036.

Carvalho, A.S., J. Silva, P. Ho, P. Teixeira, F.X. Malcata, and P. Gibbs. 2004. Effects of various sugars added to growth and drying media upon thermotolerance and survival throughout storage of freeze-dried *Lactobacillus delbrueckii* ssp. *Bulgaricus. Biotechnology Progress* 20(1):248–254. doi: 10.1021/bp034165y.

Chacon, M., J. Molpeceres, L. Berges, M. Guzman, and M.R. Aberturas. 1999. Stability and freeze-drying of cyclosporine loaded poly(D,L-lactic-glycolide) carriers. *European Journal of Pharmaceutical Science* 8:99–107.

Chanioti, S., P. Siamandoura, and C. Tzia. 2016. Evaluation of extracts prepared from olive oil by-products using microwave-assisted enzymatic extraction: Effect of encapsulation on the stability of final products. *Waste and Biomass Valorization* 7(4):831–842. doi: 10.1007/s12649-016-9533-1.

Chávez, B.E., and A.M. Ledeboer. 2007. Drying of probiotics: Optimization of formulation and process to enhance storage survival. *Drying Technology* 25(7–8):1193–1201. doi: 10.1080/07373930701438576.

Chen, Q., F. Zhong, J. Wen, D. McGillivray, and S.Y. Quek. 2013. Properties and stability of spray-dried and freeze-dried microcapsules co-encapsulated with fish oil, phytosterol esters, and limonene. *Drying Technology* 31(6):707–716. doi: 10.1080/07373937.2012.755541.

Choi, M.J., U. Ruktanonchai, S.G. Min, J.Y. Chun, and A. Soottitantawat. 2010. Physical characteristics of fish oil encapsulated by ß-cyclodextrin using an aggregation method or polycaprolactone using an emulsion-diffusion method. *Food Chemistry* 119(4):1694–1703.

Chou, S.K., and K.J. Chua. 2001. New hybrid drying technologies for heat sensitive foodstuffs. *Trends in Food Science & Technology* 12(10):359–369. doi: http://dx.doi.org/10.1016/S0924-2244(01)00102-9.

Chranioti, C., and C. Tzia. 2014. Arabic gum mixtures as encapsulating agents of freeze-dried fennel oleoresin products. *Food and Bioprocess Technology* 7(4):1057–1065. doi: 10.1007/s11947-013-1074-z.

Claussen, I.C., T.S. Ustad, I. Strommen, and P.M. Walde. 2007. Atmospheric freeze drying— A review. *Drying Technology* 25(6):947–957. doi: 10.1080/07373930701394845.

da Rosa, C.G., C.D. Borges, R.C. Zambiazi, M.R. Nunes, E.V. Benvenutti, S.R. da Luz, R.F. D'Avila, and J.K. Rutz. 2013. Microencapsulation of gallic acid in chitosan, β-cyclodextrin and xanthan. *Industrial Crops and Products* 46:138–146. doi: http://dx.doi.org/10.1016/j.indcrop.2012.12.053.

Delgado-Vargas, F., A.R. Jimenez, and O. Pardes-Lopez. 2000. Natural pigments: Carotenoids, anthocyanins and betalains-characteristics, biosynthesis, processing and stability. *Critical Reviews in Food Science and Nutrition* 40:173–289.

Desai, K.G.H., and H. Jin Park. 2005. Recent developments in microencapsulation of food ingredients. *Drying Technology* 23(7):1361–1394. doi: 10.1081/DRT-200063478.

Desobry, S.A., F.M. Netto, and T.P. Labuza. 1997. Comparison of spray-drying, drum-drying and freeze-drying for b-carotene encapsulation and preservation. *Journal of Food Science* 62(6):1158–1162.

Dianawati, D., and N.P. Shah. 2011. Enzyme stability of microencapsulated *Bifidobacterium animalis* ssp. *Lactisb B12* after freeze drying and during storage in low water activity at room temperature. *Journal of Food Science* 76(6):M463–M471. doi: 10.1111/j.1750-3841.2011.02246.x.

Duan, X., M. Zhang, A.S. Mujumdar, and R. Wang. 2010a. Trends in microwave-assisted freeze drying of foods. *Drying Technology* 28(4):444–453. doi: 10.1080/07373931003609666.

Duan, X., M. Zhang, A.S. Mujumdar, and S. Wang. 2010b. Microwave freeze drying of sea cucumber (*Stichopus japonicus*). *Journal of Food Engineering* 96(4):491–497. doi: http://dx.doi.org/10.1016/j.jfoodeng.2009.08.031.

Dube, A., K. Ng, J.A. Nicolazzo, and I. Larson. 2010. Effective use of reducing agents and nanoparticle encapsulation in stabilizing catechins in alkaline solution. *Food Chemistry* 122(3):662–667. doi: http://dx.doi.org/10.1016/j.foodchem.2010.03.027.

Ezhilarasi, P.N., D. Indrani, B.S. Jena, and C. Anandharamakrishnan. 2013a. Freeze drying technique for microencapsulation of Garcinia fruit extract and its effect on bread quality. *Journal of Food Engineering* 117(4):513–520. doi: http://dx.doi.org/10.1016/j.jfoodeng.2013.01.009.

Ezhilarasi, P.N., P. Karthik, N. Chhanwal, and C. Anandharamakrishnan. 2013b. Nanoencapsulation techniques for food bioactive components: A review. *Food and Bioprocess Technology* 6(3):628–647. doi: 10.1007/s11947-012-0944-0.

Fan, L., S. Ding, Y. Liu, and L. Ai. 2012. Dehydration of crude protein from *Ginkgo biloba* L. by microwave freeze drying. *International Journal of Biological Macromolecules* 50(4):1008–1010. doi: http://dx.doi.org/10.1016/j.ijbiomac.2012.02.027.

Fang, Z., and B. Bhandari. 2010. Encapsulation of polyphenols—A review. *Trends in Food Science & Technology* 21(10):510–523. doi: http://dx.doi.org/10.1016/j.tifs.2010.08.003.

Fang, Z., and B. Bhandari. 2012. Spray drying, freeze drying and related processes for food ingredient and nutraceutical encapsulation. In *Encapsulation Technologies and Delivery Systems for Food Ingredients and Nutraceuticals*, pp. 73–109. Woodhead Publishing.

Feng, H., and J. Tang. 1998. Microwave finish drying of diced apples in a spouted bed. *Journal of Food Science* 63(4):679–683. doi: 10.1111/j.1365-2621.1998.tb15811.x.

Flink, J., and M. Karel. 1970. Effects of process variables on retention of volatiles in freeze-drying. *Journal of Food Science* 35(4):444–447. doi: 10.1111/j.1365-2621.1970.tb00953.x.

Francis, F.J., and P.C. Markakis. 1989. Food colorants: Anthocyanins. *Critical Reviews in Food Science and Nutrition* 28(4):273–314. doi: 10.1080/10408398909527503.

Garza Sáenz, E.N. 2013. Radiant energy vacuum microwave microencapsulation of natural antimicrobials for a controlled release application in fresh-cut ambrosia apples. Electronic Theses and Dissertations University of British Columbia.

Gaukel, V., T. Siebert, and U. Erle. 2017. Microwave-assisted drying. In *The Microwave Processing of Foods*, edited by M. Regier, K. Knoerzer, and H. Schubert. Woodhead Publishing series in Food Science, Technology and Nutrition, pp. 152–178. Cambridge, UK: Elsevier.

Geankoplis, C.J. 1993. Drying of process materials. In *Transport Processes and Unit Operations*, pp. 520–583. Englewood Cliffs, NJ: Prentice-Hall, Inc.

Gharsallaoui, A., G. Roudaut, O. Chambin, A. Voilley, and R. Saurel. 2007. Applications of spray-drying in microencapsulation of food ingredients: An overview. *Food Research International* 40:1107–1121.

Goderska, K. 2012. Different methods of probiotics stabilization. In *Probiotics*, edited by E.C. Rigobelo. Rijeka, Croatia: InTech.

Gradinaru, G., C.G. Biliaderis, S. Kallithraka, P. Kefalas, and C. Garcia-Viguera. 2003. Thermal stability of *Hibiscus sabdariffa* L. anthocyanins in solution and in solid state: Effects of copigmentation and glass transition. *Food Chemistry* 83(3):423–436. doi: http://dx.doi.org/10.1016/S0308-8146(03)00125-0.

Gurak, P.D., L.-M.-C. Cabral, and M.-H. Rocha-Leão. 2013. Production of grape juice powder obtained by freeze-drying after concentration by reverse osmosis. *Brazilian Archives of Biology and Technology* 56:1011–1017.

Hammami, C., and F. René. 1997. Determination of freeze-drying process variables for strawberries. *Journal of Food Engineering* 32(2):133–154. doi: http://dx.doi.org/10.1016/S0260-8774(97)00023-X.

Heinzelmann, K., and K. Franke. 1999. Using freezing and drying techniques of emulsions for the microencapsulation of fish oil to improve oxidation stability. *Colloids and Surfaces B: Biointerfaces* 12(3–6):223–229. doi: http://dx.doi.org/10.1016/S0927-7765(98)00077-0.

Heldman, D.R. 1992. Food freezing. In *Handbook of Food Engineering*, edited by D.R. Heldman and D.B. Lund, pp. 277–315. New York: Marcel Dekker, Inc.

Hogan, S.A., B.F. McNamee, E.D. O'Riordan, and M. O'Sullivan. 2001. Emulsification and microencapsulation properties of sodium caseinate/carbohydrate blends. *International Dairy Journal* 11(3):137–144. doi: http://dx.doi.org/10.1016/S0958-6946(01)00091-7.

Indrawatia, R., H. Sukowijoyoa, Indriatmokoa, R.D.E. Wijayantib, and L. Limantara. 2015. Encapsulation of brown seaweed pigment by freeze drying: Characterization and its stability during storage *Procedia Chemistry* 14:353–360.

Jafari, S.-M., K. Mahdavi-Khazaei, and A. Hemmati-Kakhki. 2016. Microencapsulation of saffron petal anthocyanins with cress seed gum compared with Arabic gum through freeze drying. *Carbohydrate Polymers* 140:20–25. doi: http://dx.doi.org/10.1016/j.carbpol.2015.11.079.

Jiang, H., M. Zhang, A.S. Mujumdar, and R.-X. Lim. 2011. Comparison of the effect of microwave freeze drying and microwave vacuum drying upon the process and quality characteristics of potato/banana re-structured chips. *International Journal of Food Science and Technology* 46:570–576.

Kaasgaard, T., and D. Keller. 2010. Chitosan coating improves retention and redispersibility of freeze-dried flavor oil emulsions. *Journal of Agricultural and Food Chemistry* 58(4):2446–2454. doi: 10.1021/jf903464s.

Karaca, A.C., M. Nickerson, and N.H. Low. 2013. Microcapsule production employing chickpea or lentil protein isolates and maltodextrin: Physicochemical properties and oxidative protection of encapsulated flaxseed oil. *Food Chemistry* 139(1–4):448–457. doi: http://dx.doi.org/10.1016/j.foodchem.2013.01.040.

Kaushik, V., and Y.H. Roos. 2007. Limonene encapsulation in freeze-drying of gum Arabic–sucrose–gelatin systems. *LWT—Food Science and Technology* 40(8):1381–1391. doi: http://dx.doi.org/10.1016/j.lwt.2006.10.008.

Kaushik, V., and Y.H. Roos. 2008. Lipid encapsulation in glassy matrices of sugar-gelatin systems in freeze-drying. *International Journal of Food Properties* 11(2):363–378. doi: 10.1080/10942910701409278.

Khoramnia, A., N. Abdullah, S.L. Liew, C.C. Sieo, K. Ramasamy, and Y.W. Ho. 2011. Enhancement of viability of a probiotic Lactobacillus strain for poultry during freeze-drying and storage using the response surface methodology. *Animal Science Journal* 82(1):127–135. doi: 10.1111/j.1740-0929.2010.00804.x.

Kim, S.J., G.B. Park, C.B. Kang, S.D. Park, M.Y. Jung, J.O. Kim, and Y.L. Ha. 2000. Improvement of oxidative stability of conjugated linoleic acid (CLA) by microencapsulation in cyclodextrins. *Journal of Agricultural and Food Chemistry* 48(9):3922–3929. doi: 10.1021/jf991215z.

Kopelman, I.J., S. Meydav, and P. Wilmersdorf. 1977. Freeze drying encapsulation of water soluble citrus aroma. *International Journal of Food Science & Technology* 12(1):65–72. doi: 10.1111/j.1365-2621.1977.tb00086.x.

Krokida, M.K., V.T. Karathanos, and Z.B. Maroulis. 1998. Effect of freeze-drying conditions on shrinkage and porosity of dehydrated agricultural products. *Journal of Food Engineering* 35(4):369–380.

Krokida, M.K., and Z.B. Maroulis. 1997. Effect of drying method on shrinkage and porosity. *Drying Technology* 15(10):2441–2458. doi: 10.1080/07373939708917369.

Laine, P., P. Kylli, M. Heinonen, and K. Jouppila. 2008. Storage stability of microencapsulated cloudberry (*Rubus chamaemorus*) phenolics. *Journal of Agricultural and Food Chemistry* 56:11251–11261.

Land, C.M. 1991. *Industrial Drying Equipment: Selection and Application*. New York: Marcel Dekker, Inc.

Laokuldilok, T., and N. Kanha. 2016. Microencapsulation of black glutinous rice anthocyanins using maltodextrins produced from broken rice fraction as wall material by spray drying and freeze drying. *Journal of Food Processing and Preservation* 41:1–10. doi: 10.1111/jfpp.12877.

Lee, S.-W., S.-Y. Kang, S.-H. Han, and C. Rhee. 2009. Influence of modification method and starch concentration on the stability and physical properties of modified potato starch as wall materials. *European Food Research and Technology* 228(3):449–455. doi: 10.1007/s00217-008-0952-5.

Liu, H., K. Nakagawa, D.-I. Kato, D. Chaudhary, and M.O. Tadé. 2011. Enzyme encapsulation in freeze-dried bionanocomposites prepared from chitosan and xanthan gum blend. *Materials Chemistry and Physics* 129(1–2):488–494. doi: http://dx.doi.org/10.1016/j.matchemphys.2011.04.043.

Liu, X.-D., T. Atarashi, T. Furuta, H. Yoshii, S. Aishima, M. Ohkawara, and P. Linko. 2001. Microencapsulation of emulsified hydrophobic flavors by spray drying. *Drying Technology* 19(7):1361–1374. doi: 10.1081/DRT-100105293.

Lombraña, J.I., I. Zuazo, and J. Ikara. 2001. Moisture diffusivity behavior during freeze drying under microwave heating power application. *Drying Technology* 19(8):1613–1627. doi: 10.1081/DRT-100107262.

López-Leiva, M., and B. Hallström. 2003. The original Plank equation and its use in the development of food freezing rate predictions. *Journal of Food Engineering* 58(3):267–275. doi: http://dx.doi.org/10.1016/S0260-8774(02)00385-0.

Luo, Y., B. Zhang, M. Whent, L. Yu, and Q. Wang. 2011. Preparation and characterization of zein/chitosan complex for encapsulation of α-tocopherol, and its *in vitro* controlled release study. *Colloids and Surfaces B: Biointerfaces* 85(2):145–152. doi: http://dx.doi.org/10.1016/j.colsurfb.2011.02.020.

Madene, A., M. Jacquot, J. Scher, and S.A. Desobry. 2006. Flavour encapsulation and controlled release—A review. *International Journal of Food Science & Technology* 41(1): 1–21. doi: 10.1111/j.1365-2621.2005.00980.x.

Mahdavee Khazaei, K., S.M. Jafari, M. Ghorbani, and A. Hemmati Kakhki. 2014. Application of maltodextrin and gum Arabic in microencapsulation of saffron petal's anthocyanins and evaluating their storage stability and color. *Carbohydrate Polymers* 105:57–62. doi: http://dx.doi.org/10.1016/j.carbpol.2014.01.042.

Mahfoudhi, N., and S. Hamdi. 2015. Kinetic degradation and storage stability of β-carotene encapsulated by freeze-drying using almond gum and gum Arabic as wall materials. *Journal of Food Processing and Preservation* 39(6):896–906. doi: 10.1111/jfpp.12302.

Malacrida, C.R., S. Ferreira, L.A.C. Zuanon, and V.R. Nicoletti Telis. 2015. Freeze-drying for microencapsulation of turmeric oleoresin using modified starch and gelatin. *Journal of Food Processing and Preservation* 39(6):1710–1719. doi: 10.1111/jfpp.12402.

Marefati, A., M. Sjöö, A. Timgren, P. Dejmek, and M. Rayner. 2015. Fabrication of encapsulated oil powders from starch granule stabilized W/O/W Pickering emulsions by freeze-drying. *Food Hydrocolloids* 51:261–271. doi: http://dx.doi.org/10.1016/j.foodhyd.2015.04.022.

Maroulis, Z.B., and G.D. Saravacos. 2003. *Food Process Design*, 1st edn. New York: Marcel Dekker, Inc.

Marques, L.G., and J.T. Freire. 2005. Analysis of freeze-drying of tropical fruits. *Drying Technology* 23(9):2169–2184.

Marques, L.G., A.M. Silveira, and J.T. Freire. 2006. Freeze-drying characteristics of tropical fruits. *Drying Technology* 24(4):457–463.

Mehrnoush, A., C.P. Tan, M. Hamed, N.A. Aziz, and T.C. Ling. 2011. Optimisation of freeze drying conditions for purified serine protease from mango (*Mangifera indica* Cv. *Chokanan*) peel. *Food Chemistry* 128(1):158–164. doi: http://dx.doi.org/10.1016/j.foodchem.2011.03.012.

Miao, S., S. Mills, C. Stanton, G.F. Fitzgerald, Y.H. Roos, and R.P. Ross. 2008. Effect of disaccharides on survival during storage of freeze dried probiotics. *Dairy Science & Technology* 88(1):19–30. doi: 10.1051/dst:2007003.

Michailidis, P.A., and M.K. Krokida. 2014. Drying and dehydration processes in food preservation and processing. In *Conventional and Advanced Food Processing Technologies*, edited by S. Bhattacharya, pp. 1–32. Chichester, UK: John Wiley & Sons, Ltd.

Michailidis, P.A., M.K. Krokida, and M.S. Rahman. 2008. Data and models of density, shrinkage, and porosity. In *Food Properties Handbook*, edited by M.S. Rahman. New York: Taylor & Francis Group.

Mohd Nawi, N., I.I. Muhamad, and A. Mohd Marsin. 2015. The physicochemical properties of microwave-assisted encapsulated anthocyanins from *Ipomoea batatas* as affected by different wall materials. *Food Science & Nutrition* 3(2):91–99. doi: 10.1002/fsn3.132.

Montel Mendoza, G., S.E. Pasteris, M.C. Otero, and M.E. Fatima Nader-Macías. 2014. Survival and beneficial properties of lactic acid bacteria from raniculture subjected to freeze-drying and storage. *Journal of Applied Microbiology* 116(1):157–166. doi: 10.1111/jam.12359.

Morais, A.R. do V., É. do N. Alencar, F.H. Xavier Júnior, C.M. de Oliveira, H.R. Marcelino, G. Barratt, H. Fessi, E.S.T. do Egito, and A. Elaissari. 2016. Freeze-drying of emulsified systems: A review. *International Journal of Pharmaceutics* 503(1–2):102–114. doi: http://dx.doi.org/10.1016/j.ijpharm.2016.02.047.

Morgan, C.A., N. Herman, P.A. White, and G. Vesey. 2006. Preservation of micro-organisms by drying: A review. *Journal of Microbiological Methods* 66(2):183–193. doi: http://dx.doi.org/10.1016/j.mimet.2006.02.017.

Mumenthaler, M., and H. Leuenberger. 1991. Atmospheric spray-freeze drying: A suitable alternative in freeze-drying technology. *International Journal of Pharmaceutics* 72(2):97–110. doi: http://dx.doi.org/10.1016/0378-5173(91)90047-R.

Murali, S., A. Kar, D. Mohapatra, and P. Kalia. 2014. Encapsulation of black carrot juice using spray and freeze drying. *Food Science and Technology International* 21(8):604–612.

Naik, A., V. Meda, and S.S. Lele. 2014. Freeze drying for microencapsulation of α-linolenic acid rich oil: A functional ingredient from *Lepidium sativum* seeds. *European Journal of Lipid Science and Technology* 116(7):837–846. doi: 10.1002/ejlt.201300305.

Nakagawa, K., S. Surassmo, S.-G. Min, and M.-J. Choi. 2011. Dispersibility of freeze-dried poly(epsilon-caprolactone) nanocapsules stabilized by gelatin and the effect of freezing. *Journal of Food Engineering* 102(2):177–188. doi: http://dx.doi.org/10.1016/j.jfoodeng.2010.08.017.

Nastaj, J., and K. Witkiewicz. 2011. Experimental and simulation studies of the primary and secondary vacuum freeze drying at microwave heating. In *Advances in Induction and Microwave Heating of Mineral and Organic Materials*, S. Grundas. Rijeka, Croatia: In Tech.

Nireesha, G.R., L. Divya, C. Sowmya, N. Venkateshan, M. Niranjan Babu, and V. Lavakumar. 2013. Lyophilization/freeze drying—An review. *International Journal of Novel Trends in Pharmaceutical Sciences* 3(4):87–98.

Oikonomopoulou, V.P., M.K. Krokida, and V.T. Karathanos. 2011. Structural properties of freeze-dried rice. *Journal of Food Engineering* 107(3–4):326–333. doi: http://dx.doi.org/10.1016/j.jfoodeng.2011.07.009.

Onwulata, C.I., P.W. Smith, and V.H. Holsinger. 1995. Flow and compaction of spray-dried powders of anhydrous butteroil and high melting milkfat encapsulated in disaccharides. *Journal of Food Science* 60(4):836–840. doi: 10.1111/j.1365-2621.1995.tb06242.x.

Owusu-Ansah, Y.J. 1991. Advances in microwave drying of foods and food ingredients. *Canadian Institute of Food Science and Technology Journal* 24(3–4):102–107. doi: http://dx.doi.org/10.1016/S0315-5463(91)70027-8.

Plank, R. 1913. Die Gefrierdauer von Eisblöcken (Freezing times of ice blocks). *Zeitschrift für die gesamte Kälte-Industrie* 20(6):109–114.

Poshadri, A., and A. Kuna. 2010. Microencapsulation technology: A review. *Journal of Research ANGRAU* 38(1):86–102.

Pothakamury, U.R., and G.V. Barbosa-Cánovas. 1995. Fundamental aspects of controlled release in foods. *Trends in Food Science & Technology* 6(12):397–406. doi: http://dx.doi.org/10.1016/S0924-2244(00)89218-3.

Ramírez, M.J., G.I. Giraldo, and C.E. Orrego. 2015. Modeling and stability of polyphenol in spray-dried and freeze-dried fruit encapsulates. *Powder Technology* 277:89–96. doi: http://dx.doi.org/10.1016/j.powtec.2015.02.060.

Ratti, C. 2001. Hot air and freeze-drying of high-value foods: A review. *Journal of Food Engineering* 49(4):311–319.

Ratti, C. 2013. Freeze drying for food powder production. In *Handbook of Food Powders: Processes and Properties*, edited by B. Bhandari, N. Bansal, M. Zhang, and P. Schuck. Philadelphia, PA: Woodhead Publishing Limited.

Ray, S., U. Raychaudhuri, and R. Chakraborty. 2016. An overview of encapsulation of active compounds used in food products by drying technology. *Food Bioscience* 13:76–83. doi: http://dx.doi.org/10.1016/j.fbio.2015.12.009.

Reineccius, G.A. 1991. Carbohydrates for flavor encapsulation. *Food Technology* 454(45):144–147.

Rhim, J.-W., S. Koh, and J.-M. Kim. 2011. Effect of freezing temperature on rehydration and water vapor adsorption characteristics of freeze-dried rice porridge. *Journal of Food Engineering* 104(4):484–491. doi: http://dx.doi.org/10.1016/j.jfoodeng.2010.08.010.

Roos, Y.H. 1995. *Phase Transitions in Foods*, 1st edn. London, U.K.: Academic Press.

Roy, I., and M.N. Gupta. 2004. Freeze-drying of proteins: Some emerging concerns. *Biotechnology and Applied Biochemistry* 39(2):165–177. doi: 10.1042/BA20030133.

Saarela, M., I. Virkajärvi, L. Nohynek, A. Vaari, and J. Mättö. 2006. Fibres as carriers for *Lactobacillus rhamnosus* during freeze-drying and storage in apple juice and chocolate-coated breakfast cereals. *International Journal of Food Microbiology* 112(2):171–178. doi: http://dx.doi.org/10.1016/j.ijfoodmicro.2006.05.019.

Sanchez, V., R. Baeza, M.V. Galmarini, M.C. Zamora, and J. Chirife. 2013. Freeze-drying encapsulation of red wine polyphenols in an amorphous matrix of maltodextrin. *Food and Bioprocess Technology* 6(5):1350–1354. doi: 10.1007/s11947-011-0654-z.

Sanoj Rejinold, N., M. Muthunarayanan, V.V. Divyarani, P.R. Sreerekha, K.P. Chennazhi, S.V. Nair, H. Tamura, and R. Jayakumar. 2011. Curcumin-loaded biocompatible thermoresponsive polymeric nanoparticles for cancer drug delivery. *Journal of Colloid and Interface Science* 360(1):39–51. doi: http://dx.doi.org/10.1016/j.jcis.2011.04.006.

Santagapita, P.R., M.F. Mazzobre, and M.P. Buera. 2011. Formulation and drying of alginate beads for controlled release and stabilization of invertase. *Biomacromolecules* 12(9):3147–3155. doi: 10.1021/bm2009075.

Schwegman, J.J. 2009. Understanding the physical properties of freeze-dried materials. *Innovations in Parmaceutical Technology* 29(72–74):72–77.

Shahidi, F., and X.-Q. Han. 1993. Encapsulation of food ingredients. *Critical Reviews in Food Science and Nutrition* 33(6):501–547. doi: 10.1080/10408399309527645.

Shaikh, J., D.D. Ankola, V. Beniwal, D. Singh, and M.N.V.R. Kumar. 2009. Nanoparticle encapsulation improves oral bioavailability of curcumin by at least 9-fold when compared to curcumin administered with piperine as absorption enhancer. *European Journal of Pharmaceutical Sciences* 37(3–4):223–230. doi: http://dx.doi.org/10.1016/j.ejps.2009.02.019.

Soottitantawat, A., H. Yoshii, T. Furuta, M. Ohkawara, and P. Linko. 2003. Microencapsulation by spray drying: Influence of emulsion size on the retention of volatile compounds. *Journal of Food Science* 68(7):2256–2262. doi: 10.1111/j.1365-2621.2003.tb05756.x.

Sunderland, J.E. 1980. Microwave freeze drying. *Journal of Food Process Engineering* 4:195–212.

Tiyaboonchai, W., W. Tungpradit, and P. Plianbangchang. 2007. Formulation and characterization of curcuminoids loaded solid lipid nanoparticles. *International Journal of Pharmaceutics* 337(1–2):299–306. doi: http://dx.doi.org/10.1016/j.ijpharm.2006.12.043.

Tobitsuka, K., M. Miura, and S. Kobayashi. 2006. Retention of a European pear aroma model mixture using different types of saccharides. *Journal of Agricultural and Food Chemistry* 54(14):5069–5076. doi: 10.1021/jf060309n.

Toledo, R.T. 1991. *Fundamentals of Food Process Engineering*, 2nd edn. New York: Chapman & Hall.

Vega-Mercado, H., M.M. Gongora-Nieto, and G.V. Barbosa-Canovas. 2001. Advances in dehydration of foods. *Journal of Food Engineering* 49:271–289.

Walstra, P. 2003. *Physical Chemistry of Food.* New York: Macel Dekker.

Wang, J., Y.Z. Li, R.R. Chen, J.Y. Bao, and G.M. Yang. 2007. Comparison of volatiles of banana powder dehydrated by vacuum belt drying, freeze-drying and air-drying. *Food Chemistry* 104(4):1516–1521. doi: http://dx.doi.org/10.1016/j.foodchem.2007.02.029.

Wang, L., X. Yu, H. Xu, Z.P. Aguilar, and H. Wei. 2016. Effect of skim milk coated inulin-alginate encapsulation beads on viability and gene expression of *Lactobacillus plantarum* during freeze-drying. *LWT—Food Science and Technology* 68:8–13. doi: http://dx.doi.org/10.1016/j.lwt.2015.12.001.

Wang, R., M. Zhang, A.S. Mujumdar, and J.-C. Sun. 2009. Microwave freeze–drying characteristics and sensory quality of instant vegetable soup. *Drying Technology* 27(9):962–968. doi: 10.1080/07373930902902040.

Wilkowska, A., W. Ambroziak, A. Czyżowska, and J. Adamiec. 2016. Effect of microencapsulation by spray-drying and freeze-drying technique on the antioxidant properties of blueberry (*Vaccinium myrtillus*) juice polyphenolic compounds. *Polish Journal of Food and Nutrition Sciences* 66(1):11–16.

Wray, D., and H.S. Ramaswamy. 2015. Novel concepts in microwave drying of foods. *Drying Technology* 33(7):769–783. doi: 10.1080/07373937.2014.985793.

Xu, M., F. Gagné-Bourque, M.-J. Dumont, and S. Jabaji. 2016. Encapsulation of *Lactobacillus casei* ATCC 393 cells and evaluation of their survival after freeze-drying, storage and under gastrointestinal conditions. *Journal of Food Engineering* 168:52–59. doi: http://dx.doi.org/10.1016/j.jfoodeng.2015.07.021.

Ying, D.Y., J. Sun, L. Sanguansri, R. Weerakkody, and M.A. Augustin. 2012. Enhanced survival of spray-dried microencapsulated *Lactobacillus rhamnosus* GG in the presence of glucose. *Journal of Food Engineering* 109(3):597–602. doi: http://dx.doi.org/10.1016/j.jfoodeng.2011.10.017.

Zhang, J., M. Zhang, L. Shan, and Z. Fang. 2007. Microwave-vacuum heating parameters for processing savory crisp bighead carp (*Hypophthalmichthys nobilis*) slices. *Journal of Food Engineering* 79(3):885–891. doi: http://dx.doi.org/10.1016/j.jfoodeng.2006.03.008.

Zhang, M., J. Tang, A.S. Mujumdar, and S. Wang. 2006. Trends in microwave-related drying of fruits and vegetables. *Trends in Food Science & Technology* 17(10):524–534. doi: http://dx.doi.org/10.1016/j.tifs.2006.04.011.

Zhao, L.P., H. Xiong, H. Peng, Q. Wang, D. Han, C.Q. Bai, Y.Z. Liu, S.H. Shi, and B. Deng. 2011. PEG-coated lyophilized proliposomes: Preparation, characterizations and *in vitro* release evaluation of vitamin E. *European Food Research and Technology* 232(4):647–654. doi: 10.1007/s00217-011-1429-5.

Zouboulis, A., T. Karapantsios, K. Mates, and P. Mavros. 2009. *Details for Physical Processes.* Thessaloniki, Greece: Tziolas.

Zuidam, N.J., and E. Shimoni. 2010. Overview of microencapsulates for use in food products or processes and methods to make them. In *Encapsulation Technologies for Active Food Ingredients and Food Processing*, edited by N.J. Zuidam and V. Nedovic, pp. 3–29. New York: Springer.

Zuidam, N.J., E. Van Winden, R. De Vrueh, and D.J.A. Crommelin. 2003. Stability, storage and sterilization of liposomes. In *Liposomes*, edited by V.P. Torchilin and V. Weissig, pp. 149–165. Oxford, U.K.: Oxford University Press.

3 Encapsulation Methods
Spray Drying, Spray Chilling and Spray Cooling

Maciej Jaskulski, Abdolreza Kharaghani, and Evangelos Tsotsas

CONTENTS

3.1 INTRODUCTION

Encapsulation denotes the process during which very small droplets of liquid or gas or very small solid particles is enclosed within a film of soluble material. The trapped material is commonly referred to as core material, payload, actives, fill, or internal phase. The material that creates the coating is called wall material, carrier, membrane, or shell. The product obtained by this process is referred to as capsules. In a relatively simple form, a capsule is a small sphere with a uniform wall around it. Capsules can be classified into three groups based on their sizes: nanocapsules with diameter about 100–300 nm, microcapsules with diameter about 3–800 µm, and macrocapsules with diameter larger than about 1000 µm. The manufacturing technique of capsules depends on the physical and chemical properties of the material to be encapsulated, and it can be classified into three methods:

1. *Physical methods*: pan coating, air-suspension, centrifugal extrusion, vibrational nozzle, spray drying, spray chilling, spray cooling
2. *Physicochemical methods*: ionotropic gelation, coacervation-phase separation
3. *Chemical methods*: interfacial polycondensation, interfacial cross-linking, in situ polymerization, matrix polymerization

This chapter highlights the spray drying, spray chilling, and spray cooling methods, which are commonly used in the food and pharmaceutical industry to produce microcapsules (Fang and Bahandri 2010). In spray drying microencapsulation, a carrier material solidifies around a liquid core due to solvent evaporation, whereas in spray chilling and cooling, carrier temperature goes below its melting point. To better understand the spray microencapsulation processes, it is necessary to describe in detail methods of droplets creation, mechanisms of droplet solidification, mathematical principles of heat, mass and momentum transfer between particles, and surrounding agent.

In this chapter, the principles of spray microencapsulation technology, different constructions of spray equipment, and their applications in various industrial sectors are presented. The methodologies of spray drying and other spray microencapsulation techniques (spray chilling and cooling) are described. Methods to determine single droplet drying (SDD) kinetics are explained. The methodology used to scale up evaporation models from a particle-scale to an industrial-scale drying system by using both the traditional linear model and the most advanced 3D computational fluid dynamics (CFD) simulations is presented.

3.2   SPRAY DRYING MICROENCAPSULATION

The idea of using spray drying technology for the production of microcapsules was born in the 1950s. In this method, volatile liquid in the carrier material evaporates leading to solidification of capsules around the core (Re 1998). Powder produced by this method is characterized by physical or chemical properties different from components from which it was created. Some applications of spray drying microencapsulation are as follows:

- Avoid emission of the core material from the capsules. This is of great importance for the protection of the environment if the core material shows toxic or hazardous properties.
- Retain undesirable properties of the core material, for example, taste, smell, pH, or chemical activity.
- Protect desired core properties: retain aroma, protect functional ingredients from oxidation.
- Convert liquids into bulk solid materials with new unique physical or chemical properties.
- Release a core material in a controlled way.
- Regenerate microcapsules, for example, as chemical catalysts.

It is very important to properly choose materials that are used in spray drying microencapsulation. Depending on the core material and desired properties of produced capsules, a carrier material can be chosen from natural or synthetic polymers. Most industrial microencapsulation processes are based on solvent evaporation; therefore, the carrier material must be soluble in water in order to create two phase solutions with a low viscosity at high concentrations of a carrier. Polymers mostly used as carrier materials are: carbohydrates (maltodextrin, starch, cyclodextrin, cellulose, konjac glucomannan), gums (acacia, arabic, agar, sodium alginate), fats (paraffin wax), microbiological substances (enzymes living cells), or resins. The selected material cannot interfere or react with the core substance. Furthermore, it should meet the expectations of the properties of the finished products, such as a gradual release of encapsulated substances or durability of the crust.

Several studies have been conducted on the encapsulation of volatile substances, proteins, and carotenoids (Table 3.1). In these studies, seeking a carrier material satisfying all the main requirements of the ideal carrier is a difficult task due to a change of the behavior of the carrier material when it comes in contact with various substances of the core.

Microcapsules created by the spray drying technique have a shape close to spherical and sizes from a few microns to several millimeters. The shape and size of the microcapsules depend on the method of emulsion preparation and the substances from which they were created. We can distinguish six basic types of microcapsules (see Figure 3.1):

1. *Simple*: the core material is surrounded by a single layer of carrier material; the particle is spherical or similar to spherical shape.
2. *Irregular*: the core material is surrounded by a single layer of carrier material; the particle shape deviates substantially from the spherical.
3. *Multicore*: a few drops of the core material is enclosed by a carrier material.
4. *Multilayer*: the core is encased by several layers made of different carrier materials.
5. *Matrix*: the fine droplets of core material are distributed uniformly in the particle volume.
6. *Hollow*: the fine droplets of core material are closed inside the particle crust.

The final geometrical appearance of microcapsules produced by spray drying is affected by several parameters: type of carrier material, feed composition, fragmentation of the dispersed phase, feed stability, atomization parameters (type of atomizer and fluid pressure at the atomizer outlet), drying conditions (type of phase contact [co- or counter-current], mass flow rate of feed and drying agent, temperature of feed and drying medium, and residence time of particles within dryer).

3.3   SPRAY DRYING TECHNIQUE

Spray drying is a ubiquitous industrial operation used to produce dry loose powders without prior grinding. The production of powders by this method is usually economical, easy to perform and controllable. The spray drying technique started in 1872 when, for the first time, drying by liquid atomization was performed (Hunziker 1920). However, the first industrial-scale installation of this equipment appeared in the 1920s. Chester Earl Gray and Aage Jansen developed the first installation of a spray dryer in 1913 and 1914, respectively. This dryer was used for milk powder production (Fogler and Kleinschmidt 1938). In those times, spray dryers were mainly used in food and chemical industries. With the beginning of global conflicts, the demand for easy to transport and light products

TABLE 3.1

A Summary of Studies on Spray Drying Microencapsulation

Encapsulated Material (EM)	Encapsulating Agent (EA)	Drying Air Temperature (°C)	Operational Conditions	Encapsulation Efficiency (%)	Reference
Canola oil	Lentil protein Isolate/maltodextrin	Inlet: 180	Nozzle diameter: 0.7 mm	42.3–87.9	Chang et al. (2016)
	Lentil protein Isolate/maltodextrin/ lecithin	Outlet: 85	Air flow rate: 35,000 L/h		
	Lentil protein Isolate/maltodextrin/ sodium Alginate		Feed pump rate: 5–20%		
	Lentil protein Isolate/maltodextrin/ lecithin/sodium alginate		Air pressure: 5 bar Total solid content: 25% EM to EA ratio: 1:4 w/w		
Ginger essential oil	Gum arabic	Inlet: 170	Atomizer: Two fluid nozzles	48–86.5	Fernandes et al. (2016)
	Gum arabic/maltodextrin	Outlet: N/A	Atomizing air flow rate: 35 L/min		
	Gum arabic/inulin		Feed flow rate: 0.8 L/h EM to EA ratio:1:4 w/w		
Bruncea javanica oil	Gum arabic/gelatin	Inlet: 140, 160, 180	Air flow rate: 50, 65, 80 L/min	82.9	Hu et al. (2016)
		Outlet: N/A	Feed flow rate: 1, 1.5, 2 mL/min EM to EA ratio:1:6 w/w		
Curcumin	Whey protein isolate	Inlet: 110, 150	Microfluidic jet spray dryer microfluidic aerosol nozzle system	95–97.5	Liu et al. (2016)
		Outlet: N/A	Orifice diameter: 75 μm Curcumin concentration: 25 mg/100 mL WPI solution		

(Continued)

TABLE 3.1 (Continued)
A Summary of Studies on Spray Drying Microencapsulation

Encapsulated Material (EM)	Encapsulating Agent (EA)	Drying Air Temperature (°C)	Operational Conditions	Encapsulation Efficiency (%)	Reference
Fish oil	Glucose syrup	Inlet: 180 Outlet: 70	Atomizer: Rotary Atomizer speed: 22,000 rpm Pressure: 4 bar EM to EA ratio: 1:6 w/w	98	Morales-Medina et al. (2016)
Jussara pulp	Gum arabic/modified starch/whey protein concentrate Gum arabic/modified Starch/soy protein isolate	Outlet: 90–98	Atomizer: Two fluid nozzles Nozzle diameter: 7mm Aspirator rate: 90% Spraying air flow rate: 500 L/h Feed flow rate: 5 mL/min (pump rate 18%) EM to EA ratio: 1:1w/w	80.3–99.5	Santana et al. (2016)
Ginger extract (*Zingiber officinale*)	Maltodextrin (DE 9-13) Gum arabic Maltodextrin (DE 9-13)/gum arabic	Inlet: 160 Outlet: 65	EM to EA ratio: 475 mL extract/25 g encapsulating agent	33–40	Simon-Brown et al. (2016)
Flexirubin	Gum arabic κ-Carrageenan	Inlet: 140–220 Outlet: 85	Atomizer: Single standard nozzle Nozzle diameter: 0.7 mm Air flow rate: 60 m³/h Feed flow rate: 6 g/min EM to EA ratio: 1:2 w/w	18.31–70.06	Venil et al. (2016)
Fish oil	Whey protein isolate	Inlet: 160 Outlet: N/A	Spraying air flow rate: 900 L/h EM to EA ratio: 10:1, 5:1, 1:1, 1:2 w/w	13.1–93.2	Wang et al. (2016)

(Continued)

TABLE 3.1 (*Continued*)

A Summary of Studies on Spray Drying Microencapsulation

Encapsulated Material (EM)	Encapsulating Agent (EA)	Drying Air Temperature (°C)	Operational Conditions	Encapsulation Efficiency (%)	Reference
Phytostrols	Gum arabic/maltodextrin/ surfactants (Tween 20, SDS)	Inlet: 160 Outlet: <102	Atomizer: Two fluid nozzles Nozzle diameter: 0.5 mm Air flow rate: 35–38 m³/h Feed flow rate: 2 mL/min Spraying air flow rate: 601 L/h EM to EA ratio: 0.29–0.33 w/w	10–50	Battista et al. (2015)
Fish oil	Maltodextrin/fish gelatin Maltodextrin/fish gelatin/k-Carrageenan Maltodextrin/k-Carrageenan	Inlet: 180 Outlet: 80	EM to EA ratio: 1:4 w/w	~70–90	Mehrad et al. (2015)
Betalains of cactus fruit	Maltodextrin (DE 20) Maltodextrin/cladode mucilage	Inlet: 170 Outlet: 98	Nozzle diameter: 0.5 mm Air flow rate: 100 m³/h Feed flow rate: 485 mL/h Air pressure: 4 bar EM to EA ratio: 1:1, 1:0.225 w/w	16.4–25.4	Otálora et al. (2015)
Phenolics compound of olive pomace	Maltodextrin (DE 16.5-19.5)	Inlet: 130, 160 Outlet: 75–102	Air flow rate: 30 m³/h Feed flow rate: 5, 10 mL/min Concentration of dispersion: 100, 500 g/L Volume of used extract: 50 mL	51–94	Paini et al. (2015)

(*Continued*)

TABLE 3.1 (Continued)
A Summary of Studies on Spray Drying Microencapsulation

Encapsulated Material (EM)	Encapsulating Agent (EA)	Drying Air Temperature (°C)	Operational Conditions	Encapsulation Efficiency (%)	Reference
Polyphenols (Gallic acid)	Maltodextrin	Inlet: 80, 100, 120	Aspirator rate: 75% (28 m³/h)	N/A	Ramírez et al. (2015)
	Gum arabic	Outlet: N/A	Spraying air flow rate: 600 L/h		
	Maltodextrin/gum arabic		Feed flow rate: 72, 108, 144 mL/h		
			Encapsulant concentration: 10, 20, 30% w/w		
			Encapsulant ratio: 0, 50, 100		
Cinnamon infusion	Maltodextrin	Inlet: 140, 160, 180	Feed flow rate: 8, 10 mL/min	~85	Santiago-Adame et al. (2015)
		Outlet: N/A	Air pressure: 6.5 bar		
Blueberry bioactive compounds	Maltodextrin (DE 4–7)	Inlet: 125	Atomizer: Two fluid nozzles	70.33–98 (Anthocyanin retention)	Tatar et al. (2015)
	Gum arabic	Outlet: N/A	Aspirator rate: 100% (35 m³/h)		
			Spraying air flow rate: 601 L/h		
			Feed pump rate: 30%		
			Atomizer: Ultrasonic nozzle		
			Nozzle frequency: 60 kHz		
			Power outlet at nozzle: 1–15 W		
			Aspirator rate: 50% (35 m³/h)		
			Spraying air flow rate: 601 L/h		
			Feed pump rate: 8%		
Amaranthus gangeticus extract	B-cyclodextrin/maltodextrin	Inlet: 132–210	Feed flow rate: 3–15 mL/min	N/A	Chong et al. (2014)
		Outlet: N/A	Concentration of extract: 10% w/w		

(Continued)

TABLE 3.1 (Continued)
A Summary of Studies on Spray Drying Microencapsulation

Encapsulated Material (EM)	Encapsulating Agent (EA)	Drying Air Temperature (°C)	Operational Conditions	Encapsulation Efficiency (%)	Reference
Gac oil	Whey protein concentrate/gum arabic	Inlet: 148–162 Outlet: 73–87	Atomizer: Two fluid nozzles Nozzle diameter: 0.5 mm Air flow speed: 4.3 m/s Air pressure: 2 bar EM to EA ratio: 1:5 w/w	77.68–94.10	Kha et al. (2014)
Fish oil	Skim milk powder	Inlet: 140–180 Outlet: N/A	Aspirator rate: 55, 65, 75% Spraying air flow rate: 600, 800 L/h Feed pump rate: 5, 10, 15% EM to EA ratio: 1:2 w/w	59.05–85.67	Aghbashlo et al. (2013a)
Fish oil	Skim milk powder Whey protein concentrate Whey protein isolate Milk protein concentrate Sodium caseinate	Inlet: 140–180 Outlet: N/A	Aspirator rate: 65% Spraying air flow rate: 700 L/h Feed pump rate: 10% EM to EA ratio: 1:2 w/w	40.59–81.94	Aghbashlo et al. (2013b)
Flax seed oil	Maltodextrin/gum arabic Maltodextrin/whey protein concentrate Maltodextrin/modified starch (Hi-Cap, Capsule)	Inlet: 180 Outlet: 110	Nozzle diameter: 0.5 mm Air flow rate: 73 m³/h Feed flow rate: 12 g/min Compressor air pressure: 0.06 MPa Total solid content: 30% w/w Oil concentration: 20% w/w (with respect to total solids)	62.3–95.7	Carneiro et al. (2013)

(Continued)

TABLE 3.1 (*Continued*)
A Summary of Studies on Spray Drying Microencapsulation

Encapsulated Material (EM)	Encapsulating Agent (EA)	Drying Air Temperature (°C)	Operational Conditions	Encapsulation Efficiency (%)	Reference
Linseed oil	Gum arabic Gum arabic/maltodextrin Gum arabic/maltodextrin/whey protein isolate Maltodextrin/methylcellose	Inlet: 175 Outlet: 75	Feed flow rate: 15 mL/min Air pressure: 2.8 bar EM to EA ratio: 0.1, 0.55, 0.68, 0.7 w/w	25.5–91.4	Gallardo et al. (2013)
Coffee oil	Gum arabic	Inlet: 150–190 Outlet: N/A	Atomizer: Two fluid nozzles Nozzle diameter: 1.2 mm Air flow rate: 36 m³/h Feed flow rate: 0.8 L/h Total solid content: 10–30% w/w Oil concentration: 10–30% w/w (with respect to total solids)	47.93–82.57	Frascareli et al. (2012)
Pomegranate seed oil	Skim milk powder	Inlet: 150–180 Outlet: N/A	Atomizer: Two fluid nozzles Nozzle diameter: 0.5 mm Air flow rate: 17.5–22.8 m³/h Feed flow rate: 1.75 g/min Air pressure: 5 bar Total solid content: 10–30% w/w EM to EA ratio: 1:9–1:2.3 w/w	~50–95.6	Goula and Adamopoulos (2012)

(*Continued*)

TABLE 3.1 (Continued)
A Summary of Studies on Spray Drying Microencapsulation

Encapsulated Material (EM)	Encapsulating Agent (EA)	Drying Air Temperature (°C)	Operational Conditions	Encapsulation Efficiency (%)	Reference
Capsicum annuum L. (Chilli)	Gum arabic/maltodextrin	Inlet: 160 Outlet: 70	Atomizer: Pneumatic nozzle Feed flow rate: 13.3 mL/min Air pressure: 0.4 kg/cm² EM to EA ratio:1:4 w/w	84.2–86.6	Guadarrama-lezama et al. (2013)
Flax oil	Zein	Inlet: 135 Outlet: 55–60	Feed flow rate: 9 mL/min Concentration of dispersion: 6, 10, 14 g/100 mL EM to EA ratio: 1:28–1:4 w/w	75.42–93.26	Quispe-Condori et al. (2011)
Fish oil Fish oil/extra virgin olive oil	Sugar beet pectin Glucose syrup	Inlet: 180 Outlet: 80	EM to EA ratio: 1:2–1:4 w/w	90.42–97.87	Polavarapu et al. (2011)
Docosahexaenoic acid ethyl esther	Gum arabic/tween 80	Inlet: 155–215 Outlet: N/A	Aspirator rate: 90% Spraying air flow rate: 635–730 L/h Feed pump rate: 25–58% EM to EA ratio: 1:1.9–1:5.6 w/w	4.73–17.52	Rodrigues et al. (2011)

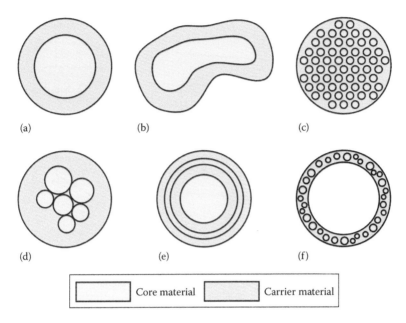

FIGURE 3.1 The most common form of microcapsules: (a) simple, (b) irregular, (c) matrix, (d) multi-core, (e) multi-layer, and (f) hollow.

with low water content, which prevents product deterioration and prolongs product stability, was rapidly increased. For military needs mainly, food, medicine, and plastic granulates were produced on a large scale by using the spray drying technology. Nowadays, this technology is preferably used for drying many thermally sensitive materials, such as foods (e.g., milk powder, coffee, spices, flavorings, starch), chemicals (e.g., catalysts, fertilizers, pigments, ceramics), and pharmaceuticals (e.g., antibiotics, vitamins, enzymes, medical ingredients, fillers) (Mujumdar 2007). Despite being widely used, it has however some disadvantages, such as a relatively low thermal efficiency of a spray drying installation, a necessity to install a complex dedusting system at the outlet of the drying medium, and a lack of possibility for heat recovery from dried materials.

Spray drying is a continuous process in which a liquid feed in the form of a solution or suspension is sprayed into the stream of a hot drying agent (e.g., air or nitrogen) by nozzles installed inside a drying chamber. As a result of rapid extension of phase contact area, the moisture evaporation is intense without leading to a significant increase in dried material temperature. This prevents powder overheating and retains its quality due to short drying time. The structures of powder particles produced in this way depend on the drying process conditions and on the materials used for the production. Nowadays, advanced studies on the structure of single particles produced by spray drying are conducted (Walton and Mumford 1999). New techniques and methods of product preparation with different, sometimes really complex, internal structures are introduced (Walton 2002). Phenomena such as crystallization, microencapsulation, and agglomeration are also combined with the spray drying technique

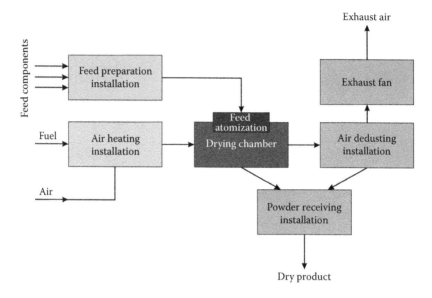

FIGURE 3.2   General scheme of the spray drying process.

to design new materials with unique properties. Generally there are four main stages in this technique (Figure 3.2):

1. Preparation of liquid feed and heating drying agent
2. Spraying liquid feed by using atomizers
3. Drying of particles due to contact with drying agent
4. Product collection and separation of powder from exhaust air

The first stage in the spray drying process is feed preparation. Mostly liquid solutions, suspensions, or slurries are used as a feed in spray drying. However, sometimes feed is specially prepared for future processing. For microcapsules produced by spray drying, an emulsion of carrier and core materials is used as feed. For this purpose, two or more mutually immiscible substances are subjected to a homogenization process. To decrease the probability of emulsion degradation and division of phases, feed is stabilized with the help of surface-active substances called emulsifiers. In spray drying, microencapsulation oil-in-water emulsions are used. Insoluble core material is dispersed in carrier solution. The size of the dispersed droplets inside the emulsion affects the final microcapsules morphology. If core material droplets are small in comparison to droplets generated during the atomization, microcapsules can contain several cores enclosed inside the carrier. In other cases, single core microcapsules are created or core material droplets are too large to close it inside the carrier (Soottitantawat et al. 2003, 2005). Control of the core material droplet size inside the emulsion is performed mostly by microscopic analysis. Emulsion quality or emulsification kinetics can be controlled by an analysis of changes in emulsion properties, such as viscosity or electrical conductivity during feed preparation.

To keep parameters such as temperature and solid concentration constant, the feed is placed in a thermostated tank with a mounted stirrer to provide uniform concentration of solids and to prevent emulsion flocculation and sedimentation. Feeds with chemically active ingredients are mixed just before atomization.

It is worth noting that in some spray drying operations, the feed is initially dehydrated before atomization in order to increase the solid concentration of the feed. This process can be performed either by a conventional evaporator system or by an advanced membrane technology. For instance, in milk powder production the initial concentration of solid is increased from around 18% to a level of 45–50%. The concentrated milk is sprayed into the tower in a form of slurry or paste. This operation can significantly reduce the size of the spray dryer.

The second stage is to spray an already prepared liquid feed into the drying chamber by atomizers, which break the liquid stream into droplets. Atomizers can be divided into four groups.

1. *Pressure atomizers*: Transform pressure energy into kinetic energy as a result of decompression. This creates a cone of liquid at the nozzle outlet comprising single droplets. Depending on the type of nozzle and process in which it is used, droplet size of up to 800 μm can be generated. The spraying is carried out at a pressure between 0.7 and 3 MPa. The outlet hole in which the depressurization takes place in the range between 0.5 mm and 4 mm. For microencapsulation process, pressure atomizers are not suitable. Because a high mechanical stress generated during the atomization may destroy the structure of microcapsules, the core material may release to the droplet surface.

2. *Pneumatic atomizers*: Spraying occurs as a result of collision of slurry and compressed gas. The kinetic energy required to break the liquid stream is accelerated by the air stream with a speed of 100–200 m/s. Spraying of liquid feeds by this method is recommended mainly for slurries with a high density and viscosity, as well as heterogeneous substances (suspensions or emulsions). The fragmentation ratio of liquid stream and diameter of droplet strongly depend on the speed of added gas. However, additional phase (atomization air) can cause aeration of the emulsion and enclose air bubbles inside the particle. This can cause oxidation of the core material or the growth of biologically active materials.

3. *Rotary atomizers*: The main part of this group of atomizers is a rotating disc usually with a diameter of 500–350 mm. The fragmentation ratio of the slurry droplets depends on the rotational speed of the disc, which varies between 4,000 and 30,000 rpm. Droplets are ejected from the atomizer by centrifugal force of the rotating disk. The initial speed of particles is relatively low in comparison with other types of atomizers. Rotary devices are widely used in spraying chemically active liquids, which can lead to nozzle corrosion or materials that can clog the nozzle outlet. Rotary atomizers are commonly used in spray drying encapsulation because of low mechanical stress at the nozzle outlet and good control on the generated droplet size.

4. *Special atomizers*: In this group, other technical solutions of atomizers are constructed for the purpose of individual installations. Examples of such systems are atomizers with a large processing capacity (cooling towers), acoustic atomizers, which use sound waves with a frequency in the range of 0.006–20 kHz, ultrasonic atomizers, which create a vibration with a frequency of up to 26 MHz, electrostatic atomizers, which use the electrostatic fields for droplet fragmentation, and microatomizers, where fine droplets are created during foam bubbles breakage. Also new constructions of monodisperse atomizers are available on the market. In these atomizers, droplets are created by the movement of a piezoelectric crystal that disturbs the liquid stream and disintegrates it into droplets. A precise control on the created droplet size allows prediction of structure and morphology of produced capsules. It also helps in the design of new material properties (Soottitantawat et al. 2005).

After atomization, the particles fall into the drying chamber in which the drying agent—mostly hot air—flows. In cases where flammable vapor is formed as a result of evaporation, inert gases such as nitrogen are used to decrease risk of explosion. From the viewpoint of phase mixing at the industrial scale, the drying process can be carried out in three ways (see Figure 3.3): co-current, counter-current, and mixed.

Co-current spray dryers are characterized by a significant height of the drying chamber. Particles flow in the same direction as drying agent. This decreases the rate of moisture removal and prevents overheating of the dry material. Additionally, parallel phase flow prevents formation of vortexes and particle recirculation resulting in a lower probability of particle agglomeration and more uniform particle size distribution. For this reason co-current spray dryers are mostly recommended for heat-sensitive materials and they are commonly used in food and pharmaceutical productions. It is worth mentioning that co-current systems can be used in new technology of flame spray drying, in which heat for the moisture evaporation is provided by the flame generated by the fuel combustion (mostly alcohols) inside the drying

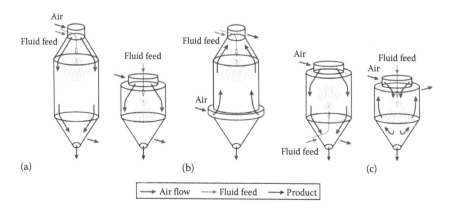

FIGURE 3.3 Flow patterns of phases used in the spray drying process: (a) co-current, (b) counter-current, and (c) mixed flow.

chamber (Piatkowski et al. 2014). This method avoids the process of air heating. However, due to high temperatures inside flame spray drying, it is recommended to be used for drying of materials with high heat resistance.

In counter-current spray dryers, particles fall down in the direction opposite to the drying agent flowing from the drying chamber bottom. In terms of mass and heat transfer between phases counter-current flow is more favorable than co-current flow geometry. However, strong particles recirculation causes the phenomenon of droplet coalescence and particle agglomeration. Additionally, dried particles have to flow through the drying medium inlet area where the temperature inside the drying chamber is very high. For this reason, counter-current spray drying systems are not recommended for heat-sensitive materials.

Spray dryers with mixed flow geometry allow for an efficient drying similar to counter-current flow, but with a lower inlet temperature of drying medium. Additionally, mixed system creates much less dust in exhaust air. Spray dryers with mixed phase flow are characterized in bigger diameters than co- and counter-current constructions and they are used in production of materials such as ceramics (fountain dryers).

The last stage in the spray drying process is powder separation. Dried material can be retrieved with the exhaust drying agent or separately. Retrieving powder with the exhaust air is the easiest way in drying chambers with a conical bottom. However, in constructions where gravitational outflow of particles is difficult or impossible to perform it is necessary to install an additional scraping device to push behindhand product. In constructions where drying agent and powder are received separately, divisions of phases occur inside the drying chamber where particles fall to the dryer bottom and the drying agent is retrieved at the top of the tower (counter-current flow). Exhaust air is dedusted by the system of cyclones and bag filters. In case of hazardous substances, wet dedusting is performed by using scrubbers.

Spray drying operation involves three variables: feed (flow rate, temperature, and concentration), drying agent supply (flow rate, inlet temperature, outlet temperature, and humidity), and atomizer selection (type, rotational velocity, nozzle diameter, and number of nozzles). An optimization approach can be used to determine effectively suitable values of these parameters (e.g., see Woo et al. 2007).

3.4 SINGLE DROPLET DRYING

Designing of a complex spray drying system, optimized for a given product, requires a thorough understanding of the kinetics of moisture evaporation and the morphological changes of the material during drying. Measurements of properties of discrete phases during the spray drying process are difficult to perform due to rapid evaporation, intensive mixing of phases, and limited access to spray drying installations. In literature, there are few descriptions of measurements of drying kinetics during the spray drying process (Zbiciński and Piątkowski 2004). To avoid those difficulties, systems for measurement of single droplet drying (SDD) were developed. SDD experiments allow to dry single droplets under controlled process conditions. Analysis of morphological changes and measurement of drying kinetics for stationary droplets can be carried out in a more controlled and accurate manner.

In terms of how the droplet is suspended during experiments, the SDD measurement systems can be categorized into three groups (Fu et al. 2012): free fall, levitation, and glass filament methods. The free fall method seems to be the best way to imitate conditions inside industrial spray drying towers. Experiments were performed by generation of a mono-disperse stream of droplets, which can freely fall inside the column (Kinzer and Gunn 1951). Falling particles can be caught to measure their mean temperature and moisture content at different positions from the droplet generator. However, measurements of the drying kinetics of the particles falling in free motion are difficult to carry out. For this reason, the free fall method is mostly used for pure solvent droplets or to measure changes in particle morphology, without concerning drying kinetics of droplets/particles (El-Sayed et al. 1990). To measure diameter changes during the drying process, falling droplets are usually recorded by a high-speed camera.

A simpler way to measure drying kinetics of a single liquid droplet is to suspend it in an acoustic or a magnetic field (Schiffter and Lee 2007). In this method, the droplet is stationary and placed at a fixed position, which allows the performing of more precise measurements. Diameter changes of a single droplet, dried in an acoustic field, can be determined by analysis of movie frames recorded during experiment (Kastner et al. 2001). Determination of drying kinetics can be divided into two stages. The evaporation rate in the first drying stage can be estimated on the basis of changing droplet volume. In the second drying stage, in which the decrease of volume is minimal, the drying curve can be estimated by changes of particle position in the acoustic field (Kastner et al. 2001) or, more precisely, by analysis of humidity in the outlet air (Groenewold et al. 2002). Apart from the lack of direct methods for measuring the rate of evaporation in the second drying stage, there are disadvantages to the acoustic levitator systems. It has been observed that acoustic waves speed up slightly the evaporation rate and decrease the critical moisture content (Yarin et al. 2002). A low temperature of the drying air (max. 80°C) is caused by the limited thermal resistance of acoustic systems. This range of temperatures does not reflect the real drying conditions inside spray drying chambers.

Glass filament method can be used for a wider range of air temperatures. In this type of SDD experiment, the droplet is suspended at the tip of a thin glass filament (Renksizbulut and Yuen 1983). Methods used to measure the drying kinetics are similar to the acoustic levitator system. Additionally, a thin thermocouple can be inserted into the droplet to measure changes in droplet/particle temperature (Lin and Chen 2002; Shamaei et al. 2016). Drying parameters determined by glass filament SDD experiments are useful for Computational Fluid Dynamics (CFD) simulations of spray drying using Reaction Engineering Approach (REA) or Characteristic Drying Curve (CDC), see Section 3.5. However, in this method the drying process is disturbed due to heat conduction in the glass filament and a relatively large droplet size (0.5–2 mm), which is several times bigger than droplets used in real spray dryers (Lin and Chen 2002). To decrease the impact of heat conduction through the glass filament on the drying process, the droplet can be suspended on a PTFE fiber (polytetrafluoroethylene) inside the drying chamber (Tran et al. 2016a). A schematic diagram of this modified glass filament method is presented in Figure 3.4.

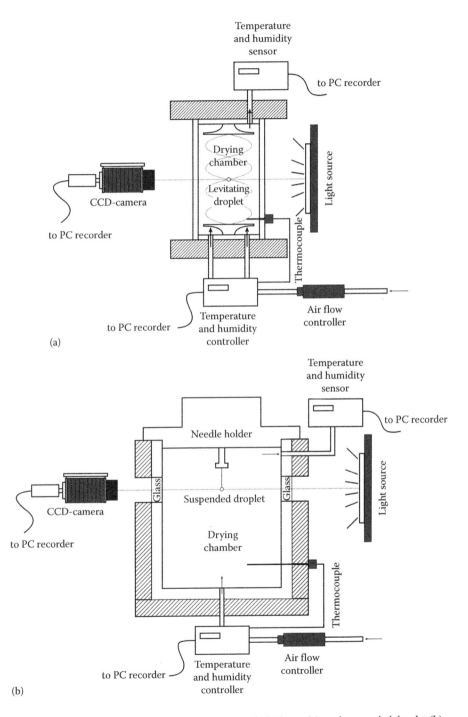

FIGURE 3.4 Single droplet drying systems: acoustic levitator (a), and suspended droplet (b).

Using an optical camera, changes in particle diameters during drying can be obtained after post-processing of images (Figure 3.5). However, to determine the evaporation rate it is necessary to use the indirect method by recalculations of changes in exhaust air humidity. For this purpose, in the suspended droplet method, humidity and mass flow rate of air at the drying chamber inlet are constant and can be controlled by the operator. Assuming that the drying chamber is hermetic and there is no influence from its surrounding, changes in exhaust air humidity are caused only by water evaporation from the droplet. The amount of water evaporated during the drying process can be calculated from the difference in inlet and outlet air moisture content, which allows determining the drying rate in the second stage of drying, when changes of particle volume do not correspond to changes in moisture content. Examples of data received from SDD measurement for skim milk droplets are presented in Figure 3.6. A detailed analysis of particle diameter changes for whey-lactose mixture was carried out for five different air temperatures between 40 and 180°C. It can be observed that particle diameter is constantly decreasing in the first stage of drying. When droplet solidifies at the critical moisture content and changes into wet particle (the so-called locking point), particle shrinkage almost stops; however, some deformations can be observed. At temperatures above 100°C, after the locking point, the wet core of the particles may begin to boil. This phenomenon can increase particle diameter rapidly.

Additionally, in Figure 3.6, changes of evaporation flux as a function of particle water mass fraction for each measured case are presented. Three characteristic

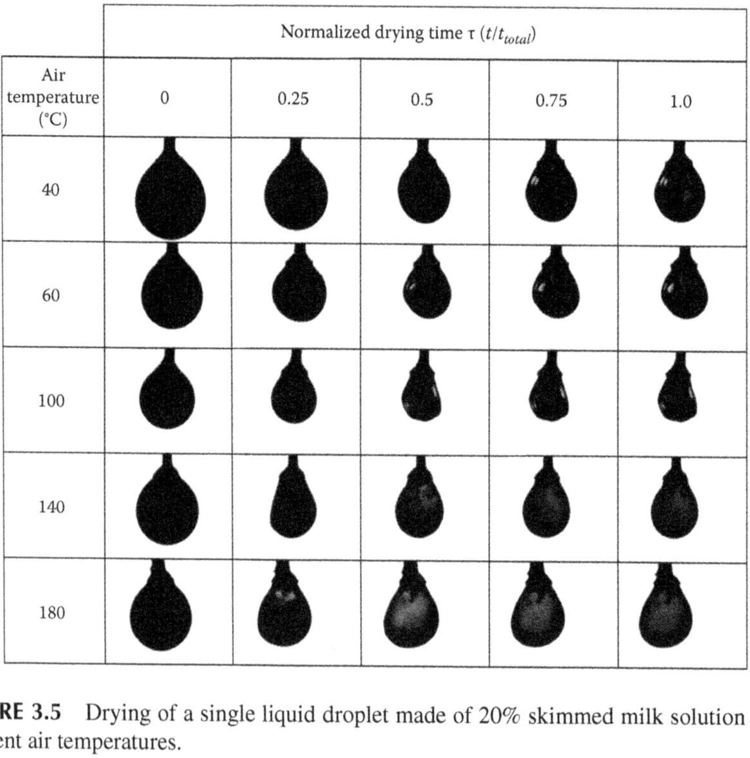

	Normalized drying time τ (t/t_{total})				
Air temperature (°C)	0	0.25	0.5	0.75	1.0
40					
60					
100					
140					
180					

FIGURE 3.5 Drying of a single liquid droplet made of 20% skimmed milk solution at five different air temperatures.

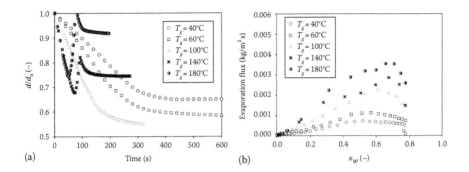

FIGURE 3.6 Changes in particle diameter (a) and evaporation flux (b) during the single droplet drying of whey (15%)-lactose (30%) solution performed for five different air temperatures. (From Tran, T.T.H., From single droplet to spray tower drying of dairy solutions, PhD thesis, Otto-von-Guericke-Universität Magdeburg, Magdeburg, Germany, 2017.)

stages of drying are observed. The initial stage, when a droplet warms up, is characterized by a rapid growth in the evaporation rate. In the first stage of drying, evaporation rate decreases slowly only due to changes in particle diameter. When the particle reaches critical moisture content, the evaporation rate starts to decrease visibly.

It is clear that the boiling period has an influence on the evaporation rate. Small disturbances can be observed in the drying rate when water vapor is generated in the particle center. This phenomenon reduces noticeably the total drying time by the immediate release of water vapor from the particle center without diffusion through the particle crust (Tran et al. 2016a). This phenomenon is very important in spray drying encapsulation. Breaking the particle crust by water vapor can release the core material. Even if the temperature remains below the boiling point, overheating of the particle may lead to thermal expansion of core material and eventually the surrounding crust may be damaged. For this reason, the microencapsulation process is very sensitive to drying parameters. The amount of energy introduced with air and feed rate of emulsion into the drying chamber should be calculated in a way that energy consumed for the drying process increases air temperature without causing any damage to the particle (Adamiec 2009). Very recently, the method for measurement of SDD described above was successfully used to determine evaporation rates of emulsion droplets (Shamaei et al. 2016).

3.5 MODELING OF THE SPRAY DRYING PROCESS

Mathematical models for simulation of the spray drying process can be divided into three groups: one-dimensional, axisymmetric, and three-dimensional models. One-dimensional models of spray drying can predict residence time of particles and changes in temperature, moisture content, and velocity of continuous and dispersed phases with the height of the drying tower. For one-dimensional models, we refer the reader to Zbiciski (1995) and Langrish (2009). In this section, axisymmetric and three-dimensional models are described briefly.

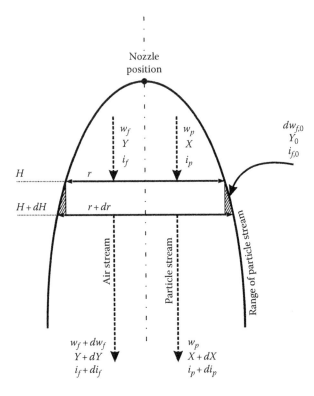

FIGURE 3.7 Heat and mass balance between the atomized droplets and the hot air stream for a co-current flow in a spray drying tower. (Adapted from Zbiciski, I., *Chem. Eng. J.*, 58, 123, 1995.)

Assuming axisymmetric flow for the atomized droplets/particles (Figure 3.7), momentum balance in axial (x), radial (r), and tangential (t) directions gives changes in particle positions with distance from the atomizer (H)

$$\frac{du_{p,x}}{dH} = \left[\left(1 - \frac{\rho_f}{\rho_p}\right)g - \frac{3}{4}\frac{\rho_f C_D u_R \left(u_{p,x} - u_{f,x}\right)}{\rho_p d_p}\right]\frac{1}{u_{p,x}}, \tag{3.1}$$

$$\frac{du_{p,r}}{dH} = \left[\frac{u_{p,t}^2}{r} - \frac{3}{4}\frac{\rho_f C_D u_R \left(u_{p,r} - u_{f,r}\right)}{\rho_p d_p}\right]\frac{1}{u_{p,r}}, \tag{3.2}$$

$$\frac{du_{p,t}}{dH} = \left[-\frac{u_{p,t}u_{p,r}}{r} - \frac{3}{4}\frac{\rho_f C_D u_R \left(u_{p,t} - u_{f,t}\right)}{\rho_p d_p}\right]\frac{1}{u_{p,t}}, \tag{3.3}$$

where u_R is the relative velocity between the particle and air is expressed by

$$u_R = \sqrt{\left(u_{p,x} - u_{f,x}\right)^2 + \left(u_{p,r} - u_{f,r}\right)^2 + \left(u_{p,t} - u_{f,t}\right)^2}. \tag{3.4}$$

Radial position of particles can be calculated from

$$\frac{dr}{dH} = \frac{u_{p,r}}{u_{p,x}}. \tag{3.5}$$

Changes in particle/droplet and air temperature with distance from the atomizer can be described by

$$\frac{dT_p}{dH} = \left[\frac{6\left(\alpha_p\left(T_f - T_p\right) - \frac{dm_p}{dt}\left(\Delta h_v + c_v T_f\right)\frac{1}{\pi d_p^2}\right)}{d_p \rho_p c_p} - \frac{u_{p,x} c_w T_p}{c_p}\frac{dX}{dH}\right]\frac{1}{u_{p,x}}, \tag{3.6}$$

$$\frac{dT_f}{dH} = \frac{1}{w_f c_H}\left[\left(h_{f,0} - h_f\right)\frac{dw_f}{dH} - w_f\left(c_v T_f + \Delta h_v\right)\frac{dY}{dH} - w_p\left(X c_w + c_s\right)\frac{dT_p}{dH} - w_p c_w T_p\frac{dX}{dH}\right], \tag{3.7}$$

where dw_f/dH is the entrainment rate, which can be determined on the basis of the increment of air stream in the dryer cross section

$$\frac{dw_f}{dH} = 2\pi r\frac{u_{f,x}\rho_f}{1+Y}\frac{dr}{dH}. \tag{3.8}$$

Changes in moisture content of particle stream and of air can respectively be calculated by

$$\frac{dX}{dH} = -6\frac{N\frac{dm_p}{dt}\left(1+X\right)}{\rho_p \pi d_p^3}\frac{1}{u_{p,x}}, \tag{3.9}$$

$$\frac{dY}{dH} = \frac{dw_f}{dH}\frac{Y_0 - Y}{w_f} - \frac{w_p}{w_f}\frac{dX}{dH}. \tag{3.10}$$

However, they do not offer any possibility of determining local changes in drying parameters, which is necessary to optimize system operation. The presence of multiple simplifications, such as the assumption of monodisperse droplets or perfect mixing of phases (no dispersion of particles, constant concentration of particles in drying chamber), does not allow for a thorough analysis of changes in the properties of the manufactured product.

In recent simulations of spray dryer operations, more advanced calculation methods have been used. The technique of Computational Fluid Dynamics (CFD), based on numerical calculations, makes it possible to analyze local changes of drying parameters and particle morphology by using 2D axisymmetric and full 3D simulations of the spray drying process.

CFD is a comprehensive calculation tool for solving problems related to the flow of fluids. The basis of the CFD method is a numerical calculation of fluid stream continuity inside a computational mesh, which involves solving the Navier–Stokes equation of momentum, mass, and energy conservation (N-S) in order to obtain local distributions of velocity, pressure, temperature, etc. (Ferziger and Perit 2002). The N-S equations in three-dimensional space are recalled here.

Momentum conservation equation:

$$\frac{\partial \rho_f}{\partial t}\left(\rho_f u_i\right) + \frac{\partial}{\partial x_i}\left(\rho_f u_j u_i\right) = -\frac{\partial p}{\partial x_i} + \frac{\partial}{\partial x_j}\left[\mu_e\left(\frac{\partial u_i}{\partial x_j} + \frac{\partial u_j}{\partial x_i}\right)\right] + \Delta g_i + u_{p,i} S_c + F_i.$$

$$i,j \in \langle 1,2,3\rangle \tag{3.11}$$

Mass conservation equation:

$$\frac{\partial \rho_f}{\partial t} + \frac{\partial}{\partial x_j}\left(\rho_f u_j\right) = S_c. \tag{3.12}$$

Energy conservation equation:

$$\frac{\partial}{\partial t}\left(\rho_f h\right) + \frac{\partial}{\partial x_j}\left(\rho_f u_j h\right) = \frac{\partial}{\partial x_j}\left(\frac{\mu_e}{\sigma_h}\frac{\partial h}{\partial x_j}\right) + S_h \tag{3.13}$$

where h (J/kg) denotes the specific enthalpy and it is calculated by:

$$h = \int_{T_{ref}}^{T} c_p \, dT. \tag{3.14}$$

In Equation 3.13, μ_e denotes the effective viscosity, which is the sum of the dynamic viscosity of liquid and the turbulent viscosity. It is calculated from the turbulence model:

$$\mu_e = \mu_f + \mu_t. \tag{3.15}$$

The N-S equations cannot usually be solved analytically. However, the N-S equations can be solved with high accuracy by numerical calculations.

Efforts to numerically solve the N-S equations started in the beginning of the twentieth century for calculations of laminar fluid flow ($Re = 20$) around cylinders (Thom 1933). The calculations were repeated by Kawaguti in 1953, who manually calculated the distribution of fluid velocity during steady flow around a cylindrical object for $Re = 40$. Kawaguti spent 18 months for calculations with the use of a mechanical calculator, applying the finite difference method (Kawaguti 1953). Based on the above work, researchers started to use numerical calculations of the N-S equations in aerodynamics (Apelt 1958). It was noted, however, that the calculation of flow around bodies for Reynolds numbers from the transient and turbulent flow range could not be performed at the steady state and required additional calculations describing flow turbulence. An important turning point in the history of CFD calculations was the development and validation of the first turbulence models for a wide range of Reynolds numbers (Launder and Spalding 1974). The ability to quickly solve complex differential equations resulted in the development of increasingly complex mathematical models of industrial-scale processes. These models required algorithms allowing a fast solution of the N-S equations. One of the first developed algorithms was SIMPLE (Semi Implicit Method for Pressure Linked Equations) (Patankar and Spalding 1972). The SIMPLE algorithm, which belongs to a group called "predictor-corrector" algorithms, calculates iteratively changes in the flow rate and pressure of fluid based on the momentum conservation equation (Equation 3.11). The calculation continues until the desired convergence of solution is achieved. The algorithms, which are most frequently used in the CFD calculations, are: SIMPLER (Semi Implicit Method for Pressure Linked Equations Revised), SIMPLEC (Semi Implicit Method for Pressure Linked Equations Consistent), and PISO (Pressure Implicit with Splitting of Operator). Here the method for solving the SIMPLE algorithm is summarized (Ferziger and Perit 2002):

1. Start calculations for time t^{n+1} using the last solution for velocity u_i^n and pressure p^n as an initial condition for the calculation of μ_i^{n+1} and p^{n+1}.
2. Calculation of velocity and pressure gradients using the momentum conservation equation.
3. Solution of the correction equation for pressure.
4. Correction of the results of local changes in velocity and pressure to obtain a convergent field of velocity distribution μ_i^n, on the basis of which the field of pressure changes p^n is calculated. For the PISO algorithm, an additional pressure-correcting equation is solved to obtain a new field of velocity and pressure distribution. For the SIMPLER algorithm, first the pressure distribution is calculated and then on this basis the velocity field is computed.
5. Return to point 2, where u_i and p^n are used as corrected initial conditions for the determination of μ_i^{n+1} and p^{n+1}. The calculation loop is performed as long as all correction factors are negligible.
6. Perform the next time step.

CFD calculation software has become a common tool for engineers. Not only fluid flow calculations are made but also calculations relating to heat and mass transfer, chemical reactions, multiphase flow, or the impact of flow on the structure

(Fluid Structure Interaction—FSI), as well as electromagnetic calculations. A range of CFD simulation programs are available on the market. The largest commercial provider of this software is ANSYS, which offers Fluent, CFX, or Polyflow (ANSYS Inc 2016). Other software, such as Comsol Multiphysics (COMSOL Multiphysics 2016), or open-source solvers such as OpenFOAM, (OpenFOAM Foundation 2016) is also available.

Fluid flow in industrial installations is usually turbulent, and laminar flows occur only in specific cases, for example, non-Newtonian fluid flow. The simulation of turbulent flows requires application of complex mathematical models that can describe swirl flow.

Although the origin of research on fluid flow turbulence dates back to a century ago, there is still no solution to this problem owing to its multiscale nature. Several methods have already been developed to solve this problem. The first method is Direct Numerical Simulation (DNS), though it requires high computing power due to the necessity of using dense computational meshes. This method is applied to solve the N-S equations for turbulent fluid flow without using additional models of turbulence (Ferziger and Perit 2002). As a result, a full instantaneous flow velocity field can be obtained. In this method, mesh elements cannot be bigger than the so-called Kolmogorov length scale, which decreases with an increase in Reynolds number. In addition, for calculations in transient state, the time step must be chosen such that fluid particles move by one mesh element in each time step, satisfying Courant–Friedrichs–Lewy (CFL) conditions (Courant number <1). The DNS is widely used to simulate the shape of turbulent eddies, the dispersion of the so-called aerodynamic noise, or calculations of hydrodynamic resistance during flow around solid bodies.

Another common method used in turbulent flow calculations is Large Eddy Simulation (LES), which also fully reflects the eddy structure of bigger eddies, but omits microturbulences. The LES method was developed in 1963 for computing air-mass movement in simulations of meteorological phenomena (Smagorinsky 1963). Later, it was implemented in engineering calculations in unlimited space, especially in aerodynamics.

The classical models of turbulence, which are commonly used in the CFD calculations, are based on the Reynolds concept, according to which turbulence may be regarded as the sum of time-averaged values and the fluctuation component that is a random function of time and space. Many models have been developed to solve the N-S equations for turbulent flow using the iterative method (RANS, Reynolds Average Navier–Stokes equations). The most common model of turbulence is the k-ε model, which allows the system of N-S equations to be solved by introducing two new variables: turbulent kinetic energy of fluid (k), and turbulence dissipation coefficient (ε). In a standard version of the k-ε model, these values are described by Launder and Spalding (1974)

$$\frac{\partial}{\partial t}\left(\rho_f k\right) + \frac{\partial}{\partial x_j} = \frac{\partial}{\partial x_j}\left(\frac{\mu_e}{\sigma_k}\frac{\partial k}{\partial x_j}\right) + G_k + G_b - \rho_f \varepsilon + S_k, \qquad (3.16)$$

$$\frac{\partial}{\partial t}\left(\rho_f \varepsilon\right) + \frac{\partial}{\partial x_j}\left(\rho_f \varepsilon u_j\right) = \frac{\partial}{\partial x_j}\left(\frac{\mu_e}{\sigma_\varepsilon}\frac{\partial \varepsilon}{\partial x_j}\right) + \frac{\varepsilon}{k}\left(C_1 G_k + C_2 \rho_f \varepsilon\right) + S_\varepsilon. \qquad (3.17)$$

In Equations 3.16 and 3.18, G_k is the source of turbulent kinetic energy connected with velocity changes. It is expressed by

$$G_k = \mu_t \left(\frac{\partial u_i}{\partial x_j} + \frac{\partial u_j}{\partial x_i} \right) \frac{\partial u_i}{\partial x_j}. \qquad (3.18)$$

G_b is the source of turbulent kinetic energy associated with fluid elasticity

$$G_b = -\beta g_i \frac{\mu_t}{Pr_t} \frac{\partial T}{\partial x_j}, \qquad (3.19)$$

where Pr_t is the turbulent Prandtl number, which is 0.85 in the standard k-ε model. β is the thermal expansion coefficient

$$\beta = -\frac{1}{\rho_f} \left(\frac{\partial \rho_f}{\partial T} \right)_p. \qquad (3.20)$$

Turbulent viscosity of fluid is calculated by

$$\mu_t = \rho_f C_\mu \frac{k^2}{\varepsilon}. \qquad (3.21)$$

Constants in the model are $C_1 = 1.44$, $C_2 = 1.92$, $C_\mu = 0.09$, $\sigma_k = 1.0$, $\sigma_\varepsilon = 1.3$, $\sigma_h = 0.9$.

The CFD solvers include many available RANS models, including modifications of the k-ε model such as RNG k-ε or Realizable k-ε, models of type k-ω and k-ω-SST, and the most advanced Reynolds Stress Model (RSM). The model of turbulence is selected individually for each problem at hand.

The method for calculation of the continuous phase properties after flowing over an element placed rigidly in space and time is called the Euler approach. The Euler method is used to calculate continuous phase or loose materials that behave like a fluid, for example, inside a fluidized bed. A system in which volume fraction of the discrete phase is less than 10%, calculations by the Euler method is impossible, since it may require generation of a mesh for every single particle. Moreover, the mesh may have to move and morph with particles. Thus, simplifications are required to reduce the computational cost. A group of discrete-phase particles can be reduced to points that move inside a computational domain. This method does not require a mesh for each particle and is called the Lagrange method. The combination of the Euler and Lagrange methods, first achieved in 1977, led to the development of the Particle Source in Cell method (PSI-Cell) (Crowe et al. 1977). In this method, streams of particles tracked inside the computational domain are treated as the sources of mass, heat, and momentum within each single calculation cell. Figure 3.8 depicts the numerical calculations by the Euler–Lagrange method for steady and unsteady conditions.

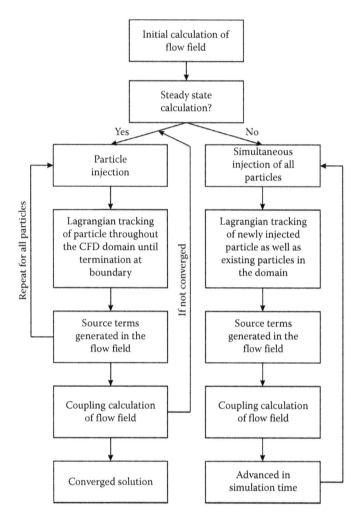

FIGURE 3.8 Numerical procedure for coupled Lagrangian particle tracking in steady and unsteady states.

For stationary calculations, the trajectory of particles is updated as long as the desired convergence of the solution is reached. For unsteady state simulations, the calculations are performed until the pseudo-steady state is reached. Convergence of the results should be obtained in each time step. Particle trajectories are described by Newton's second law

$$m_p \frac{d\vec{u}_p}{dt} = \vec{F}_{drag} + \vec{F}_{grav} + \vec{F}_{buo} + \vec{F}_{other}, \tag{3.22}$$

where F_{drag} is the hydrodynamic resistance force calculated from

$$\vec{F}_{drag} = \frac{\rho_f A_{cross}}{2} C_D |\vec{u}_f - \vec{u}_p| (\vec{u}_f - \vec{u}_p),$$ (3.23)

where A_{cross} is the cross-sectional area of a droplet. Drag coefficient (C_D) can be calculated for spherical particles from

$$C_D = \frac{24}{Re_p} \left(1 + 0.15 Re_p^{0.687}\right),$$ (3.24)

or for nonspherical particles the following model can be used (Haider and Levenspiel 1989):

$$C_D = \frac{24}{Re_p} \left(1 + b_1 Re_p^{b_2}\right) + \frac{b_3 Re_p}{b_4 + Re_p},$$ (3.25)

where

$$b_1 = \exp\left(2.3288 - 4.4581 \cdot d_r + 2.4486 \cdot d_r^2\right),$$
$$b_2 = 0.0964 - 0.5565 \cdot d_r,$$
$$b_2 = 0.0964 - 0.5565 \cdot d_r,$$
$$b_4 = \exp\left(1.4681 + 12.2584 \cdot d_r - 20.7322 \cdot d_r^2 + 15.8855 \cdot d_r^3\right),$$
$$d_r = \frac{A_s}{A_p},$$

where A_s is the surface area of a sphere of the same volume as the particle and A_p is the surface area of the particle. Re_p is the relative Reynolds number, which is defined as

$$Re_p = \frac{\rho_f d_p |\vec{u}_f - \vec{u}_p|}{\mu_f}.$$ (3.26)

F_{grav} is the gravity force

$$F_{grav} = m_p \frac{\rho_p - \rho_f}{\rho_p} \vec{g}.$$ (3.27)

F_{buo} is the buoyancy force given by

$$\vec{F}_{buo} = -\rho_f \frac{\pi d_p^3}{6} \vec{g}. \tag{3.28}$$

The particle may be affected by other forces (F_{other}), such as Saffman's lift force, Brownian motion, the Magnus effect, or electromagnetic forces. However, in the calculation of particle flow, where density of fluid is close to density of particle, additional forces arise due to the pressure gradient in the fluid and for acceleration of fluid surrounding particle (the so-called pressure and virtual mass forces).

The temperature of a particle during drying changes due to convection, moisture evaporation, or heat exchange by radiation:

$$\frac{d\left(m_p c_p T_p\right)}{dt} = \pi d_p^2 \alpha_p \left(T_f - T_p\right) + \sum \dot{M}_i \Delta h_{i,v} + \pi d_p^2 \varepsilon_p \sigma \left(T_s^4 - T_p^4\right). \tag{3.29}$$

Note that in standard calculations, thermal radiation heat transfer is usually neglected. The evaporation rate is calculated on the basis of mass transfer equation

$$\frac{dm_p}{dt} = -\sum \dot{M}_i = -\pi d_p^2 \sum \tilde{M}_i \beta_p \left(C_{i,s} - C_{i,f}\right), \tag{3.30}$$

where α_p and β_p denote the heat and mass transfer coefficients between particle surface and surrounding air, respectively. These coefficients can be calculated from Ranz–Marshall equations

$$\alpha_p = \frac{\lambda_f}{d_p}\left(2.0 + 0.6 Re_p^{1/2} \cdot Pr^{1/3}\right), \tag{3.31}$$

$$\beta_p = \frac{\delta_f}{d_p}\left(2.0 + 0.6 Re_p^{1/2} \cdot Sc^{1/3}\right), \tag{3.32}$$

where δ_v is the diffusion coefficient of water vapor in air.

The molar concentration of liquid vapor evaporating from the particle surface (based on Raoult's law) and into the surrounding air are calculated by

$$C_{i,s} = x_i \frac{P_{sat}\left(T_p\right)}{R \cdot T_p}, \quad C_{i,f} = y_i \frac{P_{op}}{R \cdot T_p}, \tag{3.33}$$

Mass transfer from the wet core of a particle through the solid crust into the surrounding air during drying is a complex process. The last step in devising a model for predicting the spray drying process is to develop or use a single droplet/particle drying model. Such a drying model accounts for moisture evaporation from a droplet surface and also the process of particle drying with reference to internal diffusion resistance. In the literature, various models of single droplet drying are presented. They can be divided into three groups (Mezhericher et al. 2010): deterministic models, empirical models based on the characteristic drying curves (CDC), and models based on the reaction engineering approach (REA).

Deterministic models take into account the description of drying kinetics and the phenomena associated with changes in particle morphology, such as the formation and growth of dry crust on particle surface. Determination of dry-layer thickness, which changes during drying, makes it possible to calculate resistance in moisture transport inside the particle. This method, however, requires moving boundary conditions. One of the deterministic models was proposed by Mezhericher (2011), who divided particle drying time into stages in reference to particle temperature.

After droplets leave the atomizer, they start to be heated until they reach a constant temperature, the so-called wet bulb temperature (Figure 3.9 points 0–1). During this stage, assuming no gradient in solid concentration and the lack of temperature distribution inside the droplet, the evaporation rate decreases only due to the particle shrinkage. All thermal energy supplied to the droplet is used for phase transition and can be described by Equations 3.29 and 3.30. A constant particle temperature is maintained until the droplet reaches a level of moisture content at which a solid crust begins to form on its surface (Figure 3.9 point 2). This point refers to the onset of the locking point. This moisture content is called critical moisture content (X_{cr}).

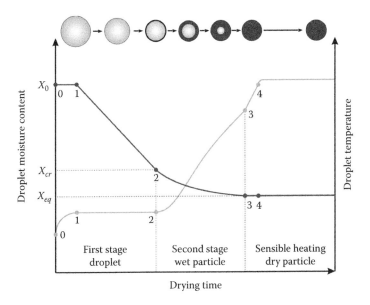

FIGURE 3.9 Changes in particle moisture content and temperature during the single droplet drying process.

Upon exceeding the critical moisture content, the particle enters the falling drying stage (Figure 3.9 points 2–3), in which its temperature begins to rise and the drying rate drops. In the course of drying, the solid crust surrounding the droplet is enlarged, thus diffusion resistance inside the crust increases. Hence, the particle can be considered as a sphere with variable wall thickness, which requires calculations of two different temperatures, that is, calculations of the wet core (T_{wc}) and the solid crust (T_{crs}) (Mezhericher 2011):

$$\frac{d\left(\rho_{wc} c_{wc} T_{wc}\right)}{dt} = \frac{1}{r^2}\frac{d}{dr}\left(\lambda_{wc} r^2 \frac{dT_{wc}}{dr}\right), \quad 0 \leq r \leq R_{crs}\left(t\right) \tag{3.34}$$

$$\frac{dT_{crs}}{dt} = \frac{\alpha_{cr}}{r^2}\frac{d}{dr}\left(r^2 \frac{dT_{crs}}{dr}\right), \quad R_{crs}\left(t\right) \leq r \leq R_p. \tag{3.35}$$

The change in the crust wall thickness is determined from

$$\frac{dR_{crs}}{dt} = -\frac{1}{\varepsilon\rho_{wc} 4\pi R_{crs}^2}\frac{dm_p}{dt}. \tag{3.36}$$

Once the particle is completely dried, its temperature reaches the ambient air temperature (Figure 3.9 points 3–4) and can be determined from Equations 3.34 and 3.35 for the full range of radius r from 0 to R_p.

The method proposed by Mezhericher is based on the thermal state of a particle, a different approach for calculation of a single particle drying was proposed by Tran et al. (2016b) where the locking point is determined from the measured critical solid concentration on the droplet surface. Changes in radial gradient of solid concentration during the drying process can be calculated from

$$\frac{dC}{dt} = \frac{1}{r^2}\frac{d}{dr}\left(r^2 \delta_{w,s}\frac{dC}{dr}\right). \tag{3.37}$$

Changes in particle size due to water evaporation can be calculated from

$$\frac{dr_p}{dt} = -\frac{\rho_f}{\rho_p} k_m \left(Y_p - Y_f\right). \tag{3.38}$$

When the solid concentration on the droplet surface reaches the critical value, particle shrinkage stops; however, water core volume still decreases. Difference between radius of water core (r_{wc}) and critical diameter gives the thickness of growing solid crust. To calculate the evaporation rate from water core to surrounding air, internal resistance for heat and mass transfer in the crust is respectively accounted for in the heat and mass transfer coefficients

$$k_m = \frac{1}{\dfrac{2r_p}{Sh \cdot \delta_{v,f}} + \dfrac{r_p^2}{\delta_{v,crs}}\left(\dfrac{1}{r_{wc}} - \dfrac{1}{r_p}\right)}, \tag{3.39}$$

$$k_h = \cfrac{1}{\cfrac{2r_p}{Nu \cdot \lambda_f} + \cfrac{r_p^2}{\lambda_{crs}}\left(\cfrac{1}{r_{wc}} - \cfrac{1}{r_p}\right)}.$$

(3.40)

The most frequently used methods of two-stage modeling of drying are based on the experimental determination of drying kinetics of the material being tested. One of these methods is the Characteristic Drying Curve model (CDC), in which drying rate decreases as a function of particle moisture content. For this purpose, drying rate (Equation 3.30) is multiplied by the additional correction factor f (drying rate retardation coefficient), which depends on mean particle water content ($\bar{X}$) (Van Meel 1958; Keey and Suzuki 1974; Kuts et al. 1996)

$$f = \frac{\dot{M}_w}{\dot{M}_{w,I}},$$

(3.41)

where
$\dot{M}_w$ is the drying rate
$\dot{M}_{w,I}$ is the drying rate in the first drying stage

In the first drying stage, the relative drying rate is equal to unity. After the critical point, due to an increase in internal mass transfer resistance in the particle, f decreases until it reaches $f = 0$ at the equilibrium point. In the second drying stage, f can be expressed as a function of normalized particle water content ϕ

$$\phi = \frac{\bar{X} - \bar{X}_{eq}}{\bar{X}_{cr} - \bar{X}_{eq}}.$$

(3.42)

When the equilibrium moisture content of the material is small or unknown, it is assumed that $X_{eq} = 0$, and Equation 3.42 is simplified to the form $\phi = \bar{X}/\bar{X}_{cr}$. Experimental analyses show that the decrease in the drying rate in the second stage can be nonlinear, and can be described by polynomial functions of instantaneous moisture content of particles (Huang et al. 2004). However, often the basic Equation 3.42 is modified as a power function whose exponent is determined experimentally (Woo et al. 2008)

$$f = \phi^n.$$

(3.43)

The last group of droplet drying models is based on the Reaction Engineering Approach (REA). In the REA method, molar concentrations of a volatile component on the particle surface are modified (Chen and Patel 2008). The molar concentration of a volatile component on the surface is a function of saturation concentration ($C_{i,sat}$)

and is expressed by the equation analogous to the description of chemical reaction kinetics

$$C_{i,s} = C_{i,sat} \exp\left(-\frac{\Delta E_v}{RT_p}\right). \qquad (3.44)$$

Parameter ΔE_v is the apparent energy of activation and is material-type dependent. It can be determined by relating it to the material moisture content

$$\frac{\Delta E_v}{\Delta E_{v,\infty}} = a \cdot e^{-b(\bar{X} - \bar{X}_{eq})^c}, \qquad (3.45)$$

where constants a, b, and c should be obtained from curves fitted to experimental data. Assuming that the continuous phase is a perfect gas, $\Delta E_{v,\infty}$ can be calculated from the relationship

$$\Delta E_{v,\infty} = -RT_g \ln(\varphi), \qquad (3.46)$$

where φ is the relative humidity of air.

Models of droplet drying based on the CDC and REA methods are most frequently implemented in CFD solvers. The advantage of the CDC and REA methods is an easy implementation in complex CFD models and fast simulation of the droplet drying process. A serious limitation of both methods is the necessity of experimental determination of the drying curve for a tested material in order to fit parameters of the two models. The CDC and REA models can be used to calculate the spray drying microencapsulation process. In a single liquid droplet with core material which is not evaporating, temperature changes due to convective heating of the core and carrier and due to evaporation from the carrier into the surrounding air. Therefore, heat balance for the microcapsule during drying can be expressed by:

$$\frac{dT_p \left(m_{crs} c_{crs} + m_{core} c_{core}\right)}{dt} = \pi d_p^2 \alpha_p \left(T_f - T_p\right) + \sum \dot{M}_i \Delta h_{i,v}, \qquad (3.47)$$

where
  m_{crs} is the mass of carrier/crust
  m_{core} is the mass of encapsulated core

However, mass of the encapsulated core depends on the dispersion rate of the core material in the droplet and on the atomization parameters (e.g., droplet particle size distribution). For this reason, it is very difficult to determine the mass of the core material inside the crust.

3.6 SPRAY COOLING AND SPRAY CHILLING MICROENCAPSULATION

Though spray drying is an easy and fast method used for powder production, even a small overheating of the powder can damage or destroy thermosensitive core materials. For this reason, low temperature spray encapsulation methods are developed. Spray cooling and chilling use the physical phenomenon of phase change as a result of reducing the carrier material temperature. In this process, solid particle creation is not the consequence of moisture evaporation but physical change of phase due to decreasing temperature of droplet below the solidification or crystallization point. Similar to spray drying, there are four main stages in spray cooling/chilling techniques (Figure 3.10):

1. Preparation of feed and cooling medium
2. Spraying the slurry by using different types of atomizers inside the cooling chamber
3. Particle solidification by decreasing temperature of the carrier material
4. Product receiving and separation of the powder from the exhaust air; air recirculation is optional

Because in spray cooling/chilling method there is no water evaporation, feed is prepared in a different way. Core material can be emulsified in molten carrier material. Spray cooling and chilling can be distinguished with regard to the melting point of

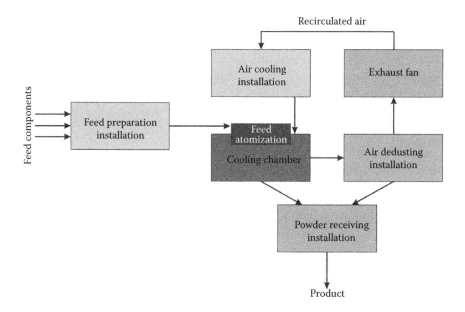

FIGURE 3.10 General scheme of the spray cooling/chilling process.

the carrier material. In spray chilling carrier materials with a relatively low melting temperature are used (32–42°C), whereas in spray cooling melting point of carrier is in the range between 45 and 122°C (Madene et al. 2006; Okuro et al. 2013). To keep the feed in liquid state, carrier material needs to be warmed up to a temperature over the melting point and constantly mixed. For this reason, preparation tanks are thermostated and have a mounted stirrer.

A relatively low melting temperature of carrier is a criterion for the selection of material that will create the capsule wall. In spray cooling, mostly natural polymers, polysaccharides, waxes, or fats are used as a molten matrix material. For chilling encapsulation methods, materials such as vegetable oils, fatty acids, or waxes are used as the carrier material. Low temperature encapsulation is mostly used to capsulate thermosensitive, volatile, or biologically active materials such as vitamins, proteins, enzymes, acidulants, probiotics, or living bacteria (Pedroso et al. 2012).

Spraying of the feed inside the cooling chamber can be performed by using the same methods and by the same atomizers as in spray drying microencapsulation. However, here atomizers and pipes, which connect to the feed tank and cooling chamber, need to be warmed up and thermostated to prevent premature carrier solidification before atomization. During the atomization inside the cooling chamber, the temperature of carrier material decreases due to the contact with the cooling medium. In spray cooling microencapsulation, mostly cold air is used, but in chilling processes when much lower temperatures are required, cooling can be performed by spraying liquid nitrogen into the chamber or using a carbon-dioxide ice bath (Madene et al. 2006).

No mass transfer (evaporation) between droplets and cooling air results in a lack of thermal stress during droplet solidification, therefore particles have an almost perfect spherical shape. Microcapsules produced by spray cooling/chilling are mostly insoluble due to the lipids content in the capsule crust. For this reason, spray cooling/chilling microencapsulation is applied in many manufacturing sectors like pharmaceutical, cosmetic, agriculture, and food industry. Separation of the powder from the exhaust cooling agent can be performed similar to the spray drying process using cyclones and bag filters. In spray cooling systems, exhaust air can be recycled for reuse. Unlike spray drying, exhaust air is dry and can be cooled and used again after purification from the powder.

Though spray cooling and chilling microencapsulations are easy to perform, there are several disadvantages. First, cooling of the chamber in these techniques can be more expensive than heating the drying agent in spray drying. Second, the powder needs to be stored under specific conditions as otherwise carrier material may melt and thus release an encapsulated core material. Table 3.2 presents state-of-the-art research on the spray cooling and spray chilling microencapsulation.

TABLE 3.2

A Summary of Studies on Spray Cooling and Spray Chilling Microencapsulation

Encapsulated Material (EM)	Encapsulating Agent (EA)	Operational Conditions	Encapsulation Efficiency (%)	Reference
Ascorbic acid	Interesterified fat produced from fully hydrogenated palm oil and palm kernel oil (melting point: 43°C) Emulsifier: soy lecithin powder	Atomizer: twin fluid Nozzle diameter: 1.2 mm Temperature of cooled chamber: 13°C Air pressure: 216 kPa Feed flow rate: 50 mL/min EM to EA ratio: 1:7, 1:10	59.2–72.5	Matos-Jr et al. (2017)
Ascorbic acid	Stearic acid (SA)/hydrogenated vegetable fat (HVF) (SA/HVF: 1:1 w/w, melting point: 49.5°C)	Nozzle diameter: 0.7 mm Temperature of cooled chamber: Inlet: 5°C, Outlet: 15°C Feed flow rate: 12 mL/min Aspiration rate: 24.5 m³/h	97.8	Alvim et al. (2016)
Gallic acid	Fully hydrogenated soybean oil (FHSO)/soybean oil (SO) (FHSO/SO: 60:40, 70:30, 80:20, 90:10 w/w, melting point: 74°C) Surfactant: PGRP	Atomizer: double fluid Nozzle diameter: 0.7 mm Temperature of cooled chamber: Inlet: 7°C, Outlet: 11°C Feed flow rate: 0.530 L/h Cooling air flow rate: 35,000 L/h Atomizing air flow rate: 667 L/h EM to EA ratio: 30:70, 20:80	54.14–83.48	Consoli et al. (2016)

(Continued)

TABLE 3.2 (*Continued*)

A Summary of Studies on Spray Cooling and Spray Chilling Microencapsulation

Encapsulated Material (EM)	Encapsulating Agent (EA)	Operational Conditions	Encapsulation Efficiency (%)	Reference
Ginger oleoresin	Palmitic acid/oleic acid/palm fat	Atomizer: double fluid Nozzle diameter: 2 mm Temperature of cooled chamber: 7°C Feed flow rate: 0.7 kg/h Cooling air flow rate: 35,000 L/h Atomizing air flow rate: 1,052 L/h EM to EA ratio: 10:90	89.8–97.3	Oriani et al. (2016)
Vitamin D$_3$	Vegetable fat (melting point: 49°C) Emulsifier: soy lecithin, beeswax	Atomizer: double fluid pressurized Nozzle diameter: 1.2 mm Temperature of cooled chamber: 13°C Air pressure: 2.2 kgf/cm^2 Feed flow rate: 50 mL/min EM/EA: 0.1:99.9	NR	Paucar et al. (2016)
Proanthocyanidins-rich-cinnamon	Vegetable fat (melting point: 48°C)	Nozzle diameter: 1.2 mm Temperature of cooled chamber: 13°C Feed flow rate: 45 mL/min EM/EA: 4.8:75.2, 8:72	~87	Tulini et al. (2016)
Ascorbic acid	Vegetable glycerol monostearate 90 (melting point: 79.15°C) Fully hydrogenated palm oil (melting point: 64.02°C)	Atomizer: double fluid Temperature of cooled chamber: 22°C Air pressure: 4 bar Atomization temperature: 100°C Ascorbic acid concentration: 15, 25, 40 %w/w	74.66–84.5	Matos-Jr et al. (2015)

(Continued)

TABLE 3.2 (*Continued*)
A Summary of Studies on Spray Cooling and Spray Chilling Microencapsulation

Encapsulated Material (EM)	Encapsulating Agent (EA)	Operational Conditions	Encapsulation Efficiency (%)	Reference
Lycopene	Shortening composed of hydrogenated and interesterified cottonseed, soy and palm oil (melting point: 51°C)	Nozzle diameter: 1.2 mm Temperature of cooled chamber: 13°C Air pressure: 1 kgf/cm² Feed flow rate: 40 mL/min EM/EA: 15:60, 18:60, 25:60, 30:60	NR	Pelissari et al. (2016)
Ascorbic acid	Lauric acid (LA)/oleic acid (OA) (LA/OA: 70:30, 80:20, 90:10 w/w, melting points: 40–48°C) Surfactant: PGRP	Atomizer: double fluid pressurized Nozzle diameter: 0.7 mm Temperature of cooled chamber: Inlet: 6°C, Outlet: 9.5°C Feed flow rate: 5.28e–4 m³/h Cooling air flow rate: 35 m³/h Atomizing air flow rate: 0.66 m³/h EM to EA ratio: 20:80, 25:75, 30:70	58–88	Sartori et al. (2015)
Phytosterol	Low trans hydrogenated vegetable fat/stearic acid	Atomizer: Two fluid pressurized Nozzle diameter: 0.7 mm Temperature of cooled chamber: 5°C Atomizing air flow rate: 500–600 L/h	NR	Alvim et al. (2013)
Glucose solution	Stearic acid (SA)/oleic acid (OA) (SA/OA: 70:30, 80:20, 90:10, 100:0 w/w, melting points: 46.9–58.4°C) Emulsifier: soy lecithin	Atomizer: double fluid Nozzle diameter: 0.7 mm Temperature of cooled chamber: 0°C Air pressure: 1.25 kgf/cm² Atomization temperature: 60–65°C EM/EA: 25:75	75.1–96.9	Ribeiro et al. (2012)

(*Continued*)

TABLE 3.2 (*Continued*)

A Summary of Studies on Spray Cooling and Spray Chilling Microencapsulation

Encapsulated Material (EM)	Encapsulating Agent (EA)	Operational Conditions	Encapsulation Efficiency (%)	Reference
Bovine serum albumin	Glyceryl palmitostearate (melting point: 58°C)	Atomizer: wide pneumatic nozzle (an innovative external-mix two fluid atomizer)	Above 90	Sabatino et al. (2012)
	Trimirystin (melting point: 57°C)	Temperature of cooled chamber: room temperature		
	Tristearin (melting point: 72°C)	Air pressure: 2.5 bar		
		Atomization temperature: 100°C		
		Bovine serum albumin concentration: 10, 20%w/w		
α-Tocopherol	Interesterified fat without trans isomerase produced by cotton seed oil with fully hydrogenated palm oil (IF)/fully hydrogenated soy bean oil (FHSO) (IF/FHSO: 70:30 w/w)	Atomizer: double fluid	90–95.8	Gamboa et al. (2011)
		Nozzle diameter: 0.7 mm		
		Temperature of cooled chamber: 10°C		
		Air pressure: 0.25 MPa		
		Atomization temperature: 65°C		
		EM/EA: 20:80, 15:85, 10:90, 5:95		

(Continued)

TABLE 3.2 (*Continued*)

A Summary of Studies on Spray Cooling and Spray Chilling Microencapsulation

Encapsulated Material (EM)	Encapsulating Agent (EA)	Operational Conditions	Encapsulation Efficiency (%)	Reference
Milled theophylline	Compritol	Atomizer: wide pneumatic nozzle	~100	Passerini et al. (2010)
		Orifice diameter: 4.5 mm		
		Temperature of cooled chamber: ambient temperature		
		Air pressure: 2 bar		
		Atomization temperature: 120°C		
		EM to EA ratio: 30:70, 20:80, 10:90		
		Atomizer: conventional air pressure nozzle	<100	
		Temperature of cooled chamber: ambient temperature		
		Air pressure: 1.5 bar		
		EM to EA ratio: 30:70, 20:80, 10:90		
Econazole nitrate	Gelucire 53/10 (melting point: 55°C)/chitosan	Atomizer: wide pneumatic nozzle	21.03–35.23 (EM content)	Albertini et al. (2009)
	Gelucire 53/10 (melting point: 55°C)/sodium carboxymethylcellulose	Air pressure: 1.5 bar		
	Gelucire 53/10 (melting point: 55°C)/poloxamers (Lutrol F68 and F127)	Econazole nitrate concentration: 20, 30 %w/w		

(Continued)

TABLE 3.2 (*Continued*)
A Summary of Studies on Spray Cooling and Spray Chilling Microencapsulation

Encapsulated Material (EM)	Encapsulating Agent (EA)	Operational Conditions	Encapsulation Efficiency (%)	Reference
Glimepiride	Gelucire 50/13 Poloxamer 188 PEG 6000	Atomizer: duale fluid atomizer Nozzle diameter: 0.7 mm Nozzle cap: 1.5 mm Temperature of cooled chamber: 0–5°C Air pressure: 1.4, 1.8, 2 bar Feed flow rate: 6 mL/min Glimepiride: 1.7% w/w	~100	Ilić et al. (2009)
Glucose solution	Stearic acid (SA)/oleic acid (OA)/hydrogenated vegetable oil (HVO) (SA/OA/HVO: 2:1:1 w/w, melting point: 49.3°C) Surfactant: Lecithin	Atomizer: double fluid atomizer Nozzle diameter: 0.7 mm Temperature of cooled chamber: 10°C Air pressure: 1 kgf/cm² EM to EA ratio: 10:90, 25:75	78.3–97.8	Leonel et al. (2010)
Glucose, Casein protein hydrolyzed	Stearic acid (SA)/Lauric acid (LA) (SA/LA: 57:43 w/w, melting point: 56.8°C) Stearic acid (SA)/oleic acid (OA) (SA/OA: 50:50 w/w, melting point: 56.8°C)	Atomizer: double fluid atomizer Nozzle diameter: 0.7 mm Temperature of cooled chamber: 10°C Air pressure: 1 kgf/cm² Atomization temperature: 70°C EM to EA ratio: 30:70, 20:80, 10:90	76.4–100	Chambi et al. (2008)

(Continued)

TABLE 3.2 (*Continued*)

A Summary of Studies on Spray Cooling and Spray Chilling Microencapsulation

Encapsulated Material (EM)	Encapsulating Agent (EA)	Operational Conditions	Encapsulation Efficiency (%)	Reference
Insulin	Glycerol tripalmitate	Atomizer: pressure nozzle Nozzle orifice: 300 μm Spraying pressure: 5, 5.5, 6 bar Atomization temperature: 70, 75, 80°C	~100	Maschke et al. (2007)
Verapamil HCL	Crocrystalline wax Stearyl alcohol Crocrystalline wax/stearyl alcohol Surfactant: soya lecithin	Atomizer: ultrasound atomizer Frequency: 25 kHz Power output: 2.2 kW Feed temperature: 70°C	49–100	Passerini et al. (2003)

NOMENCLATURE

A	area (m^2)
C	molar concentration (mol/m)
c	specific heat (J/kg/K)
C_D	drag coefficient (—)
d	diameter (m)
ΔE_v	apparent activation energy (J/mol)
f	drying rate retardation coefficient (—)
F	force (N)
g	gravity acceleration (m/s^2)
h	specific enthalpy (J/kg)
H	height (m)
Δh	latent heat of evaporation (J/kg)
k	turbulent kinetic energy ($m^{-2}\ s^{-2}$)
k_h	overall heat transfer coefficient ($W/m^2/K$)
k_m	overall mass transfer coefficient (m/s)
$\dot{M}$	evaporation rate (kg/s)
m	mass (kg)
Nu	Nusselt number (—)
p	pressure (Pa)
Pr	Prandtl number (—)
r	radius (m)
R	universal gas constant (J/mol/K)
Re	Reynolds number (—)
Sc	Schmidt number (—)
T	temperature (K, °C)
t	time (s)
u	velocity (m/s)
w	mass flux (kg/s)
X	moisture content (kg/kg)
x	space coordinate (m)
x	mole fraction in liquid (—)
Y	gas moisture content (kg/kg)
y	mole fraction in gas (—)

GREEK LETTERS

α	heat transfer coefficient ($W/m^2/K$)
β	mass transfer coefficient (m/s)
δ	droplet void fraction (—)
δ	diffusion coefficient (m^2/s)
ε	turbulent dissipation rate (m^2/s^3)
ε	emissivity (—)
λ	thermal conductivity (W/m/K)
μ_e	effective viscosity (kg/m/s)
ρ	mass density (kg/m^3)

σ   mechanical stress (Pa)

σ   Stefan–Boltzmann constant ($W/m^2/K^4$)

φ   normalized water content (—)

φ   relative air humidity (—)

Subscripts

‾   average

0   initial

cr   critical

crs   crust

eq   equilibrium

f   fluid

h   heat

in   inlet

m   mass

op   operating

out   outlet

p   particle

r   radial direction

r   relative

s   surface

sat   saturation

t   tangential direction, turbulent

v   vapor

w   water

wc   water core

x   axial direction

REFERENCES

Adamiec, J. 2009. Drying of microorganisms for food applications, in C. Ratti (ed.), *Advances in Food Dehydration*. Boca Raton, FL: CRC Press, pp. 315–354.

Aghbashlo, M., H. Mobli, A. Madadlou, and S. Rafiee. 2013a. Fish oil microencapsulation as influenced by spray dryer operational variables. *International Journal of Food Science and Technology* 48:1707–1713.

Aghbashlo, M., H. Mobli, A. Madadlou, and S. Rafiee. 2013b. Influence of wall material and inlet drying air temperature on the microencapsulation of fish oil by spray drying. *Food and Bioprocess Technology* 6(6):1561–1569.

Albertini, B., N. Passerini, M.D. Sabatino, B. Vitali, P. Brigidi, and L. Rodriguez. 2009. Polymer–lipid based mucoadhesive microspheres prepared by spray-congealing for the vaginal delivery of econazole nitrate. *European Journal of Pharmaceutical Sciences* 36:591–601.

Alvim, I.D., F.S. Souza, I.P. Koury, T. Jurt, and F.B.H. Dantas. 2013. Use of the spray chilling method to deliver hydrophobic components: Physical characterization of microparticles. *Ciência e Tecnologia de Alimentos* 33(1):34–39.

Alvim, I.D., M.A. Stein, I.P. Koury, F.B.H. Dantas, and C.C.V. Cruz. 2016. Comparison between the spray drying and spray chilling microparticles contain ascorbic acid in a baked product application. *LWT—Food Science and Technology* 65:689–694.

ANSYS, Inc. 2016. ANSYS Customer Portal. (http://www.ansys.com/). Accessed May 06, 2017.

Apelt, C.J. 1958. The steady flow of a viscous fluid past a circular cylinder at Reynolds Numbers 40 and 44. Ministry of Aviation A.R.C. Technical Report R. & M. No. 3175. University of Oxford, UK.

Battista, C.A.D., D. Constenla, M.V. Ramírez-Rigo, and J. Piña. 2015. The use of arabic gum, maltodextrin and surfactants in the microencapsulation of phytosterols by spray drying. *Powder Technology* 286:193–201.

Carneiro, H.C.F., R.V. Tonon, C.R.F. Grosso, and M.D. Hubinger. 2013. Encapsulation efficiency and oxidative stability of flaxseed oil microencapsulated by spray drying using different combinations of wall materials. *Journal of Food Engineering* 115:443–451.

Chambi, H.N.M., I.D. Alvim, D. Barrera-Arellano, and C.R.F. Grosso. 2008. Solid lipid microparticles containing water-soluble compounds of different molecular mass: Production, characterization and release profiles. *Food Research International* 41:229–236.

Chang, C., N. Varankovich, and M.T. Nickerson. 2016. Microencapsulation of canola oil by lentil protein isolate-based wall materials. *Food Chemistry* 212:264–273.

Chen, X.D. and K.C. Patel. 2008. The reaction engineering approach to estimate surface properties of aqueous droplets during convective drying, in *Proceedings of 16th International Drying Symposium*, Hyderabad, India, pp. 235–241.

Chong, P.H., Y.A. Yusof, M.G. Aziz, N.M. Nazli, N.L. Chin, and S.K. Syed Muhammad. 2014. Effects of spray drying conditions of microencapsulation of *Amaranthus gangeticus* extract on drying behaviour. *Agriculture and Agricultural Science Procedia* 2:33–42.

COMSOL Multiphysics. 2016. COMSOL Multiphysics customer portal. (http://www.comsol. com/). Accessed May 6, 2017.

Consoli, L., R. Grimaldi, T. Sartori, F.C. Menegalli, and M.D. Hubinger. 2016. Gallic acid microparticles produced by spray chilling technique: Production and characterization. *Food Science and Technology* 65:79–87.

Crowe, C.T., M.P. Sharma, and D.E. Stock. 1977. The particle-source-in-cell (PSI-Cell) model for gas-droplet flows. *Journal of Fluid Engineering* 99(2):325–332.

El-Sayed, T.M., D.A. Wallack, and C.J. King. 1990. Changes in particle morphology during drying of drops of carbohydrate solutions and food liquids, Part 1. Effect of composition and drying conditions. *Industrial & Engineering Chemistry Research* 29(12):2346–2354.

Fang, Z. and B. Bahandri. 2010. Encapsulation of polyphenols—A review. *Trends in Food Science & Technology* 21:510–523.

Fernandes, R.V.B., S.V. Borges, E.K. Silva, Y.F. Silva, H.J.B. Souza, E.L. Carmo, C.R. Oliveira, M.I. Yoshida, and D.A. Botrel. 2016. Study of ultrasound-assisted emulsions on microencapsulation of ginger essential oil by spray drying. *Industrial Crops and Products* 94:413–423.

Ferziger, J.H. and M. Perit 2002. *Computational Methods for Fluid Dynamiks*, 3rd edn. New York: Springer.

Fogler, B.B. and R.V. Kleinschmidt. 1938. Spray drying. *Industrial and Engineering Chemistry* 30:1771–1785.

Frascareli, E.C., V.M. Silva, R.V. Tonon, and M.D. Hubinger. 2012. Effect of process conditions on the microencapsulation of coffee oil by spray drying. *Food and Bioproducts Processing* 90:413–424.

Fu, N., M.W. Woo, and X.D. Chen. 2012. Single droplet drying technique to study drying kinetics measurement and particle functionality: A review. *Drying Technology* 30:1771–1785.

Gallardo, G., L. Guida, V. Martinez, M.C. López, D. Bernhardt, R. Blasco, R. Pedroza-Islas, and L.G. Hermida. 2013. Microencapsulation of linseed oil by spray drying for functional food application. *Food Research International* 52:473–482.

Gamboa, O.S., L.G. Gonçalves, and C.F. Grosso. 2011. Microencapsulation of tocopherols in lipid matrix by spray chilling method. *Procedia Food Science* 1:1732–1739.

Goula, A.M. and K.G.A. Adamopoulos. 2012. Method for pomegranate seed application in food industries: Seed oil encapsulation. *Food and Bioproducts Processing* 90(4):639–652.

Groenewold, C., C. Möser, H. Groenewold, and E. Tsotsas. 2002. Determination of single-particle drying kinetics in an acoustic levitator. *Chemical Engineering Journal* 86:217–222.

Guadarrama-Lezama, A.Y., L. Dorantes-Alvarez, M.E. Jaramillo-Flores, C. Perez-Alonso, K. Niranjan, G.F. Gutierrez-Lopez, and L. Alamilla-Beltran. 2013. Preparation and characterization of non-aqueous extracts from chilli (*Capsicum annuum* L.) and their microencapsulates obtained by spray-drying. *Journal of Food Engineering* 112:29–37.

Haider, A. and O. Levenspiel. 1989. Drag coefficient and thermal velocity of spherical and nonspherical particles. *Powder Technology* 58:63–70.

Hu, L., J. Zhang, Q. Hu, N. Gao, S. Wang, Y. Sun, and X. Yang. 2016. Microencapsulation of brucea javanica oil: Characterization, stability and optimization of spray drying conditions. *Journal of Drug Delivery Science and Technology* 36:46–54.

Huang, L., K. Kumar, and A.S. Mujumdar 2004. Computational fluid dynamic simulations of droplet drying in a spray dryer, in *Drying 2004—Proceedings of the 14th International Drying Symposium*, Sao Paulo, Brazil, pp. 326–332.

Hunziker, O.F. 1920. *Condensed Milk and Milk Powder*. La Grange, IL.

Ilić, I., R. Dreu, M. Burjak, M. Homar, J. Kerč, and S. Srčič. 2009. Microparticle size control and glimepiride microencapsulation using spray congealing technology. *International Journal of Pharmaceutics* 381:176–183.

Kastner, O., G. Brenn, D. Rensink, and C. Tropea. 2001. The acoustic tube evitator: A novel device for determining the drying kinetics of single droplets. *Chemical Engineering and Technology* 24(4):335–339.

Kawaguti, M. 1953. Numerical solution of the NS equations for the flow around a circular cylinder at Reynolds number 40. *Journal of Physical Society of Japan* 8(6):747–757.

Keey, R.B. and M. Suzuki. 1974. On the characteristic drying curve. *International Journal of Heat and Mass Transfer* 17:1455–1464.

Kha, T.C., M.H. Nguyen, P.D. Roach, and C.E. Stathopoulos. 2014. Microencapsulation of Gac oil: Optimisation of spray drying conditions using response surface methodology. *Powder Technology* 264:298–309.

Kinzer, G.D. and R. Gunn. 1951. The evaporation, temperature and thermal relaxation-time of freely falling waterdrops. *Journal of Meteorology* 8(2):71–83.

Kuts, P.S., C. Strumillo, and I. Zbicinski. 1996. Evaporation kinetics of single droplets containing dissolved biomass. *Drying Technology* 14(9):2041–2060.

Langrish, T.A.G. 2009. Multi-scale mathematical modelling of spray dryers. *Journal of Food Engineering* 93:218–228.

Launder, B.E. and D.B. Spalding. 1974. The numerical computation of turbulent flows. *Computer Methods in Applied Mechanics* 3(2):269–289.

Leonel, A.J., H.N.M. Chambi, D. Barrera-Arellano, H.O. Pastore, and C.R.F. Grosso. 2010. Production and characterization of lipid microparticles produced by spray cooling encapsulating a low molar mass hydrophilic compound. *Food Science and Technology* 30(1):276–281.

Lin, S.X.Q. and X.D. Chen. 2002. Improving the glass-filament method for accurate measurement of drying kinetics of liquid droplets. *Chemical Engineering Research and Design* 80(A):401–410.

Liu, W., X.D. Chen, Z.Z. Cheng, and C. Selomulya. 2016. On enhancing the solubility of curcumin by microencapsulation in whey protein isolate via spray drying. *Journal of Food Engineering* 169:189–195.

Madene, A., M. Jacquot, J. Scher, and S. Desobry. 2006. Flavour encapsulation and controlled release—A review. *International Journal of Food Science and Technology* 41:1–21.

Maschke, A., C. Becker, D. Eyrich, J. Kiermaier, T. Blunk, and A. Göpferich. 2007. Development of a spray congealing process for the preparation of insulin-loaded lipid microparticles and characterization thereof. *European Journal of Pharmaceutics and Biopharmaceutics* 65:157–187.

Matos-Jr, F.E., T.A. Comunian, M. Marcelo Thomazini, and C.S. Favaro-Trindade. 2017. Effect of feed preparation on the properties and stability of ascorbic acid microparticles produced by spray chilling. *LWT—Food Science and Technology* 75:251–260.

Matos-Jr, F.E., M.D. Sabatino, N. Passerini, C.S. Favaro-Trindade, and B. Albertini. 2015. Development and characterization of solid lipid microparticles loaded with ascorbic acid and produced by spray congealing. *Food Research International* 67:52–59.

Mehrad, B., B. Shabanpour, S.M. Jafari, and P. Pourashouri. 2015. Characterization of dried fish oil from Menhaden encapsulated by spray drying. *AACL Bioflux* 8(1):57–69.

Mezhericher, M. 2011. *Theoretical Modelling of Spray Drying Process*. Vol. 1. Drying Kinetics, Two and Three Dimensional CFD Modelling. Saarbrucken, Germany: Lambert Academic Publishing GmbH & Co. KG.

Mezhericher, M., A. Levy, and I. Borde. 2010. Theoretical models of single droplet drying kinetics: A review. *Drying Technology* 28(2):278–293.

Morales-Medina, R., F. Tamm, A.M. Guadix, E.M. Guadix, and S. Drusch. 2016. Functional and antioxidant properties of hydrolysates of sardine (*S. pilchardus*) and horse mackerel (*T. mediterraneus*) for the microencapsulation of fish oil by spray-drying. *Food Chemistry* 194:1208–1216.

Mujumdar, A.S. 2007. *Handbook of Industrial Drying*. Boca Raton, FL: CRC Press.

Okuro, P.K., F.E. Matos Jr, and C.S. Favaro-Trindade. 2013. Technological challenges for spray-chilling encapsulation of functional food ingredients. *Food Technology and Biotechnology* 51:171–183.

OpenFOAM Foundation. 2016. OpenFOAM customer portal. (http://www.openfoam.org). Accessed May 06, 2017.

Oriani, V.B., I.D. Alvim, L. Consoli, G. Molina, G.M. Pastore, and M.D. Hubinger. 2016. Solid lipid microparticles produced by spray chilling technique to deliver ginger oleoresin: Structure and compound retention. *Food Research International* 80:41–49.

Otálora, M.C., J.G. Carriazo, L. Iturriaga, M.A. Nazareno, and C. Osorio. 2015. Microencapsulation of betalains obtained from cactus fruit (*Opuntia ficus-indica*) by spray drying using cactus cladode mucilage and maltodextrin as encapsulating agents. *Food Chemistry* 187:174–181.

Paini, M., B. Aliakbarian, A.A. Casazza, A. Lagazzo, R. Botter, and P. Perego. 2015. Microencapsulation of phenolic compounds from olive pomace using spray drying: A study of operative parameters. *LWT—Food Science and Technology* 62(1):177–186.

Passerini, N., B. Perissutti, B. Albertini, D. Voinovich, M. Moneghini, and L. Rodriguez. 2003. Controlled release of verapamil hydrochloride from waxy microparticles prepared by spray congealing. *Journal of Controlled Release* 88:263–275.

Passerini, N., S. Qi, B. Albertini, M. Grassi, L. Rodriguez, and D.Q.M. Craig. 2010. Solid lipid microparticles produced by spray congealing: Influence of the atomizer on microparticle characteristics and mathematical modeling of the drug release. *Journal of Pharmaceutical Sciences* 99(2):916–931.

Patankar, S.V. and D.B. Spalding. 1972. A calculation procedure for heat, mass and momentum transfer in three-dimensional parabolic flows. *International Journal of Heat and Mass Transfer* 15:1787–1806.

Paucar, O.C., F.L. Tulini, M. Thomazini, J.C.C. Balieiro, E.M.J.A. Pallone, and C.S. Favaro-Trindade. 2016. Production by spray chilling and characterization of solid lipid microparticles loaded with vitamin D3. *Food and Bioproducts Processing*. 100(Part A):344–350.

Pedroso, D.L., M. Thomazini, R.J.B. Heinemann, and C.S. Favaro-Trindade. 2012. Protection of *Bifdobacterium lactis* and *Lactobacillus acidophilus* by microencapsulation using spray-chilling. *International Dairy Journal* 183:133–143.

Pelissari, J.R., V.B. Souza, A.A. Pigoso, F.L. Tulini, M. Thomazini, C.E.C. Rodrigues, A. Urbano, and C.S. Favaro-Trindade. 2016. Production of solid lipid microparticles loaded with lycopene by spray chilling: Structural characteristics of particles and lycopene stability. *Food and Bioproducts Processing* 98:86–94.

Piatkowski, M., M. Taradaichenko, and I. Zbiciński. 2014. Flame spray drying. *Drying Technology* 15(10):1343–1351.

Polavarapu, S., C.M. Oliver, S. Ajlouni, and M.A. Augustin. 2011. Physicochemical characterisation and oxidative stability of fish oil and fish oil–extra virgin olive oil microencapsulated by sugar beet pectin. *Food Chemistry* 127:1694–1705.

Quispe-Condori, S., M.D.A. Saldaña, and F. Temelli. 2011. Microencapsulation of flax oil with zein using spray and freeze drying. *LWT—Food Science and Technology* 44:1880–1887.

Ramírez, J.M., G.I. Giraldo, and C.E. Orrego. 2015. Modeling and stability of polyphenol in spray-dried and freeze-dried fruit encapsulates. *Powder Technology* 277:89–96.

Re, M.I. 1998. Microencapsulation by spray drying. *Drying Technology* 16:1195–1236.

Renksizbulut, M. and M.C. Yuen. 1983. Experimental study of droplet evaporation in a high-temperature air stream. *Journal of Heat Transfer* 105:384–397.

Ribeiro, M.D.M.M., D.B. Arellano, and C.R.F. Grosso. 2012. The effect of adding oleic acid in the production of stearic acid lipid microparticles with a hydrophilic core by a spray-cooling process. *Food Research International* 47:38–44.

Rodrigues, R.A., M.V. Rodrigues, T.I. Oliveira, C.Z. Bueno, I.M. Souza, A. Sartoratto, and M.A. Foglio. 2011. Docosahexaenoic acid ethyl esther (DHAEE) microcapsule production by spray-drying: Optimization by experimental design. *Ciência e Tecnologia de Alimentos, Campinas* 31:589–596.

Sabatino, M.D., B. Albertini, V.L. Kett, and N. Passerini. 2012. Spray congealed lipid microparticles with high protein loading: Preparation and solid state characterization. *European Journal of Pharmaceutical Sciences* 46:346–356.

Santana, A.A., D.M. Cano-Higuita, R.A. de Oliveira, and V.R.N. Telis. 2016. Influence of different combinations of wall materials on the microencapsulation of jussara pulp (*Euterpe edulis*) by spray drying. *Food Chemistry* 212:1–9.

Santiago-Adame, R., L. Medina-Torres, J.A. Gallegos-Infante, F. Calderas, R.F. González-Laredo, N.E. Rocha-Guzmán, L.A. Ochoa-Martínez, and M.J. Bernad Bernad. 2015. Spray drying microencapsulation of cinnamon infusions (*Cinnamomum zeylanicum*) with maltodextrin. *LWT—Food Science and Technology* 64(2):571–577.

Sartori, T., L. Consoli, M.D. Hubinger, and F.C. Menegalli. 2015. Ascorbic acid microencapsulation by spray chilling: Production and characterization. *LWT—Food Science and Technology* 63(1):353–360.

Schiffter, H. and G. Lee. 2007. Single-droplet evaporation kinetics and particle formation in an acoustic levitator, Part 1: Evaporation of water microdroplets assessed using boundary-layer and acoustic levitation theory. *Journal of Pharmaceutical Sciences* 96(9):2274–2283.

Shamaei, S., A. Kharaghani, S.S. Seiiedlou, M. Aghbashlo, F. Sondej, and E. Tsotsas. 2016. Drying behavior and locking point of single droplets containing functional oil. *Advanced Powder Technology* 27(4):1750–1760.

Simon-Brown, K., K.M. Solval, A. Chotiko, L. Alfaro, V. Reyes, C. Liu, B. Dzandu et al. 2016. Microencapsulation of ginger (*Zingiber officinale*) extract by spray drying technology. *LWT—Food Science and Technology* 70:119–125.

Smagorinsky, J. 1963. General circulation experiments with the primitive equations. *Monthly Weather Review* 91(3):99–164.

Soottitantawat, A., F. Biegeard, H. Yoshii, T. Furuta, M. Ohkawara, and P. Linko. 2005. Influence of emulsion and powder size on the stability of encapsulated D-limonene by spray drying. *Innovative Food Science and Emerging Technologies* 6:107–114.

Soottitantawat, A., H. Yoshii, T. Furuta, M. Ohkawara, and P. Linko. 2003. Microencapsulation by spray drying: Influence of emulsion size on the retention of volatile compounds. *Journal of Food Science* 68:2256–2262.

Tatar, F., A. Cengiz, and T. Kahyaoglu. 2015. Evaluation of ultrasonic nozzle with spray drying as a novel method for the microencapsulation of blueberry's bioactive compounds. *Innovative Food Science and Emerging Technologies* 32:136–145.

Thom, A. 1933. The flow past circular cylinders at low speeds. *Proceedings of the Royal Society* 141(845):651–666.

Tran, T.T.H. 2017. From single droplet to spray tower drying of dairy solutions. PhD thesis, Otto-von-Guericke-Universität Magdeburg, Magdeburg, Germany.

Tran, T.T.H., J. Avila-Acevedo, and E. Tsotsas. 2016a. Enhanced methods for experimental investigation of single droplet drying kinetics and application to lactose/water. *Drying Technology* 34(10):1185–1195.

Tran, T.T.H., M. Jaskulski, J.G. Avila-Acevedo, and E. Tsotsas. 2016b. Model parameters for single droplet drying of skim milk and its constituents at moderate and elevated temperatures. *Drying Technology* 35(4):444–464.

Tulini, F.L., V.B. Souza, M.A. Echalar-Barrientos, M. Thomazini, E.M.J.A. Pallone, and C.S. Favaro-Trindade. 2016. Development of solid lipid microparticles loaded with a proanthocyanidin-rich cinnamon extract (*Cinnamomum zeylanicum*): Potential for increasing antioxidant content in functional foods for diabetic population. *Food Research International* 85:10–18.

Van Meel, D.A. 1958. Adiabatic convection batch drying with recirculation of air. *Chemical Engineering Science* 9:36–44.

Venil, C.K., A.R. Khasim, C.A. Aruldass, and W.A. Ahmad. 2016. Microencapsulation of flexirubin-type pigment by spray drying: Characterization and antioxidant activity. *International Biodeterioration & Biodegradation* 113:350–356.

Walton, D.E. 2002. Spray-dried particle morphologies. *Asia-Pacific Journal of Chemical Engineering* 10(3–4):323–348.

Walton, D.E. and C.J. Mumford. 1999. The morphology of spray-dried particles: The effect of process variables upon the morphology of spray-dried particles. *Chemical Engineering Research and Design* 77(5):442–460.

Wang, Y., W. Liu, X.D. Chen, and C. Selomulya. 2016. Micro-encapsulation and stabilization of DHA containing fish oil in protein-based emulsion through mono-disperse droplet spray dryer. *Journal of Food Engineering* 175:74–84.

Woo, M.W., W.R.W. Daud, A.S. Mujumdar, Z.H. Wu, M.Z.M. Talib, and S.T. Tasirin. 2008. CFD evaluation of droplet drying models in a spray dryer fitted with rotary atomizer. *Drying Technology* 26:1180–1198.

Woo, M.W., W.R.W. Daud, S.M. Tasirin, and M.Z.M. Talib. 2007. Optimization of the spray drying operating parameters—A quick trial-and-error method. *Drying Technology* 25:1741–1747.

Yarin, A.L., G. Brenn, O. Kastner, and C. Tropea. 2002. Drying of acoustically levitated droplets of liquid-solid suspensions: Evaporation and crust formation. *Physics of Fluids* 14(7):2289–2298.

Zbiciński, I. and M. Piątkowski. 2004. Spray drying tower experiments. *Drying Technology* 22(6):1325–1349.

Zbiciski, I. 1995. Development and experimental verification of momentum, heat and mass transfer model in spray drying. *Chemical Engineering Journal* 58:123–133.

4 Electro-Hydrodynamic Processes (Electrospinning and Electrospraying)
Nonthermal Processes for Micro- and Nanoencapsulation

M.A. Busolo, S. Castro, and J.M. Lagaron

CONTENTS

4.1 INTRODUCTION: BACKGROUND AND DRIVING FORCES

Encapsulation techniques were developed approximately 60 years ago as an approach to modify or protect sensitive molecules and substances to preserve their activity until their final application. The encapsulation process involves the entrapment of a substance (active agent) by the formation of a wall around it and ensuring that undesired leakage does not occur, but that the release does take place under specific conditions (Fang and Bhandarri, 2010; Nedovic et al., 2011). Typically, encapsulation processes comprise spray drying, spray cooling, extrusion, emulsion techniques, fluidized bed coating, coacervation, liposome entrapment, inclusion complexation, etc. (Nedovic et al., 2011; Pérez-Masiá et al., 2015; Katouzian and Jafari, 2016; Paximada et al., 2017). However, the application of these techniques is limited because they mostly require heating and pressure, and the use of strong and toxic organic solvents or expensive equipment. In this regard, electrohydrodynamic processing (EHD, referring to the dynamics of electrically charged fluids) has emerged as an advantageous alternative technology for encapsulation that needs neither temperature nor expensive equipment, and the use of organic solvents can be avoided by adjusting some processing conditions (i.e., use of molten polymers) (Pérez-Masiá et al., 2015; Brown et al., 2016). EHD constitutes the basis for electrospinning (ES) and electrospray (ESP). This chapter will provide an overview of the EHD process for micro- and nano-encapsulation of active substances and sensitive compounds for multisectorial applications.

4.2 ELECTROSPINNING

The electrospinning process, (from "electrostatic spinning") was patented in 1934 by Formhals as a technique for the fabrication of polymer filaments using electrostatic forces. Later studies enabled to file more than 60 patents related to polymer melt and solutions (Bhardwaj and Kundu, 2010; Newsome, 2013). It was not until the 1990s that the ES process regained attention due to developments in the field of nanotechnology that holds tremendous potential for different industrial fields, such as food safety and quality, delivery of micronutrients and bioactives (Aceituno-Medina et al., 2015; Paaver et al., 2016), biomedical applications (tissue engineering, drug delivery, biosensors, and immunoassays) (Sill and von Recum, 2008; Jiang et al., 2014; Kai et al., 2014; Rogina, 2014; Repanas et al., 2015), active and intelligent food packaging (Busolo et al., 2009; Fabra et al., 2013, 2016; Chalco-Sandoval et al., 2015, Cerqueira et al., 2016), in other important sectors of technology (fuel cells, luminescence, supercapacitors, lithium ion batteries, and catalysts) (Chang et al., 2013; Qu et al., 2013; Aruna et al., 2017) and engineering as well (filtration, automotive, environmental remediation) (Cacciotti et al., 2015; Stanishevsky et al., 2016; Malwal and Gopinath, 2017).

Electrospinning (ES) is an efficient technique and currently the only one that allows the production of continuous and uniform polymeric fibers with diameters ranging from the nano to micron scales. Nano and microfibers can be generated stably thanks to the several physical processes that do not occur in other methods, such as fiber extrusion followed by mechanical stretching. Branching and occurrence of undulated fibers are the other exclusive phenomena of the ES (Greiner and Wendorff, 2007). There is a wide range of polymers tested in ES that are capable of forming fine nanofibers (Gopal et al., 2006) involving synthetic polymers and biopolymers, as well as blends that have been reported for use in different applications, depending on the desired final characteristics of the product. ES of copolymers, on the other hand, has been observed to enhance several properties of polymeric materials, such as barrier properties and mechanical strength or thermal stability (Kundu and Bhardwaj, 2010). The versatility of this technique is well-known and lies in its applicability to every soluble and fusible polymer, and even to those tailored polymers containing additives of complex fillers to achieve the desired results from properties and functionality. Spun nanofibers offer several advantages, such as an extremely high surface-to-volume ratio, tunable porosity, and malleability to conform to a wide variety of sizes and shapes, and the possibility to control their composition.

Because of their very large area to volume ratio and relative ease of preparation, electrospun ultrathin fibers are being considered in different areas such as filtration, nanotube reinforcement, sensors, catalysis, protective clothing, biomedicine, space applications, semiconductors, micro- or nanoelectronics, and in food packaging and encapsulation (Ramakrishna et al., 2003; Lelkes et al., 2005; Torres-Giner et al., 2009; Baji et al., 2010).

4.2.1 ES PRINCIPLES AND BASIC SETUP

The unique approach of ES is the use of electrostatic forces to produce nanosized fibers from molten polymers or polymer solutions. Typically, spinning methods are classified into three groups according to the mechanism of solidification of the extruded material: solution spinning (evaporation or coagulation), emulsion spinning (phase separation), or melt blowing/spinning (cooling) (Brown et al., 2016). The technique typically involves a high voltage power supply, a syringe with a needle or capillary tube of small diameter, a pumping system, and a grounded collector screen. The mechanism of ES involves the collaborative effects of electrostatic repulsion by the accumulated charges on the surface of polymer solution and the Coulombic force exerted by the external electric field (Gong and Wu, 2012). The hemispherical droplet at the tip of the needle stretches when the electric force overcomes the surface tension of the viscous solution and a conical jet projection of fibers emerges (Taylor cone). These fibers stretch and elongate causing most of the polymer solution evaporation; thereby solidification occurs before they are collected on the screen in the form of nonwoven mats (Aruna et al., 2017).

Starting from the original single needle arrangement, other ES configurations can be used for obtaining nanofibers, which together with ES parameters will define the fiber morphology, dimensions, and properties. By changing the nozzle configuration, some limitations in polymer processability can be overcome (single, single with emulsion, side-by-side, or coaxial nozzles) (Sill and von Recum, 2008; Persano et al., 2013).

Solution ES is particularly suitable for polymers that are thermally unstable and degrade while melting. The solvent can be removed either by a dry process (by evaporation of the solvent during the jet fly or blowing hot air on the extruded filament), or wet process (the fiber is extruded in a viscous coagulation bath consisting of a liquid that is miscible with the spinning solvent, but is a nonsolvent for the polymer) (Persano et al., 2013).

In melt ES the molten polymer is used instead of a solution. This technique eliminates the use of organic solvents—many of them expensive and toxic—as well as the problem of inadequate solvent evaporation between the tip and the collector, which offers advantages in terms of productivity and environmental impact (Greiner and Wendorff, 2007; Sill and von Recum, 2008). So far, different oil-based polymers and biopolymers have been processed by this method: polyethylene (PE), polypropylene (PP), polyamide (PA), poly(ethylene terephthalate) (PET), PCL, and poly(lactic) acid (PLA) (Greiner and Wendorff, 2007; Bölgen and Vaseashta, 2015; Qin et al., 2015; Wang et al., 2015a; Brown et al., 2016; Doustgani and Ahmadi, 2016; Lee et al., 2016).

Emulsion ES has gained interest in the fabrication of core-shell structured nanofibers from water-in-oil (w/o) emulsions. The bioactive agent, generally included in the aqueous phase, can be encapsulated into the core section rather than uniformly dispersed inside the fibers. This approach provides several advantages, such as overcoming the poor solubility and instability of bioactive agents in organic solvents, and extended release due to the encapsulated agent needing to pass through the core-shell fiber matrix before reaching the medium (Wang et al., 2015b).

Coaxial electrospinning (CoES) is a robust technique for one-step encapsulation of delicate agents into core-shell nanofibers through coaxial needle geometry. CoES process eliminates the damaging effects due to direct contact of the agents with organic solvents or harsh conditions during the emulsification or suspension process (Jiang et al., 2014). When compared to the single needle ES method, the CoES approach provides greater control on encapsulant layering, morphology, bioactive loading capacity, and retention (Yao et al., 2016). CoES involves two concentrically arranged nozzles that are connected to the power source for electrospinning two separate solutions with different polymers and solvents. In this way, it is possible to simultaneously spin fibers with different outer shell-inner core materials that can have various applications in the fields of drug delivery, tissue engineering, and wound healing (Repanas et al., 2015). The system needs an optimal assembling precision to avoid nonconcentric syringes and fiber deformations. At the same time, however, it allows for a certain degree of versatility in fiber morphologies by simply "playing" with the surface tension and rheology of the two immiscible solutions. Furthermore, the inner electrospun solution does not necessarily need to be polymeric; it can simply be a liquid that finally forms a cavity. The hollow fiber so produced promises to be a versatile drug delivery system (Migliaresi et al., 2012; Chang et al., 2013). In this chapter, we will focus on single and coaxial ES processes of polymer solutions.

The scale-up of nanofibers through single jet is not economically and technically feasible and most of the electrospun fiber applications require large quantities of material (up to several kilograms). High volumes are especially desirable for biomedical applications such as tissue engineering, drug delivery, and wound-healing materials (Persano et al., 2013). Depending on the solution properties, the throughput of single-jet setup varies from 10 µL/min to 10 mL/min, which is a limiting factor to industrial scale-up where mass production is needed (Theron et al., 2005; Bhardwaj and Kundu, 2010; Nagy et al., 2015). A company called Bioinicia S.L. (www.bioinicia.com) has scaled up the process by a multinozzle technology to generate the mass production of fibers and capsules. Scaling up by a multinozzle technology is then feasible but requires high electro-hydrodynamics engineering skills, since ill-defined jets in an arrangement can interact, undergo bending instabilities, and lead to repulsion between jets because of Coulombic forces. These jet repulsions can be overcome, paradoxically, by introducing more jets in the arrangement (Theron et al., 2005). Chalco-Sandoval et al. produced polycaprolactone (PCL) nanofibers as an encapsulating matrix for a blend of paraffin waxes, using a Fluidnatek™ LE500 multinozzle injector from Bioinicia S.L., a high throughput pilot processing tool (Chalco-Sandoval et al., 2015). Nagy et al. (2015) obtained drug-loaded copovidone nanofibers by applying high speed ES setup, consisting of a spinneret with sharp edges and a spherical cap connected to a high-speed motor. They found that the final fiber morphology was very similar to that obtained in single-needle ES. A modified ES process with auxiliary grounded electrode was also developed and proven to improve the production rate of nanofibers between 7- and 10-fold regarding the traditional process (Liu et al., 2015). Other more sophisticated strategies have been tried to increase nanofiber production, such as vibration ES, corona-ES, hollow tube ES, and alternating current ES, among others. Ultrasonic vibration coupled to ES allows obtaining thinner fibers than the standard method, even from those high-viscosity polymer solutions (Si et al., 2014).

A device for launching multiple jets from a hollow, porous, cylindrical tube, including a wire electrode inserted inside the tube to maintain an equal electrical potential in each hole, was proven to produce polyvinylpyrrolidone (PVP) mats at mass rates greater than what can be obtained from a single needle setup (Varabhas et al., 2008). A patented corona-ES involves the continuous supply of the polymeric solution though a narrow, long, circular-shaped gutter bounded by a metal electrode ring (the *corona*). The spinneret rotates along a symmetry line in order to homogeneously disperse the solution, and the highest electrical charge density is formed along the edge, resulting in many self-assembled Taylor cones. The diameter and morphology of the resulting nanofibers are close to those that were processed by the classical setup (Molnar and Nagy, 2016). The alternating current (AC) ES is characterized by significantly reduced fiber jet instability and residual fiber charge, as well as much higher fiber collection rates per unit of area when compared to direct current (DC) ES. Nanocrystalline alumina from PVP precursor fibers were collected exhibiting suitable fiber diameter, surface morphology, and structure according to alumina fibers produced by other ES methods (Stanishevsky et al., 2016).

4.2.2   ES PARAMETERS

Nearly all soluble and fusible polymers are susceptible to be electrospun, and the final morphology and dimensions of nanofibers will depend on the proper setting of several parameters that can be divided by categories such as polymer properties, solution properties, processing conditions, and operating conditions (Tan et al., 2005; Sawicka and

TABLE 4.1

Parameters That Affect the Electrospun Fibers Morphology

Polymer parameters	Solubility
	Molecular weight
	Melting point
	Glass-transition temperature
	Crystallization velocity
Solution parameters	Viscosity
	Viscoelasticity
	Concentration
	Surface tension
	Solvent properties (vapor pressure)
	Electrical conductivity
Processing conditions	Applied voltage
	Needle to collector distance
	Collector composition and geometry
	Volume feed rate
	Needle diameter
Operating conditions	Temperature
	Relative humidity
	Atmospheric pressure

Gouma, 2006; Greiner and Wendorff, 2007; Torres-Giner et al., 2008, 2009; Ghorani and Tucker, 2015; Aruna et al., 2017). These influencing parameters are listed in Table 4.1.

Besides pure fiber structures, many other morphologies of products can be achieved by ES by tuning operating and processing conditions, for example, pearl-necklace-like beaded fiber, highly porous fiber, grafted fiber, hollow interior micro tubes, and twisted fibers.

CoES depends on the same parameters as those in conventional ES. However, additional phases introduce an interface interaction (miscibility or stress due to vis-cosity difference) and the individual physical behavior of different phases also complicates CoES (Qu et al., 2013).

4.2.2.1 Polymer Parameters

Polymer type and molecular weight are crucial parameters that have been taken into account to develop a semi-empirical theory about the entanglement number for a solution. This number is defined to indicate whether electrospraying (beads) or ES (fibers) will take place (Shenoy et al., 2005). In principle, molecular weight reflects the entanglement of polymer chains in solutions. At a fixed concentration, lowering the molecular weight of the polymer tends to form beads rather than smooth fiber. By increasing the molecular weight smooth fiber will be obtained (Li and Wang, 2013).

4.2.2.2 Solution Parameters

The ability of fibers to be formed and whether they are smooth or porous depends on the choice of a suitable solvent. As the solution jet travels from the needle tip to the collector, a phase separation occurs before the solid polymer fibers are deposited, a process that is greatly influenced by the volatility of the solvent (Sill and von Recum, 2008). Electrospun fibers from very volatile solvents can show a high density of pores, while fibers spun from less volatile solutions are usually smooth (Megelski et al., 2002).

Polymer concentration and molecular weight strongly influence the viscosity and surface tension of the spinnable solution. The ideal values of concentration, resulting viscosities, and electrical conductivities vary considerably with the polymer-solvent system. However, working in a suitable range of viscosity, typically from 800 to 4000 cP, is possible to collect nanofibers between 100 and 300 nm, as it has been proven, because initial jet stabilization and subsequent jet thinning are supported (Greiner and Wendorff, 2007; Chakraborty et al., 2009). The effect of viscosity and concentration occurs in the same way: low viscosity disables continuous fiber formation without droplets, while high viscosity causes drying polymer solution at needle tip (Rogina, 2014).

The ideal solution is the one that allows the highest possible percentage of polymer while still having a low enough viscosity to permit jet formation. Below this critical value, electrospray of beads will occur due to insufficient overlap between molecular chains (Ghorani and Tucker, 2015). As polymer concentration increases, a mixture of beads and fibers will be obtained, and beyond critical concentration, high viscosity of solution will impede the continuous formation of fibers (Ghorani and Tucker, 2015). The productivity of an ES process from 1 to 20 wt.% solutions is moderate. An alternative might be ES from the melt, but excepting few low-melting polymers, the average fiber diameter achieved are considerably greater than 1 μm as a result of the high viscosity of high melt polymers (Greiner and Wendorff, 2007).

The surface energy of the solvent is a key factor in the ability of the applied electrical potential to shear the polymer into fibers. The polymers that are solved in low-tension surface solvents can be electrospun more readily compared to those solved in high-tension surface ones (i.e., water) (Chakraborty et al., 2009). However, surface tension effect could be dominant with decreased solution viscosity and beaded fibers can be consequently produced (Tan et al., 2005). The high surface tension of solvents also causes instability resulting in broad fiber diameter.

The electrical conductivity of a solution reflects a charge density on a jet, thus solutions with higher electrical conductivity may cause elongation of a jet along its axis producing fibers with smaller diameters. The surface of droplets from low conductivity solutions will not have enough charge to form a Taylor cone and the ES process will not take place. It is necessary to increase the solution conductivity to a critical value in order to induce charges in the droplet surface, but conductivity should not be increased beyond a critical value, where there is a balance between the Coulomb forces in the surface of fluids and the force due to the external electric field (Tan et al., 2005; Haider et al., 2015). The conductivity of a polymer solution can be controlled by the addition of salts or organic acids, resulting in a higher charge density on the surface of the jet during the spinning, thus elongation forces are imposed on the jet in the electrical field and the production rate is increased (Rogina, 2014; Ghorani and Tucker, 2015; Haider et al., 2015).

4.2.2.3 Processing Conditions

Electrical gradient is the governing force of the ES process. The applied voltage may affect some factor such as mass of polymer fed out from the tip of a needle, elongation level of a jet by an electrical force, morphology of the jet (single or multiple jets), etc. At very high voltages, about 40 kV, the Taylor cone formation destabilizes, thereby increasing the diameter of nanofibers. Smaller diameters are achieved at high electrical field, but as voltage is increased nonuniform diameter fibers can be formed. In addition, higher voltages accelerate electrostatic forces of charged solution and prevent complete solvent evaporation. On the other hand, low voltages cause bead formation because paucity of the electrical field does not compensate surface forces of the polymer drop (Chakraborty et al., 2009; Rogina, 2014; Aruna et al., 2017). Electrospun fibers from polylactic acid (PLA) seemed not to be dramatically affected with variations in applied voltage (Tan et al., 2005).

There is a commitment between feed rate and fiber formation. Beads can result from the deformation of the Taylor's cone due to a greater feeding rate of polymer solution. If higher voltage is applied to compensate the feed rate, nonuniform fiber diameter would be produced and complete solvent evaporation could be prevented. On the other hand, at lower feed rate, the polymer solution could evaporate at the needle tip (Rogina, 2014).

In order to obtain uniform fibers, it is necessary to give them enough time to dry before reaching the collector by adjusting the needle tip-to-collector distance. The length of the distance will depend on the applied electric field and the concentration of the polymer solution (Rogina, 2014). Typically, the needle tip-to-collector distance is between 5 and 20 cm; however, a greater distance will lead to significant fiber loss to the surroundings as the polymer jet seeks the nearest grounded surface (Doshi and Reneker, 1995). At an inappropriately short distance, the electrospun fibers are more susceptible to fusion as there can be residual solvent during deposition and the nanoweb collection-zone diameter decreases (Hekmati et al., 2013).

4.2.2.4 Operating Conditions

Recently, it has been reported that environmental factors, such as relative humidity and temperature, also affect the diameter and morphology of the nanofibers. Besides the nature of polymer, humidity can induce porosity and affect the nanofiber diameter by controlling the solidification process of the charged jet. The different evaporation rates of solvents cause condensation of water into droplets that settle on the fibers. If water is miscible with the solvent, the two mix well with each other on the inner and surface of fibers. The complete evaporation of solvent and the water droplets from the fibers results in the porous morphology (Haider et al., 2015). Humidity also affects the solidification process of the charged jet, thus higher fiber diameter will be achieved at higher humidity conditions (Panda and Sahoo, 2013).

Temperature has a direct influence in the average diameter of nanofibers with two opposing effects: it increases the evaporation rate of solvent and at the same time decreases the viscosity of solution.

4.2.3   ES as Encapsulation Method

Core-shell nanofibers exhibit advantages in delivery of delicate drug or active substances, preventing decomposition or fast degradation of labile compounds, and have been investigated with several chemical drugs, bioactive substances, proteins, and macromolecules with a clear advantage over bulk material due to their large specific surface and short diffusion path lengths (Sill and von Recum, 2008; Persano et al., 2013; Kai et al., 2014). CoES and emulsion ES are mostly applied as loading method.

4.2.3.1   Active Food Packaging

Over the past 20 years, it has had a growing trend in the development of active food packaging materials containing active agents that, once they interact with the food products or the packaging atmosphere, extend the shelf life of food (Appendini and Hotchkiss, 2002). Active food packaging is an innovative approach to address the changing demands of the food industry and consumers, as well as the increasing regulatory and legal requirements about food safety (Realini and Marcos, 2014). Active packaging provides several functions that do not exist in conventional systems, such as oxygen scavenging, antimicrobial activity, or free radical trapping, among others (Ahvenainen, 2002; Brody et al., 2008; Restuccia et al., 2010).

The active agents can be directly incorporated into the packaging material, which is an advantage with regard to the potential cost saving due to increased production efficiency (Brody et al., 2008), and the objective here is that active substance can act or be delivered in a controlled manner. Nowadays, the food industry has shown increasing interest in the development of polymeric films with antioxidant and antimicrobial properties to prevent free radicals' attack on the unsaturated molecules present in foods, and to limit the microbial growth responsible for food spoilage (Ahvenainen, 2002; Lee et al., 2004; Brody et al., 2008; Arrua et al., 2010; Realini and Marcos, 2014). Antioxidants can be incorporated into food packaging materials, remain stable there, and finally be released from the packaging material in a controlled manner to inhibit the oxidation of sensitive components (Busolo and Lagaron, 2015), while volatile antimicrobial agents must be coated on a matrix that acts as a carrier (Sunil, 2012). However, many active agents are thermally sensitive and cannot be directly incorporated during the typical processing methods (Fabra et al., 2016). The immobilization by encapsulation of susceptible active components aims to protect them from oxidation, heat, or extreme conditions during the processing, or to mask component attributes as undesirable flavors to meet consumer request for organoleptic quality and functionality. In an optimized design, the capsules or carrier must provide a controlled release of active compounds during the storage (Fernandez et al., 2009). ES is a suitable alternative technology for encapsulating active agents because it does not involve any severe conditions of temperature, pressure, and chemistry with capability for continuous fabrication, versatility, and facile operating process (Ghorani and Tucker, 2015). This method involves the solubilization or dispersion of the active substance into the polymer solution for the final embedment of the substance within the electrospun fibers (Echegoyen et al., 2016).

Some studies involve the use of essential oils, volatile compounds, and phenolic compounds for producing active electrospun fibers which, once incorporated onto a polymer layer, can be integrated to a multilayer system with controlled antimicrobial properties for potential use as antimicrobial food packaging. Emulsion ES was successfully applied to encapsulate (R)-(+)-limonene in a poly(vinyl alcohol) (PVA) fibrous matrix (Camerlo et al., 2013). The release profile of the fragrance from the electrospun nanofibers over a 15-day range shows that this type of nanofibrous matrices has great potential for applications in various fields, such as cosmetics or food packaging. Cerqueira et al. (2016) electrospun zein ultrathin fibers containing cinnamaldehyde as antimicrobial agent onto polyhydroxybutyrate-co-valerate (PHBV), and developed active multilayer structures by intercalation of zein-cinnamaldehyde nanofibers with PHBV or alginate layers. The results showed that the controlled release achieved by placing the active layer between PHBV layers, the cinnamaldehyde is preserved from volatilization and burst release. An inclusion complex made of cinnamon essential oil and β-cyclodextrin was prepared and then incorporated into PLA nanofibers via solution ES to produce an active nanofilm (Wen et al., 2016). The resulting nanofilm showed strong antimicrobial activity against *Staphilococcus aureus* and *Escherichia coli*, and effectively prolonged the shelf life of meat products, suggesting that it has potential for application in food packaging. Similar antimicrobial activity against these pathogenic bacteria was found in PCL microfibers loaded with an extract of rosemary, prepared by melt ES (Bhullar et al., 2015).

Some encapsulated antioxidant compounds in hydrocolloid-based fibers have been reported as potential systems for the development of bioactive packaging materials. Neo et al. (2013) applied solution ES to obtain gallic acid loaded zein nanofibers. The antioxidant activity of gallic acid increased after being loaded into zein nanofibers, not only because this compound was protected against oxidation but also due to a small antioxidant effect of zein. Another strong natural antioxidant molecule, α-tocopherol, was encapsulated in whey protein isolate, zein and soy protein isolate nanofibers, which were directly electrospun as a coating onto wheat gluten films (Fabra et al., 2016). It was found that hydrocolloid matrices were able to protect the antioxidant compound from degradation during a typical sterilization process, and that the release in aqueous media depended on the encapsulation matrix.

Besides organic substances, some nanosized inorganic agents have been assessed for potential antimicrobial food packaging applications due to their processing convenience (high thermal resistance), strong antimicrobial performance, and high surface to volume ratio that allows enhancing of the antimicrobial effect at low concentration. There is no regulatory framework for the use of engineered nanomaterials as food contact materials in the EU, and the European Food Safety Authority (EFSA) scientific committees sustain that a case-by-case approach is necessary to make the risk assessment of specific nano-products. Zinc oxide (ZnO) has recently acquired the positive opinion from the European Food Safety Authority (EFSA) for its use as a UV light absorber for food contact materials (EFSA, 2016), thus a future authorization for its use as active agent could be possible. ZnO nanoparticles were incorporated into PHBV by solution ES and subsequently used as coating of PHBV films

(Castro-Mayorga et al., 2017). A post-annealing step was applied to form a continuous film. The authors confirmed that ZnO was properly dispersed and distributed within the PHBV fibers, and that the continuous film showed a loss of transparency regarding neat PHBV film. However, the fiber-loaded film exhibited suitable release of zinc, which fitted with bactericidal effect against *L. monocytogenes* obtained after several washing steps, but zinc concentration did not exceed the current specific migration limit of 25 mg/kg food simulant imposed by the EU regulation (European Commission, 2011).

4.2.3.2 Drug Delivery

Providing drugs in the most feasible physiological way, with smaller size and suitable coating material, improves its ability to be digested or absorbed by the targeted site (Haider et al., 2015). Nanofibers for controlled release of drugs are of great interest for pain and tumor therapy, and these nanofibers should protect the drug from decomposing in the bloodstream and allow its controlled release over a chosen time period at a release rate that is as constant as possible (Greiner and Wendorff, 2007). The drug release kinetics can be tailored, depending on the final application, by adjusting the main properties of the nanofiber, such as fiber diameter, porosity, drug binding mechanisms, and shell material (Kai et al., 2014). Biodegradable materials are preferred for drug delivery because explantation is not needed at the end of the treatment. Due to the fact that a drug may be released by diffusion as well as degradation, in some cases high local drug concentration can occur reaching toxic levels (Sill and von Recum, 2008).

For organo-soluble drugs, direct ES of drug/polymer mixed solution is a suitable method for loading the drug into the fibers. However, direct ES of water-soluble drug suspensions in organic solvents always results in a serious burst release besides the denaturing of the agent (Jiang et al., 2014). The emulsion ES was developed to overcome these issues, and it was applied to obtain bovine serum albumin (BSA) in poly(ethylene-co-vinyl-acetate) (EVA) from a water-in-oil suspension of aqueous BSA droplets in EVA/dichloromethane solution (Sanders et al., 2003). The emulsion ES was adopted for encapsulating bioactive compounds in biodegradable fibers. Li et al. (2009) prepared nanofibrous mats by emulsion ES of a water-in-oil suspension of sorbitan monooleate (Span80) in poly(L-lactide-co-caprolactone) (PLLACL)/chloroform. They found that the surfactants (not on the surface of nanofibers) would help to stabilize the nanofibers' core part, which is a desirable condition to encapsulate drugs or proteins. Emulsion ES was also used for encapsulation of a water-soluble protein into nonbiodegradable but biocompatible hydrophobic polystyrene (PS) using a natural organic L-limonene as solvent (Wang et al., 2015b). *In vitro* release profiles showed an increased release rate with PS molecular weight, which is attributed to the association of release rate with fiber inner structure and protein distribution in matrix. The water–oil emulsions made of water-soluble doxorubin in PLA-co-PGA copolymer in chloroform were electrospun to obtain nanofibers with linear release kinetics (Xu et al., 2005). Besides control of initial burst release, the delivery of multiple drugs without interfering with the release kinetics of each other is another issue to be solved (Kai et al., 2014). Wang et al. (2014) developed a novel drug delivery system with the swelling core for differential release of multiple drugs (rhodamine B and BSA) by emulsion ES, in which the aqueous phase is composed of polyvinyl alcohol and the oil phase consists of poly(ε-caprolactone). An *in vitro* drug release study demonstrated that this core-sheath structure could significantly alleviate the initial drug burst release and provide a differential diffusion pathway to delivery.

DNA was incorporated in the block copolymer PLA-PEG-PLA, but also in poly-ethylene oxide (PEO), and then released structurally intact from the nonwoven matrices (Schiffman and Schauer, 2008).

CoES has been widely applied to encapsulate biomolecules inside the polymer with a core-shell structure, leading to a reduced initial burst release and a longer release period (Kai et al., 2014). Repanas et al. studied the effect of the CoES method on ace-tylsalicylic acid (ASA) delivery from a core-shell structure made from polycaprolac-tone (PCL) comparatively with similar nanofibers created by blend electrospinning and found that coaxially spun fibers showed a more sustained ASA release profile because the drug was encapsulated in the core of the fibers (Repanas et al., 2015). The same trend had been found by Ji et al., when they incorporated BSA into PCL nanofibers by blending and CoES (Ji et al., 2010). Moreover, the addition of polyethyleneglycol (PEG) into PCL nanofibers accelerated the protein release and helped to preserve up to 5% of its initial biological activity in the coaxial nanofibers. Many other studies related to CoES for controlled release of proteins have been conducted over the past 10 years. In most cases, biodegradable crystalline polymers were used as shell materials, such as PCL, poly(L-lactide-co-glycolide) (Jiang et al., 2014; Zhang et al., 2006; Sill and von Recum, 2008; Nguyen et al., 2012; Repanas and Glasmacher, 2015). López-Rubio et al. (2009) encapsulated bifidobacteria in a core-shell nanofiber and found beneficial effects on the product storage stability instead of bacterial damaging.

Coaxial melt ES has also been widely proposed as a promising method for the microencapsulation and controlled release of growth factors and drugs (Brown et al., 2016).

4.2.3.3   Food Nanotechnology

Electrospinning of biopolymers and the encapsulation of food ingredients, enzymes, probiotics, and other active compounds related to food industry have been intensely studied over the last years, being the most widely investigated topic for the preserva-tion of active compounds (Nedovic et al., 2011; Lopez-Rubio et al., 2012; Ghorani and Tucker, 2015). For carrier of proteins or bioactive compounds purposes, the encapsulation material must be natural, biodegradable, edible, able to form a suitable barrier between the internal phase and its surroundings, and the ES process should not require the use of toxic solvents (Nedovic et al., 2011; Lopez-Rubio et al., 2012). Natural biopolymers, proteins, and carbohydrates are commonly used as carriers not only because of their sustained release behavior of the incorporated compound and biocompatibility but also because they are often in themselves valuable dietary sup-plements and functional food enhancers (Ghorani and Tucker, 2015). Common natu-ral hydrocolloids encapsulating materials for food-grade electrospun fibers are whey protein isolate, soy protein isolate, egg albumen, gelatin, collagen, zein, and casein (Lopez-Rubio et al., 2012; Neo et al., 2013; Fabra et al., 2016). In this sense, encap-sulation of β-carotene in zein nanofibers by using solvent ES allowed a remarkable protection against oxidation when fibers were exposed to UV-visible radiation, so they have a good potential in potential bioactive food packaging as well as nutraceutical formulations (Fernandez et al., 2009). Tannin from *S. adstringens* bark was also encapsulated in zein nanofibers obtained by solution ES in order to protect the compound antimicrobial activity for potential food applications (De Oliveira Moric et al., 2014). Quercetin and ferulic acid were processed by solution ES and incorporated into pullulan: amaranth protein ultrathin fibers. Amaranth is a traditional Mexican crop with high nutritional value (Aceituno-Medina et al., 2015). A sustained release of both antioxidants from the fibers was observed, which contributed to the improved antioxidant activity of the bioactives in comparison with free compounds.

Viability and stability of probiotic bacteria during food processing and storage, as well as their survival while passing through the upper gastrointestinal tract, is a challenge for the food industry (Ghorani and Tucker, 2015). *Lactobacillus acidophilus* was nanoencapsulated into hydrocolloids recovered from soybean and oil palm waste (Bushani and Anandharamakrishna, 2014). Suitable bacteria survival was obtained in the viability studies after the ES process and retained during the storage.

4.2.3.4   Other Applications

ES as an encapsulation technique has also been applied in other biomedical fields such as tissue engineering (scaffolds, dressings for wound healing, contact lenses, implants), engineering (filtration, energy generation, sensors, membranes, water treatment), and cosmetics among other applications with lesser industrial impact (Sawicka and Gouma, 2006; Sill and von Recum, 2008; Teo and Ramakrishna, 2009; Bhardwaj and Kundu, 2010; Panda and Sahoo, 2013; Kai et al., 2014; Haider et al., 2015; Stanishevsky et al., 2016).

4.3   ELECTROSPRAYING

Electrospraying (ESP) as an EHD process presents many advantages over conventional spraying systems for entrapping active compounds inside micro- and nanosized particles as it is a flexible and simple technique and does not require heat, pressure, or multiple processing (Lopez-Rubio and Lagaron, 2012; Xie et al., 2015; Eltayeb et al., 2016). Charge of droplets produced by ESP makes droplet agglomeration and coagulation impossible and a high deposition efficiency is achieved (Faridi Esfanjani and Jafari, 2016). Similar to ES, the ESP process involves a high voltage to a solution delivered through a spinneret. At critical voltage, the solution droplet at the tip of the needle distorts forming noncontinuous structures or particles. The morphology and diameter of the particles are affected by the process and operation parameters, solution and polymer properties (Teo and Ramakrishna, 2009; Lopez-Rubio and Lagaron, 2012; Pérez-Masiá et al., 2015).

4.3.1   ESP PRINCIPLES AND SETUP

In general, the principle of ESP is similar to that of ES, apart from the fact that the jet breaks down into droplets. Typically this is the consequence of using lower polymer concentration than that used in electrospinning (Chakraborty et al., 2009). Thus, the basic setup of ESP involves a syringe pump, a metal needle serving as nozzle, a high

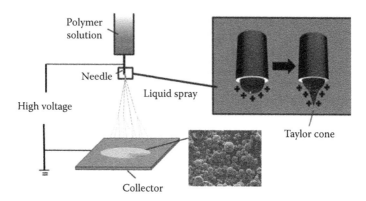

FIGURE 4.1 Schematic diagram of solution ESP setup. (Adapted from Gong, G. and Wu, J., Novel polyimide materials produced by electrospinning. In *High Performance Polymers: Polyimides Based—From Chemistry to Applications*, Abadie, M., ed., InTech, Rijeka, Croatia, DOI: 10.5772/53459, available from: http://www.intech-open.com/books/high-performance-polymers-polyimides-based-from-chemistry-to-applications/novel-polyimide-materials-produced-by-ES, 2012.)

voltage power source, and a grounded collector (Figure 4.1). Some setups employ a closed chamber with air or nitrogen flow to reduce the solvents' evaporation and facilitating the formation of smaller particles with smoother surface morphology, but also to help collect the particles into filters in some cases (Xie et al., 2015). In general, the average diameter range of electrosprayed particles increases with the polymer concentration and with the decreasing surface tension of the solvent. A broad size of particles could be achieved if the solvent does not completely evaporate during the outflow from the needle and until reaching the collector. Moreover, a smaller diameter is obtained as flow rate decreased, at higher electric field (Faridi Esfanjani and Jafari, 2016).

Various morphologies, shapes, and secondary structures can be achieved using different spraying nozzles: single, coaxial, tri-coaxial, two-channel, three-channel, side-by-side, and bipolar. In this sense, different particle morphologies have been produced depending on solution conditions, polymer properties, and used setup, such as spherical, elongated, tailed, and porous structures (Xie et al., 2015).

Commercially available one-needle ESP machines have been used for pilot scale production, such as Fluidnatek LE10 (Pérez-Masiá et al., 2015; Gomez-Mascaraque et al., 2016; Paximada et al., 2017) and MECC NF-102 (Fukui et al., 2010). Mass production of ESP capsules cannot be simply achieved by increasing the flow rate of spraying solution, but by increasing the number of needles in parallel (higher amount of Taylor cones).

4.3.2 ESP AS ENCAPSULATION METHOD

4.3.2.1 Drug Delivery

ESP is a promising technology for encapsulating both hydrophobic and hydrophilic drugs into polymeric particles. Hydrophobic drugs can be simply mixed with polymers and be further dissolved in mild organic solvents to obtain the starting

ESP solution. On the other hand, hydrophilic drugs can be firstly solved in an aqueous solution, then mixed with an organic polymer solution to get an emulsion that can be electrosprayed to form drug-loaded polymeric particles or core-shell particles (drug as core material and polymer in the shell) (Xie et al., 2015). Doxorubicine (20% loading) has been encapsulated inside temperature-sensitive polypeptide, and the particle size and delivery performance seemed not to have altered by increasing the loading to above 20% (Wu et al., 2009). Fukui et al. investigated the encapsulation of protein, dextran, and cells in polyelectrolyte microcapsules (Fukui et al., 2010). Fluorescent microspheres, albumin, dextran, and yeast cells as core substrates were solved in polyelectrolyte-alginate solutions, which were later electrosprayed onto chitosan solutions to obtain narrow-sized microcapsules. Cisplatin is a widely used anticancer agent, but its therapeutic effects are limited due to its significant toxicity. ESP process was used to fabricate cisplatin-loaded poly(lactic-co-glycolic acid) (PLGA) for controlled release of this drug and as a way to enhance its clinical response (Parhizkar et al., 2016). They found high drug encapsulation efficiency and that drug delivery in this case is based on diffusion and shell degradation mechanisms.

4.3.2.2 Food Nanotechnology

The encapsulation of nutraceutical and functional ingredients by ESP enables their protection against degradation during their incorporation into food matrices, and consequently the formulation of functional food products with health benefits. But also, encapsulation in food industry may bring the opportunity of manufacturing products with desirable texture and tastes, producing low-calorie food and beverage products with the aim of changing lifestyles into healthy ones (Katouzian and Jafari, 2016). Pérez-Masiá et al. compared the traditional spray drying technique with ESP for the encapsulation of folic acid in two different hydrocolloid matrices: whey protein concentrate (WPC) and a commercial resistant starch (Pérez-Masiá et al., 2015). Production of spherical micron and submicron particles and the encapsulation efficiency were comparable between both techniques, but ESP enabled control over the size distribution and enhanced the stability of folic acid. Paximada et al. (2017) applied emulsion ESP as one-step technique to produce epigallocathechin gallate (EGCG) loaded bacterial submicron particles. They studied the stability of EGCG at different storage conditions and found that the capsules tend to collapse at high relative humidity, but encapsulation provides EGCG great stability under neutral and basic conditions, and at 37°C/60°C, as compared to nonencapsulated EGCG. *Sarcopoterium spinosum* extract is a phenolic-rich extract and its bioavailability and stability was enhanced by encapsulation in zein nanoparticles produced by ESP (Faridi Esfanjani and Jafari, 2016).

ESP in chocolate, colorants, or other food ingredients processing could be an interesting approach, taking into account that new techniques prevail in the *haute cuisine* and can be a strategy to help sector professionals to innovate in a highly demanding market. Chocolate microspheres were obtained through the ESP process and the effects of process parameters such as sugar concentration, addition of electrolytes, flow rate, applied voltage, and collection distance on the production and morphology of as-sprayed chocolate particles were studied (Luo et al., 2012). Cocoa butter microcapsules containing an aqueous solution or an oil-in-water emulsion containing sugar, colorants, and emulsifiers were produced via ESP (Bocanegra et al., 2005). Parameters, such as thickness of the shell or number of inner cores in the microcapsules were easily controlled by properly adjusting the flow rates. ESP was used to encapsulate curcumin into zein nanoparticles for tuning chromaticities of milk-based products through the addition of various amounts of these nanoparticles (Gomez-Estaca et al., 2012). ESP processing was used for stearic acid nanoparticles encapsulating ethylvanillin (EV), both substances are representatives of typical hydrophobic coating and hydrophilic flavor component of the food industry (Eltayeb et al., 2016). The resulting nanoparticles exhibited a core-shell structure and the thickness of the outer layer depended on the concentration of lipid in the processed solution. The release of EV was consistent with the diffusion through a lipid membrane model.

4.3.2.3   Other Applications

ESP has been used to produce contrast agents for imaging applications. Magnetic and superparamagnetic nanoparticles have been encapsulated into polysorbate and PLGA microspheres; miscellar nanocomposites containing quantum dots have been assessed for potential uses in fluorescence imaging (Xie et al., 2015). In the field of regenerative medicine, ESP has shown great potential for cell delivery, immunotherapies, and engineering tissues (Chakraborty et al., 2009; Chang et al., 2012; Gasperini et al., 2013; Ma et al., 2013). Metal oxides have been used as gas sensing materials, such as iron oxyhydroxide, tungsten oxide, and indium oxide. Hybrid polypyrrole nanoparticles with various diameters containing sensing inorganic compounds were fabricated using a dual-nozzle electrospray and vapor deposition polymerization (Lee et al., 2014).

4.4   CHALLENGES AND FUTURE PROSPECTS OF ELECTROSPINNING AND ELECTROSPRAYING AS ENCAPSULATION METHODS

The advantages of the nonthermal EHD processes are extensively proven and documented to encapsulate both natural and synthetic sensitive (bio) active compounds for different applications in the fields of biomedicine, food technology, and engineering, among many others.

In the field of food nanotechnology, future research must also focus on performing toxicity experiments, residual solvent analysis, and studying the biological fate of nanoparticles and nanofibers during digestion, absorption, and excretion of carriers. Fortunately, the technologies of electrospinning and electrospraying are both currently scaled up by companies such as Bioinicia S.L., and products are currently entering into the market. The future will see this technology applied in many encapsulation applications, starting from niche areas to eventually reach commodities as production capacity increases and costs drop.

REFERENCES

Aceituno-Medina, M.; Mendoza, S.; Rodríguez, B.; Lagaron, J.M.; López-Rubio, A. (2015) Improved capacity of quercetin and ferulic acid during in-vitro digestion through encapsulation within food-grade electrospun fibers. *Journal of Functional Foods*, 12, 332–341.

Ahvenainen, R. (2002) Active and intelligent packaging: An introduction. In *Novel Food Packaging Techniques*, Ahvenainen, R., ed. Woodhead Publishing Limited: Boca Raton, FL, pp. 3–19.

Appendini, P.; Hotchkiss, J.H. (2002) Review of antimicrobial food packaging. *Innovative Food Science & Emerging Technologies*, 3, 113–126.

Aruna, S.T.; Balaji, L.S.; Kumar, S.S.; Prakash, B.S. (2017) Electrospinning in solid oxide fuel cells—A review. *Renewable and Sustainable Energy Reviews*, 67, 673–682.

Arrua, R.D.; Strumia, M.C.; Nazareno, M.A. (2010) Immobilization of caffeic acid on a polypropylene film: Synthesis and antioxidant properties. *Journal of Agricultural and Food Chemistry*, 58, 9228–9234.

Baji, A.; Mai, Y.W.; Wong, S.C.; Abtahi, M.; Chen, P. (2010) Electrospinning of polymer nanofibers: Effects on oriented morphology, structures and tensile properties. *Composites Science and Technology*, 70, 703–718.

Bhardwaj, N.; Kundu, S.C. (2010) Electrospinning: A fascinating fiber fabrication technique. *Biotechnology Advances*, 28, 325–347.

Bhullar, S.K.; Kaya, B.; Jun, M.B.G. (2015) Development of bioactive packaging structure using melt electrospinning. *Journal of Polymers and the Environment*, 23, 416–423.

Bocanegra, R.; Gaonkar, A.; Barrero, A.; Loscertales, I.G.; Pechack, D.; Marquez, M. (2005) Production of cocoa butter microcapsules using an electrospray process. *Journal of Food Science*, 70, 492–497.

Bölgen, N.; Vaseashta, A. (2015) Nanofibers for tissue engineering and regenerative medicine. In *Third International Conference on Nanotechnologies and Biomedical Engineering, ICNBME 2015*, Chisinau, Moldova, September 23–26, 2015. IFMBE Proceedings, Vol. 55, pp. 319–322.

Brody, A.L.; Bugusu, B.; Han, J.H.; Koelsch Sand, C.; McHugh, T.H. (2008) Innovative food packaging solutions. *Journal of Food Science*, 73 (8), 108–116.

Brown, T.D.; Dalton, P.D.; Hutmacher, D.W. (2016) Melt electrospinning today: An opportune time for an emerging polymer process. *Progress in Polymer Science*, 56, 116–166.

Bushani, A.; Anandharamakrishna, C. (2014) Electrospinning and electrospraying techniques: Potential food based applications. *Trends in Food Science & Technology*, 38, 21–33.

Busolo, M.A.; Lagaron, J.M. (2015) Antioxidant polyethylene films based on a resveratrol containing Clay of Interest in Food Packaging Applications. *Food Packaging and Shelf Life*, 6, 30–41.

Busolo, M.A.; Torres-Giner, S.; Lagaron, J.M. (2009) Enhancing the gas barrier properties of polylactic acid by means of electrospun ultrathin zein fibers. In *ANTEC, Proceedings of the 67th Annual Technical Conference*, Chicago, IL, pp. 2763–2768.

Cacciotti, I.; House, J.N.; Mazzuca, C.; Valentini, M.; Madau, F.; Palleschi, A.; Straffi, P.; Nanni, F. (2015) Neat and GNPs loaded natural rubber fibers by electrospinning: Manufacturing and characterization. *Materials and Design*, 88, 1109–1118.

Camerlo, A.; Vebert-Nardin, C.; Rossi, R.M.; Popa, A.M. (2013) Fragrance encapsulation in polymeric matrices by emulsion electrospinning. *European Polymer Journal*, 49, 3806–3813.

Castro-Mayorga, J.L.; Fabra, M.J.; Pourrahimi, A.M.; Olsoon, R.T.; Lagaron, J.M. (2017) The impact of zinc oxide particle morphology as an antimicrobial and when incorporated in poly(3-hydroxybutyrate-co-3-hydroxyvalerate) films for food packaging and food contact surfaces applications. *Food and Bioproducts Processing*, 101, 32–44.

Cerqueira, M.A.; Fabra, M.J.; Castro-Mayorga, J.L.; Bourbon, A.; Pastrana, L.; Vicente, A.A.; Lagaron, J.M. (2016) Use of electrospinning to develop antimicrobial biodegradable multilayer systems: Encapsulation of cinnamaldehyde and their physicochemical characterization. *Food Bioprocess Technology*, 9, 1874–1884.

Chakraborty, S.; Liao, I.C.; Adler, A.; Leong, K.W. (2009) Electrohydrodynamics: A facile technique to fabricate drug delivery systems. *Advanced Drug Delivery Reviews*, 61, 1043–1054.

Chalco-Sandoval, W.; Fabra, M.J.; López-Rubio, A.; Lagaron, J.M. (2015) Development of polysyrene-based films with temperature buffering capacity for smart food packaging. *Journal of Food Engineering*, 164, 55–62.

Chang, P.C.; Chung, M.C.; Lei, C.; Chong, L.Y.; Wang, C.H. (2012) Biocompatibility of PDGF-simvastatin double-walled PLGA (PDLLA) microspheres for dentoalveolar regeneration: A preliminary study. *Journal of Biomedical Materials Research*, 100, 2970–2978.

Chang, W.; Xu, F.; Mu, X.; Ji, L.; Ma, G.; Nie, J. (2013) Fabrication of nanostructured hollow TiO$_2$ nanofibers with enhanced photocatalytic activity by coaxial electrospinning. *Material Research Bulletin*, 48, 2661–2668.

De Oliveira Moric, C.; Almeida dos Passos, N.; Oliveira, J.E.; Capparelli Mattoso, L.H.; Akira Moric, F.; Carvalho, A.C.; De Souza Fonseca, A.; Denzin Tonoli, G.H. (2014) Electrospinning of zein/tannin bio-nanofibers. *Industrial Crops and Products*, 52, 298–304.

Doshi, J.; Reneker, D.H. (1995) Electrospinning process and applications of electrospun fibers. *Journal of Electrostatics*, 35, 151–160.

Doustgani, A.; Ahmadi, E. (2016) Melt electrospinning process optimization of polylactic acid nanofibers. *Journal of Industrial Textiles*, 45, 626–634.

Echegoyen, Y.; Fabra, M.J.; Castro-Mayorga, J.L.; Cherpinski, A., Lagaron, J.M. (2017) High throughput electro-hydrodynamic processing in food and food packaging applications: Viewpoint. *Trends in Food Science & Technology*, 60, 71–79.

Eltayeb, M.; Stride, E.; Edirisinge, M.; Harker, A. (2016) Electrosprayed nanoparticle delivery system for controlled release. *Materials Science and Engineering C*, 66, 138–146.

European Commission (2011) Commission Regulation (EU) No 10/2011 of 14 January 2011 on plastic materials and articles intended to come into contact with food. Official Journal of the European Union, Brussels, Belgium. http://eur-lex.europa.eu/LexUriServ/LexUriServ. do?uri=OJ:L:2011:012:0001:0089:EN:PDF10/21/2012. Accessed July 11, 2017.

European Food Safety Authority (EFSA) (2016) Safety assessment of the substance zinc oxide, nanoparticles, for use in food contact materials. *EFSA Journal*, 14, 4408–4416.

Fabra, M.J.; Busolo, M.A.; Lopez-Rubio, A.; Lagaron, J.M. (2013) Nanostructured biolayers in food packaging. *Trends in Food Science & Technology*, 31, 79–87.

Fabra, M.J.; López-Rubio, A.; Lagaron, J.M. (2016) Use of the electrohydrodynamic process to develop active/bioactive bilayer films for food packaging applications. *Food Hydrocolloids*, 55, 11–18.

Fang, Z.; Bhandarri, B. (2010) Encapsulation of polyphenols—A review. *Trends in Food Science & Technology*, 21, 510–523.

Faridi Esfanjani, A.; Jafari, S.M. (2016) Biopolymer nano-particles and natural nano-carriers for nano-encapsulation of phenolic compounds. *Colloids and Surfaces B: Biointerfaces*, 146, 532–543.

Fernandez, A.; Torres-Giner, S.; Lagaron, J.M. (2009) Novel route to stabilization of bioactive antioxidants by encapsulation in electrospun fibers of zein prolamine. *Food Hydrocolloids*, 23, 1427–1432.

Fukui, Y.; Mauyama, T.; Iwamatsu, Y.; Fujii, A.; Tanaka, T.; Ohmukai, Y.; Matsuyama, H. (2010) Preparation and encapsulation of monodispered polyelectrolyte microcapsules with high encapsulation efficiency by an electrospray technique. *Colloids and Surfaces A: Physicochemical and Engineering Aspects*, 370, 28–34.

Gasperini, L.; Mniglio, D.; Migliaresi, C. (2013) Microencapsulation of cells in alginate through an electrohydrodynamic process. *Journal of Bioactive and Compatible Polymers*, 28, 413–425.

Ghorani, B.; Tucker, N. (2015) Fundamentals of electrospinning as a novel delivery vehicle for bioactive compounds in food nanotechnology. *Food Hydrocolloids*, 51, 227–240.

Gomez-Estaca, J.; Balaguer, M.P.; Gavara, R.; Hernandez-Munoz, P. (2012) Formation of zein nanoparticles by electrohydrodynamic atomization: Effect of the main processing variables and suitability for encapsulating the food coloring and active ingredient curcumin. *Food Hydrocolloids*, 28, 82–91.

Gomez-Mascaraque, L.G.; Cruz Morfin, R.; Pérez-Masiá, R.; Sanchez, G.; Lopez-Rubio, A. (2016) Optimization of electrospraying conditions for the microencapsulation of probiotics and evaluation of their resistance during storage and *in-vitro* digestion. *LWT-Food Science and Technology*, 69, 438–446.

Gong, G.; Wu, J. (2012) Novel polyimide materials produced by electrospinning. In *High Performance Polymers: Polyimides Based—From Chemistry to Applications*, Abadie, M., ed. InTech: Rijeka, Croatia. DOI: 10.5772/53459. Available from: http://www. intechopen.com/books/high-performance-polymers-polyimides-based-from-chemistry-to-applications/novel-polyimide-materials-produced-by-ES. Accessed July 11, 2017.

Gopal, R.; Kaur, S.; Ma, Z.; Chan, C.; Ramakrishna, S.; Matsuura, T. (2006) Electrospun nanofibrous filtration membrane. *Journal of Membrane Science*, 281, 581–586.

Greiner, A.; Wendorff, J.G. (2007) Electrospinning: A fascinating method for the preparation of ultrathin fibers. *Angewandte Chemie International Edition*, 45, 5670–5703.

Haider, A.; Haider, S.; Kan, I.K. (2015) A comprehensive review summarizing the effect of electrospinning parameters and potential applications of nanofibers in biomedical and biotechnology. *Arabian Journal of Chemistry*. https://dx.doi.org/10.1016/j.arabjc.2015.11.015.

Hekmati, A.H.; Rashidi, A.; Ghazisaeidi, R.; Drean, J.Y. (2013) Effect of needle length, electrospinning distance, and solution concentration on morphological properties of polyamide-6 electrospun nanowebs. *Textile Research Journal*, 83, 1452–1466.

Ji, W.; Yang, F.; van den Beucken, Z.A.; Bian, Z.W.; Fan, M.W.; Chen, J.A. (2010) Fibrous scaffolds loaded with protein prepared by blend or coaxial electrospinning. *Acta Biomaterialia*, 6, 4199–4207.

Jiang, H.; Wang, L.; Zhu, K. (2014) Coaxial electrospinning for encapsulation and controlled release of fragile water-soluble bioactive agents. *Journal of Controlled Release*, 193, 296–303.

Kai, D.; Liow, S.S.; Loh, X.J. (2014) Biodegradable polymers for electrospinning: Towards biomedical applications. *Materials Science and Engineering C*, 45, 659–670.

Katouzian, I.; Jafari, S.M. (2016) Nano-encapsulation as a promising approach for targeted delivery and controlled release of vitamins. *Trends in Food Science & Technology*, 53, 34–48.

Kundu, S.C.; Bhardwaj, N. (2010) Electrospinning: A fascinating fiber fabrication technique. *Biotechnology Advances*, 28, 325–347.

Lee, C.H.; Soon An, D.; Lee, S.C.; Park, H.J.; Sun Lee, D. (2004) A coating for use and antimicrobial and antioxidative packaging material incorporating nisin and α-tocopherol. *Journal of Food Engineering*, 62, 323–329.

Lee, J.K.; Ko, J.; Jun, M.B.G.; Lee, P.C. (2016) Manufacturing and characterization of encapsulated microfibers with different molecular weight poly(ε-Caprolactone) (PCL) Resins using a melt electrospinning technique. *Materials Research Express*, 3, 025301.

Lee, J.S.; Jun, J.; Shina, D.H.; Jang, J. (2014) Urchin-like polypyrrole nanoparticles for highly sensitive and selective chemiresistive sensor application. *Nanoscale*, 6, 4188–4194.

Lelkes, P.I.; Li, M.; Mondrinos, M.J.; Gandhi, M.R.; Ko, F.K.; Weiss, A.S. (2005) Electrospun protein fibers as matrices for tissue engineering. *Biomaterials*, 26, 5999–6008.

Li, X.; Su, Y.; Zhou, X.; Mo, X. (2009) Distribution of sorbitan monooleate in poly(L-lactide-co-ε-caprolactone) nanofibers from emulsion electrospinning. *Colloids and Surfaces B: Biointerfaces*, 69, 221–224.

Li, Z.; Wang, C. (eds.) (2013) Chapter 2: Effects of working parameters on electrospinning. *One-Dimensional Nanostructures*, SpringerBriefs in Materials. Springer: Berlin, Germany. DOI: 10.1007/978-3-642-36427-3_2.

Liu, Y.; Zhang, L.; Sun, X.F.; Liu, J.; Fan, J.; Huang, D.W. (2015) Multi-jet electrospinning via auxiliary electrode. *Material Letters*, 141, 153–156.

Lopez-Rubio, A.; Lagaron, J.M. (2012) Whey protein capsules obtained from electrospraying for the encapsulation of bioactives. *Innovative Food Science & Emerging Technologies*, 13, 200–206.

López-Rubio, A.; Sanchez, E.; Sanz, Y. (2009) Encapsulation of living bifidobacteria in ultra-thin PVOH electrospun fibers. *Biomacromolecules*, 10, 2823–2829.

Lopez-Rubio, A.; Sanchez, E.; Wilkanowicz, S.; Sanz, Y.; Lagaron, J.M. (2012) Electrospinning as a useful technique for the encapsulation of living bifidobacteria in food hydrocolloids. *Food Hydrocolloids*, 1, 159–167.

Luo, C.J.; Loh, S.; Stride, E.; Edirisinghe, M. (2012) Electrospraying and electrospinning of chocolate suspensions. *Food and Bioprocess Technology*, 5, 2285–2300.

Ma, M.; Chiu, A.; Sahay, G.; Doloff, J.C.; Dholakia, N.; Thakrar, R.; Cohen, J. et al. (2013) Core–shell hydrogel microcapsules for improved islets encapsulation. *Advanced Healthcare Materials*, 2, 667–672.

Malwal, D.; Gopinath, P. (2017) Efficient adsorption and antibacterial properties of electrospun CuO-ZnO composite nanofibers for water remediation. *Journal of Hazardous Materials*, 321, 611–621.

Megelski, S.; Stephens, J.S.; Chase, D.B.; Rabolt, J.F. (2002) Micro- and nanostructured surface morphology on electrospun polymer fibers. *Macromolecules*, 22, 8456–8466.

Migliaresi, M.; Ruffo, G.A.; Volpato, F.Z.; Zeni, D. (2012) Advanced electrospinning setups and special fibre and mesh morphologies. In *Electrospinning for Advanced Biomedical Applications and Therapies*, Neves, N.M., ed. Smithers Rapra Technology: NY. DOI: 10.5772/53459. Available from: http://s3.amazonaws.com/academia.edu.documents/32624557/Chapter_2___Advanced_Electrospinning_Setups_and_Special_Fibre_and_Mesh_Morphologies.pdf?AWSAccessKeyId=AKIAJ56TQJRTWSMTNPEA&Expires=1479224673&Signature=AAJ0Z%2FumrG9qPJJUUKmGohGpqhM%3D&response-content-disposition=inline%3B%20filename%3DChapter_2_Advanced_Electrospinning_Setup.pdf. Accessed July 11, 2017.

Molnar, K.; Nagy, Z. (2016) Corona-ES: Needleless method for high-throughput continuous nanofiber production. *European Polymer Journal*, 74, 279–286.

Nagy, Z.K.; Balogh, A.; Démuth, B.; Pataki, H.; Vigh, T.; Szabó, B.; Molnár, K. et al. (2015) High speed electrospinning for scaled-up production of amorphous solid dispersion of itraconazole. *International Journal of Pharmaceutics*, 480, 137–142.

Nedovic, V.; Kalusevic, A.; Manojlovic, V.; Levic, S.; Bugarski, B. (2011) An overview of encapsulation technologies for food applications. *Procedia Food Science*, 1, 1806–1815.

Neo, Y.P.; Ray, S.; Jin, J.; Gizdavic-Nikolaidis, M.; Niewwoudt, M.; Liu, D.; Quek, S.Y. (2013) Encapsulation of food grade antioxidant in natural biopolymer by electrospinning technique: A physicochemical study based on zein-gallic acid system. *Food Chemistry*, 136, 1013–1021.

Newsome, R. (2013) IFT International Food Nanoscience Conference: Proceedings. *Comprehensive Reviews in Food Science and Food Safety*, 13, 190–228.

Nguyen, T.T.T.; Ghosh, C.; Hwang, S.; Chanunpanich, N.; Park, J.S. (2012) Porous core/sheath composite nanofibers fabricated by coaxial electrospinning as a potential mat for drug release system. *International Journal of Pharmaceutics*, 439, 296–306.

Paaver, U.; Laidmäe, I.; Santo, H.A.; Yliruusi, J.; Aruväli, J.; Kogermann, K.; Heinämäki, J. (2016) Development of a novel electrospun nanofibrous delivery system for poorly water-soluble β-sitosterol. *Asian Journal of Pharmaceutical Sciences*, 11, 500–506.

Panda, P.K.; Sahoo, B. (2013) Chapter 14: Synthesis and applications of electrospun nanofibers: A review. In *Nanotechnology: Fundamental and Applications*, Vol. 1, Navani, N.K., Sinha, S., and Govil, J.N., eds. Studium Press LLC, Houston, TX, pp. 399–416.

Parhizkar, M.; Reardon, P.; Knowles, J.C.; Browning, R.J.; Stride, E.; Barbara, P.R.; Harker, A.H.; Edirisinghe, M. (2016) Electrohydrodynamic encapsulation of cisplatin in poly(lacti-co-glycolic acid) nanoparticles for controlled drug delivery. *Nanomedicine: Nanotechnology, Biology and Medicine*, 12, 1919–1929.

Paximada, P.; Echegoyen, Y.; Koutinas, A.A.; Mandala, I.; Lagaron, J.M. (2017) Encapsulation of hydrophilic and lipophilized catechin into nanoparticles through emulsion electrospray. *Food Hydrocolloids*, 64, 123–132.

Persano, L.; Camposeo, A.; Tekmen, C.; Pisignano, P. (2013) Industrial upscaling of electrospinning and applications of polymer nanofibers: A review. *Macromolecular Materials and Engineering*, 298, 504–520.

Pérez-Masiá, R.; López-Nicolás, R.; Peroago, M.J.; Ros, G.; Lagaron, J.M. (2015) Encapsulation of folic acid in food hydrocolloids through nanospray drying and electrospray for nutraceutical applications. *Food Chemistry*, 168, 124–133.

Qin, C.C.; Duan, X.P.; Wang, L.; Zhang, L.H.; Yu, M.; Dong, R.H.; Yan, X.; He, H.H.; Long, Y.Z. (2015) Melt electrospinning of poly(lactic acid) and polycaprolactone microfibers by using a hand-operated Wimshurst generator. *Nanoscale*, 7, 16611–16615.

Qu, H.; Wei, S.; Guo, Z. (2013) Coaxial electropun nanostructures and their applications. *Journal of Materials Chemistry A*, 1, 11513–11528.

Ramakrishna, S.; Huang, Z.; Zhang, Y.Z.; Kotaki, M. (2003) A review on polymer nanofibers by ES and their applications in nanocomposites. *Composites Science and Technology*, 63, 2223–2253.

Realini, C.E.; Marcos, B. (2014) Active and intelligent packaging systems for a modern society. *Meat Science*, 98, 404–419.

Repanas, A.; Glasmacher, B. (2015) Dipyridamole embedded in Polycaprolactone fibers prepared by coaxial electrospinning as a novel drug delivery system. *Journal of Drug Delivery Science and Technology*, 29, 132–142.

Repanas, A.; Wolkers, W.F.; Gryshkov, O.; Kalozoumis, P.; Mueller, M.; Zernetsch, H.; Korossis, S.; Glasmacher, B. (2015) Coaxial electrospinning as a process to engineer biodegradable polymeric scaffolds as drug delivery systems for anti-inflammatory and anti-thrombotic pharmaceutical agents. *Clinical and Experimental Pharmacology and Physiology*, 5, 192. DOI: 10.4172/2161-1459.1000192.

Restuccia, D.; Spizzirri, U.G.; Parisi, O.I.; Cirillo, G.; Curcio, M.; Iemma, F.; Puoci, F.; Vinci, G.; Picci, N. (2010) New EU regulation aspects and global market of active and intelligent packaging for food applications. *Food Control*, 21, 1425–1435.

Rogina, A. (2014) Electrospinning process: Versatile preparation method for biodegradable and natural polymers and biocomposite systems applied in tissue engineering and drug delivery. *Applied Surface Science*, 296, 221–230.

Sanders, E.H.; Kloefkorn, R.; Bowlin, G.L.; Simpson, D.G.; Wnek, G.E. (2003) Two-phase electrospinning from a single electrified jet: Microencapsulation of aqueous reservoirs in poly(ethylene-co-vinyl acetate) fibers. *Macromolecules*, 36, 3803–3805.

Sawicka, K.; Gouma, P. (2006) Electrospun composite nanofibers for functional applications. *Journal of Nanoparticle Research*, 8, 769–781.

Schiffman, J.D.; Schauer, C. (2008) A review: Electrospinning of biopolymers nanofibers and their applications. *Polymer Reviews*, 48, 317–352.

Shenoy, S.L.; Bates, W.D.; Wnek, G. (2005) Correlation between electrospinnability and physical gelation. *Polymer*, 46, 8990–9004.

Si, N.; Xu, L.; Wang, M.Z.; Liu, F. (2014) Effect of ultrasonic vibration of electrospun poly(vinyl alcohol) (PVA) nanofibers. *Advanced Materials Research*, 843, 1–8.

Sill, T.J.; von Recum, H.A. (2008) Electrospinning: Applications in drug delivery and tissue engineering. *Biomaterials*, 29, 1989–2006.

Stanishevsky, A.; Brayer, W.A.; Pokorny, P.; Kalous, T.; Lukás, D. (2016) Nanofibrous alumina structures fabricated using high-yield alternating current electrospinning. *Ceramics International*, 42, 17154–17161.

Sunil, M. (2012) Antimicrobial food packaging to enhance food safety: Current developments and future challenges. *Journal of Food Processing and Technology*, 3, e103. DOI: 10.4172/2157-7110.1000e103.

Tan, S.H.; Inai, R.; Kotaki, M.; Ramakrishna, S. (2005) Systematic parameter study for ultrafine fabrication via process. *Polymer*, 46, 6128–6134.

Teo, W.E.; Ramakrishna, S. (2009) Electrospun nanofibers as a platform for multifunctional, hierarchically organized nanocomposite. *Composites Science and Technology*, 69, 1804–1817.

Theron, S.A.; Yarin, A.L.; Zussman, E.; Kroll, E. (2005) Multiple jets in electrospinning: Experiment and modeling. *Polymer*, 46, 2889–2899.

Torres-Giner, S.; Gimenez, E.; Lagaron, J.M. (2008) Characterization of the morphology and thermal properties of zein prolamine nanostructures obtained by electrospinning. *Food Hydrocolloids*, 22, 601–614.

Torres-Giner, S.; Ocio, M.J.; Lagaron, J.M. (2009) Novel antimicrobial ultrathin structures of zein/chitosan blends obtained by electrospinning. *Carbohydrate Polymers*, 77, 1427–1432.

Varabhas, J.; Chase, G.; Reneker, D. (2008) Electrospun nanofibers from a porous hollow tube. *Polymer*, 49, 4226–4229.

Wang, H.; Xu, Y.; Song, M.; Du, Z.; Wei, Q. (2015a) Formation of layering structure of poly(ethylene terephthalate) fiber film by melt-electrospinning. *Polymeric Materials Science and Engineering*, 31, 114–118.

Wang, X.; Yuan, Y.; Huang, X.; Yue, T. (2015b) Controlled release of protein from core-shell nanofibers prepared by emulsion electrospinning based on green chemical. *Journal of Applied Polymer Science*, 132, DOI: 10.1002/APP.41811.

Wang, Y.; Li, Z.; Shao, P.; Hao, S.; Wang, W.; Yang, Q.; Wang, B. (2014) A novel multiple drug release system in vitro based on adjusting swelling core of emulsion electrospun nanofibers with core–sheath structure. *Materials Science and Engineering C*, 44, 109–116.

Wen, P.; Zhu, D.H.; Feng, K.; Liu, F.J.; Lou, W.Y.; Li, N.; Zong, M.H.; Wu, H. (2016) Fabrication of electrospun polylactic acid nanofilm incorporating cinnamon essential oil/b-cyclodextrin inclusion complex for antimicrobial packaging. *Food Chemistry*, 196, 996–1004.

Wu, Y.; MacKay, J.A.; McDaniel, J.C.; Chilkoti, A.; Clark, R.I. (2009) Fabrication of elastine-like polypeptide nanoparticles for drug delivery by electrospraying. *Biomolecules*, 10, 19–24.

Xie, J.; Jiang, J.; Davoodi, P.; Srinivasan, M.P.; Wang, C.H. (2015) Electrohydrodynamic atomization: A two-decade effort to produce and process micro-/nanoparticulate materials. *Chemical Engineering Science*, 125, 32–57.

Xu, X.; Yang, L.; Xu, X.; Wang, X.; Chen, X.; Liang, Q.; Zeng, J.; Jing, X. (2005) Ultrafine medicated fibers electrospun from W/O emulsions. *Journal of Controlled Release*, 108, 33–42.

Yao, Z.C.; Wei, M.; Ahmad, Z.; Li, J.S. (2016) Encapsulation of rose hip seed oil into fibrous zein films for ambient and on demand food preservation via coaxial electrospinning. *Journal of Food Engineering*, 191, 115–123.

Zhang, Y.Z.; Feng, Y.; Huang, Z.M.; Ramakrishna, S.; Lim, C.T. (2006) Fabrication of porous electrospun nanofibres. *Nanotechnology*, 17, 11–15.

5 Extrusion for Microencapsulation

Andriana Lazou and Magdalini K. Krokida

CONTENTS

5.1 INTRODUCTION

In recent years, great emphasis has been given globally in the study, development, and commercialization of functional foods. It should be noted that there are a number of such products with the potential of acquiring a strong position in domestic and international food markets, due to their excellent nutritional and sensorial characteristics. Furthermore, it should be noted that a great interest exists in consumers for nutraceuticals or "functional foods," which leads to the development of "natural" and "healthy" food products. Moreover, there is a growing interest to include natural functional ingredients in diets (Ruiz-Gutiérrez et al. 2015). Though different, functional foods are sometimes considered along with dietary supplements (they include products such as vitamins, minerals, herbs, etc.) and medicinal foods (they include health bars with added medications and transgenic cows and plants) (Viswanath and Krishna 2015). They are one of the fastest growing sectors of the food industry due to increasing demand from consumers for foods that promote health and well-being (Mollet and Lacroix 2007).

Processing could seriously affect functional components of foods and hence, in developing functional foods and in many cases, specific technologies that form a structure to prevent the deterioration of physiologically active compounds, such as microencapsulation, edible films and coatings and vacuum impregnation should be applied (Betoret et al. 2011). Furthermore, food product development and production needs food ingredients with complex properties, such as controlled and/or delayed release, stability, thermal protection, and suitable sensorial profile. This could be achieved by microencapsulation (Đorđević et al. 2015). The most common encapsulation technologies used in the flavor industry comprise of spray drying, spray coating, and extrusion (Castro et al. 2016).

Extrusion is a process, which can be used for the production of such products via incorporation or encapsulation of various ingredients and natural products. By this way, an improvement in nutritive value, color, and flavor of products could be achieved. Extrusion encapsulation has the potential to provide solutions in the development of functional foods, in particular the inclusion of active compounds either with or without affecting the sensory properties of the food.

5.2   EXTRUSION PROCESS

Food extrusion is a food processing technology that combines a number of unit operations such as mixing, heating, kneading, shearing, shaping, and forming. During processing, a mixture of ingredients is enforced through a die especially designed for the product produced, and at the exit it is cut to a specified size by a set of blades (Harper and Clark 1979, Bordoloi and Ganguly 2014, Navale et al. 2015). Extrusion processing of food materials has become an increasingly important manufacturing method. The extruder consists of one or two rotating screws tightly fitting within a barrel, at the end of which is the die. The principles of operation for all extruders are similar and include raw material feeding through a feeder into the barrel and screw(s) push it toward die and cutter. More specific extruders are comprised of six parts, namely, preconditioning system, feeding system, screw(s), barrel, die, and cutting mechanism. Further, they can vary with respect to the screw, barrel, and die configuration, the selection of which depends on raw materials used and the desired final product (Harper 1989, Riaz 2000, Ramachandra Rao and Thejaswini 2015). Depending on the application, extruders may be designed either simply to form the raw mixture of ingredients or to mix and knead the ingredients in order to convert them into a plastic mass with or without heating. The latter brings about several desired physicochemical changes that contribute to the characteristics and quality of the extrudate and the extruded products. Extruders can be categorized as cold and hot extruders (extrusion cooking). Extrusion cooking is a thermomechanical process in which heat transfer, mass transfer, pressure differences, and shear are combined, leading to effects such as cooking, sterilization, drying, melting, kneading, texturizing, conveying, puffing, mixing, forming, etc. It is used in the food industries for the development of new products such as cereal-based snacks, including dietary fiber, breakfast cereals, and modified starch from cereals. Extrusion-cooking is a high temperature short time process that has an important influence on product quality, while concurrently leads to a reduction in microbial population and enzyme inactivation. Extruded products have low water activity (<0.4) and are characterized by high expansion, specific texture, prolonged shelf life, and specific color and flavor (Riaz 2000, Steel et al. 2012). In addition, extrusion offers the possibility of modifying the functional properties of food ingredients, as well as their texture (Lazou and Krokida 2010, Lazou et al. 2010, Bisharat et al. 2015). Among its beneficial effects are gelatinization of starch, destruction of antinutritional factors, increased soluble dietary fibers, reduction of lipid oxidation and contaminating microorganisms, and retention of natural colors and flavors of foods (Singh et al. 2007, Camire 2011, Navale et al. 2015, Nikmaram et al. 2015). The advantages of food extrusion systems, leading to their expanded role in the food processing industry, include versatility, high productivity, low cost, product characteristics, high productivity and automation, high product quality, energy efficiency, production of new foods, no effluents, and the extruder can operate as a continuous reactor (Mans 1982, Darrington 1987, Riaz 2000, Bouvier and Campanella 2014).

Changes occurring in extruded products depend upon process conditions, such as extruder type, feed materials, moisture content, barrel temperature, and screw speed (Kumar et al. 2010). However, it should be emphasized that extrusion is a complex process including several process variables that are referred either to extruder variables or to feed ingredient variables. Figure 5.1 shows the effect of feed material variables and process parameters on both the process and product quality.

From Figure 5.1, the complexity of the extrusion process is evident. Among extruder variables, screw rotational speed, barrel temperature, screw and barrel configuration, die opening, and feed rate are important for extruded product performance, while feed composition, moisture content, and particle size have the greatest effect on extrusion. It should be noted that the typical composition of most ingredient mixture consists of starch, protein, lipids, and fiber, all of which contribute to product quality (Lazou and Krokida 2010, Lazou et al. 2010, Navale et al. 2015). These operation factors are related with residence time distribution, which is a useful tool

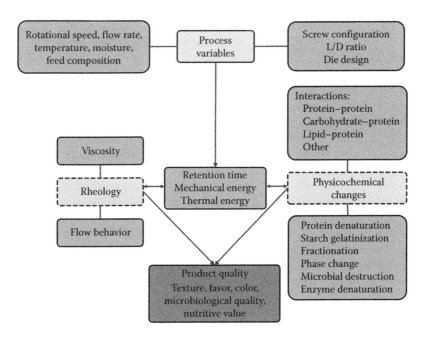

FIGURE 5.1 Extrusion variables and their effects on the quality of the extruded products.

to study physicochemical changes in feed materials and can also be used to scale up operations (Dhanasekharan and Kokini 2003, Lazou et al. 2007).

Relatively recent studies on extrusion has been focused on the development of functional products using various plant materials such as legumes, cereals and pseudocereals, fruits, and vegetables. Though some constituent degradation occurs, studies showed that in some cases good retention of quality attributes was received. Extrusion has also been applied to both the use of encapsulated powders in the extrusion process to enhance the retention of various components and in the encapsulation of a variety of bioactive and other constituents (Brennan et al. 2011).

5.3 DELIVERING BIOACTIVES THROUGH FOODS

Among of the challenges faced by the food processing industry is the delivering of high-quality food products with an added functionality, aiming to prevent some lifestyle-related diseases such as obesity, diabetes, heart disease, cancer, etc. Functional foods contain bioactive components and may provide desirable health benefits, especially in the direction of the prevention of such diseases (Wang and Bohn 2012). Encapsulation is an enabling technology that could fulfill all technical requirements for the effective delivery of various bioactives through novel or functional food product development for enhanced nutritional value (Li and Diosady 2011). Though encapsulation was originally introduced in the area of biotechnology, in recent years, it is an effective tool for the food industry, helping in the addition of functional compounds in various products. These compounds are usually highly susceptible to environmental, processing, and/or gastrointestinal conditions, and therefore, encapsulation has imposed an approach for effective protection of those (Nedovic et al. 2011). Functional compounds are used to control flavor, color, texture, or preservation properties, while bioactives with various potential health benefits are also included. Encapsulation aims to preserve the stability of the bioactive compounds during processing and storage and to prevent undesirable interactions with food matrix (Nedovic et al. 2011).

Bioactive components that may be found in functional foods are vitamins, peptides, minerals, fatty acids, phytosterols, lycopene, antioxidants, flavorings, enzymes, and living cells such as probiotics. Encapsulation of food ingredients is made (Alexe and Dima 2014) to protect the bioactive components against some physicochemical agents (temperature, pH, moisture, enzymes, oxygen, redox potential, UV light) during the storage, to prevent the reaction of bioactive components with other components in food products, for masking the bad taste or smell, to prevent the evaporation and degradation of volatile active components, to promote the conversion of liquid active compounds into a powder, and to assure the controlled release of biocompounds.

Though an increase in the production of encapsulated food ingredients is expected, many challenges still remain. Among them might be the incorporation of water-sensitive ingredients in high moisture foods (Poncelet et al. 2011). Further, it should be noted that the success of an encapsulation process is often linked to the know-how of the formulation or of the chemistry to achieve stabilization. This is often faced in the food industry as the number of acceptable materials is very limited. Another area is the engineering aspects, which are often neglected. Engineering aspects constitute a substantial part of success to maintain the integrity of active compounds, provide the right properties to the microcapsules, and reduce the cost that is directly related to the design and operating parameters (Poncelet et al. 2011). In any case, the development of encapsulated products is a challenge, requiring a multidisciplinary and integrated approach. Although many encapsulation technologies exist, many of them are still at the development stage. Once fully tested and validated, these new methods will undoubtedly broaden the spectrum of possible applications for this versatile technology (Poncelet et al. 2011).

5.3.1 Encapsulation Processes

Encapsulation is a technique that involves the incorporation of a chemically sensitive compound in a matrix or sealed capsule, protecting it against adverse reaction, preventing its degradation, and increasing its shelf life (Desai and Jin Park 2005). In other words, encapsulation is defined as the process of incorporation of a substance in the matrix of another substance, followed by the preparation of particles having a size in the nanometer (nanoencapsulation), micrometer (microencapsulation), or millimeter scale. The substance that is subjected to encapsulation forms the core of the capsule and is called active substance. The substance, which ensures the immobilization, is called a matrix or a carrier. The basic requirement for both substances is that they are permitted for use in the food industry by the food regulations. The carrier should also provide a stable barrier between the biologically active substances and the food product (Zuidam and Nedović 2010, Zuidam and Shimoni 2010, Burgain et al. 2011, Nedovic et al. 2011, Rathore et al. 2013, Lakkis 2016, Kostov et al. 2016). Besides protecting it from the harsh processing conditions and adverse storage environment, the encapsulation of bioactive compounds can also achieve targeted delivery and controlled release of entrapped substances to the specific site. In addition, encapsulation may also be useful for taste and odor-masking purposes, since some active compounds often have very strong flavors.

Depending on the encapsulation method, two types of capsules—a reservoir and a matrix type—could be obtained. The reservoir type provides a barrier (shell) between the medium and the active substance. This type is also called capsule—single-core, mono-core, or poly-core, which are capsules with several separate reservoir chambers. In the matrix type capsules, the encapsulated material is dispersed in the carrier material. It can be in the form of relatively small droplets or more homogeneously distributed over the encapsulate, while active agents are in general also present at the surface (Zuidam and Shimoni 2010, Rathore et al. 2013, Kostov et al. 2016).

The selection of the appropriate encapsulation method is based on the physical and chemical properties of the compound to encapsulate and the coating material (or matrix), as well as the final application (Desai and Jin Park 2005). It should be noted that the selection of the encapsulation method and matrix materials is interdependent.

The current encapsulation methods include spray drying, fluidized bed coating/granulation, spray chilling/cooling, (melt) extrusion, co-extrusion, freeze or vacuum drying, liposome or alginate entrapment, rotational or centrifugal suspension separation, coacervation, co-crystallization, molecular inclusion, interfacial polymerization, or rapid expansion of a supercritical fluid (Gibbs et al. 1999, Gouin 2004,

Desai and Jin Park 2005, Madene et al. 2006, Gharsallaoui et al. 2007, Zuidam and Shimoni 2010, Kostov et al. 2016). These processes could be applied for food and nonfood applications. Many of these methods start with the preparation of droplets of the active substance (gas, liquid, or powder) and then these droplets are covered by the carrier material in a gaseous or liquid medium by applying one of the above physicochemical processes. In some cases, the active substance and the carrier are mixed to form the gel phase and the solid particles are formed by physical, chemical, or combined transformations (Zuidam and Shimoni 2010). On a large-scale encapsulation in amorphous matrices, two main technologies, namely, spray drying and extrusion, are commonly used. In spray drying, the active is entrapped within the porous membranes of hollow spheres, while in extrusion the goal is to entrap the active in a dense, impermeable glass (Lakkis 2016). Spray drying is widely used in large-scale production of encapsulated substances, such as additives, vitamins, polyphenols, and flavorings, among others. This is due to the availability of equipment, low process cost, wide choice of carrier solids, good retention of volatiles, good stability of the finished product, and large-scale production in continuous mode (Gibbs et al. 1999, Gharsallaoui et al. 2007).

5.3.2 Extrusion Encapsulation

Extrusion used in the encapsulation of food ingredients is not the same as the extrusion cooking to produce puffed and/or texturized food products. It has been used almost exclusively for the encapsulation of volatile and unstable flavors in glassy carbohydrate matrices. As applied to flavor encapsulation, extrusion is a relatively low-temperature process, which involves forcing a core material in a molten carbohydrate mass through a series of dies into a bath of dehydrating liquid. The pressure and temperature employed are typically <689 kPa and seldom 115°C (Reineccius 1989, Desai and Jin Park 2005). The main advantage of this process is the very long shelf life, especially of oxidation-prone flavoring compounds. This is due to the very slow diffusion of atmospheric gases through the hydrophilic glassy matrix, thus providing an almost impermeable barrier against oxygen. Shelf lives of up to 5 years have been reported for extruded flavor oils, compared to 1 year for spray-dried flavors. Carbohydrate matrices in the glassy states have very good barrier properties and extrusion is a convenient process enabling the encapsulation of flavors in such matrices (Zasypkin and Porzio 2004, Poshadri and Aparna 2010). During processing, the coating material hardens on contacting the liquids, forming an encapsulating matrix to entrap the core material. Then the extruded filaments are separated from the liquid bath, dried, and sized (Shahidi and Han 1993, Shahidi and Pegg 2007). The carrier used may be composed of more than one ingredient, such as sucrose, maltodextrin, glucose syrup, glycerine, and glucose (Desai and Jin Park 2005).

Encapsulation into an amorphous matrix via extrusion has gained wide popularity in the past two decades with applications to different types of flavors, vitamins, and minerals (Lakkis 2016). An advantage of the extrusion technology is that it is in most cases a true encapsulation procedure instead of an immobilization technology (de Vos et al. 2010). The limitations of extrusion include its relatively high cost, low flavor loading, low solubility in cold water, and high process temperature. Its processing costs are estimated to be almost double in comparison to spray drying. The extruded product is not readily soluble in cold water and not stable in beverage application because of its large particle size. Furthermore, the compound to be extruded must be able to tolerate temperatures of 110°C –120°C. In addition, the large size of the granules (500–1000 µm), formed by extrusion, limits their use in food applications due to its negative effect on mouthfeel (Gouin 2004, Emin 2013, Lakkis 2016). Table 5.1 shows the pros and cons of extrusion processes for microencapsulation (Chokshi and Zia 2004, Bhaskaran and Lakshmi 2010, Zuidam and Shimoni 2010, Nedovic et al. 2011, Singhal et al. 2011, Jagtap et al. 2012, Maniruzzaman et al. 2012, Kleinebudde 2013, Maniruzzaman et al. 2014, Castro et al. 2016, Lakkis 2016, Patil et al. 2016).

Generally, extrusion microencapsulation includes three processes: (1) melt injection, (2) melt extrusion, and (3) centrifugal extrusion (co-extrusion) (Bakry et al. 2016). Figure 5.2 shows the process steps of the extrusion technologies used for microencapsulation of food bioactive compounds. The melt extrusion process is similar to that of melt injection. The main difference between these two processes is that melt extrusion is a horizontal screw process, while the melt injection is a vertical screwless process with surface-washed particles (Bakry et al. 2016). Figure 5.3 summarizes the load rate and the encapsulate particle size that are obtained by the main extrusion technologies.

5.3.3 Co-Extrusion

Co-extrusion process is mostly used to encapsulate flavor oils. Co-extrusion technology is used to produce core-shell particles, which consists of a concentric feed tube through which wall and core materials are pumped separately and then extruded through a concentric nozzle(s), with the payload dispersion being extruded through the inner nozzle and the wall materials being extruded from the outer nozzle (Desai and Jin Park 2005, Oxley 2012, 2014). The nozzles may be stationary, rotating, or vibrating. Further, the nozzle might also be submerged into a moving carrier and cooling fluid or it could be a dual-feed spraying nozzle in combination with ultrasonic atomization. This permits spray drying immediately in the air after atomization takes place (Zuidam and Shimoni 2010). The formed microcapsules can be collected on a moving bed of fine-grained starch (absorbs unwanted coating moisture). The capsule size depends on the rotational speed. Typical wall materials include starch, maltodextrins, gelatin, and polyethylene glycol (Oxley 2012). Particles produced by this method have diameter ranging from 80 to 2000 µm. It can be utilized to prepare spherical microbeads with a hydrophobic core of active agent and a hydrophilic or hydrophobic shell produced by interfacial gelling (Zuidam and Shimoni 2010). The co-extrusion has been used in the encapsulation of olive oil and caffeic acid, essential oils, proteins, and confectionery products (Sun-Waterhouse et al. 2011, Dolçà et al. 2015). Stability, low surface oil, and prolonged shelf life are the main advantages of extrusion encapsulation of oils (Gouin 2004). Moreover, this technique helps to reduce the evaporation rate of essential oils (Soliman et al. 2013). Co-extrusion is more expensive than spray drying. Further, it produces rather large particles, which limit the use of extruded essential oils in various applications (Desai and Jin Park 2005, Bakry et al. 2016).

TABLE 5.1
Pros and Cons of Some Extrusion Processes for Microencapsulation

Advantages	Disadvantages
Hot melt extrusion	
Solvent-free method	High processing temperatures
Limited amount of water	Difficult for heat-sensitive materials
Continuous, fast, and economic operation	Better process knowledge
Less unit operations	Equipment: requires training
High throughput	High energy input
Excellent mixing	Type and amount of plasticizers
Closed process unit	Flow properties of matrix material
Short residence time	Probably limited number of available matrix materials
Online monitoring	Difficulty in predicting physical stability
Less downstream processing	
Small-scale equipment	
Volume and scale-up flexibility	
Incorporation of bioactives at different points of the extrusion process	
Disperses bioactive in a matrix at molecular level	
Better uniformity in bioactive content in extrudates	
Better bioactive stabilization	
Medium to high load of bioactive	
Bioavailability enhancement	
Controlled release	
Taste masking	
Shaped final product	
Production of various geometries	
More uniform particle size	
Good stability	
Extrusion spheronization/co-extrusion	
Spheroids with high loading capacity of active ingredient	Wet mass extrusion
Particles of uniform size with narrow size distribution	Process parameters influence pellet quality
Successful coating	Different extruders influence final microsphere size
Pellets of different bioactives can be blended and formulated in single unit	Shear rate and shear stress in extruder leads in differences in spheroid uniformity
Pellets contribute in free spheroid dispersion and controlled release	Stability and release profile are affected by morphology, size distribution, porosity, sphericity, etc.
Improved safety and efficiency of active ingredient	Release profile is affected by presence or absence fillers, surface active agents, pH adjusters, and bioactive load
Increased bioavailability of active substance	Challenges in large-scale production

(Continued)

TABLE 5.1 (*Continued*)
Pros and Cons of Some Extrusion Processes for Microencapsulation

Advantages	Disadvantages
Low costs	Emulsion technique is more expensive
Mild operation conditions	Needs drying
Can provide smaller microspheres	Microcapsules range from 150 to 2000 μm
	Lack of suitable solvents
	Solvent residue removal process
	Handling large volumes of solvents
	Possibility of explosion hazard
	Solvent disposal
Solid lipid extrusion	
Solvent-free process	Lipids remain solid during processing, storage and application
No need for drying	Difficulties in dissolving bioactives
Low temperature process	Solid lipid extrudates not suitable to increase the bioavailability
Physically stable extrudates	In some cases, electrostatic charging during the extrusion process
Wide range of dissolution profiles	
Continuous process	

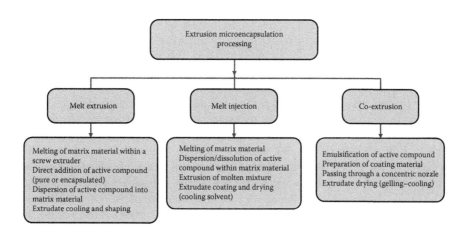

FIGURE 5.2 Extrusion microencapsulation processes. (Based on Zuidam, N.J. and Shimoni, E., Overview of microencapsulates for use in food products or processes and methods to make them, in: *Encapsulation Technologies for Active Food Ingredients and Food Processing*, edited by Zuidam, N.J. and Nedovic, V., New York, Springer, 2010, pp. 3–29.)

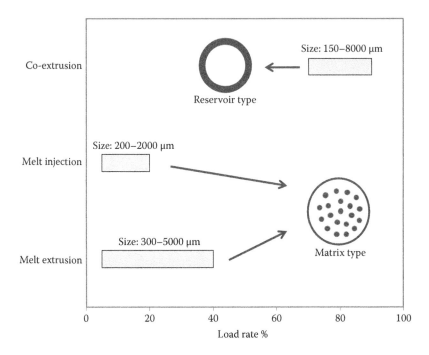

FIGURE 5.3 Load rate, encapsulate particle size, and characteristics of encapsulates obtained from the main extrusion microencapsulation processes.

5.3.4 ELECTROSTATIC EXTRUSION

Electrostatic extrusion (electrospraying or electrostatic atomization) is a method for the production of polymeric microspheres using, during polymer dipping into a hardening solution, electrostatic charges (Low and Lim 2014). This method can produce particle sizes within the range of 1000 and 5 μm (Yeo et al. 2005, Liu et al. 2010). However, it should be noted that the size of polymer microspheres is a complex function of several operating parameters, the system properties, and the properties of the polymer solution (Bugarski et al. 2004). Principally, the method includes the application of electrostatic forces in order to disrupt the liquid filament at the tip of a capillary or needle, forming thus a charged stream of small droplets (Bugarski et al. 2005). This leads to the formation of very small particles, of which their presence does not affect textural and sensorial properties when added to foods. For example, alginate spherical microbeads of 170 μm in diameter were obtained at a potential difference of 6 kV with a 26-gauge needle and an electrode distance of 2.5 cm (Bugarski et al. 2004). By increasing the applied potential, a large fraction (30%–40%) of microbeads with a mean diameter of 50 μm was obtained (Bugarski et al. 2004). Manojlovic et al. (2008) applied electrostatic extrusion for the encapsulation of ethyl vanillin in alginate gel microbeads. They obtained microbeads with ~10% w/w of ethyl vanillin encapsulated in about 2% w/w alginate; the size of microspheres was about 450 μm. Stojanovic et al. (2012) also applied electrostatic extrusion for the encapsulation of *Thymus serpyllum* L. aqueous extract within calcium alginate beads. The bead size was about 730 μm, while the encapsulation efficiency varied in the range of 50%–80% depending on the encapsulation method. Belščak-Cvitanović et al. (2011) encapsulated raspberry leaf, hawthorn, ground ivy, yarrow, nettle, and olive leaf extracts in alginate–chitosan microbeads using electrostatic extrusion. They obtained high encapsulation efficiency (80%–89%) for all extract encapsulating microbeads. It was also reported that the microbeads deliver significant biological activity and antioxidant potential. This could potentially contribute to an increase in the daily intake of antioxidants when microbeads are applied in food products.

5.4 HOT MELT EXTRUSION TECHNIQUE

Hot melt extrusion (HME) finds wide application in the plastic, rubber, pharmaceutical, and food industry. Interest in the food applications is growing rapidly and a number of patents and publications appeared. HME technology is currently being explored and could be used in food processing because it offers several advantages over traditional processing methods. HME is used for mixing, melting, and the reaction of materials, thereby combining several separate batch operations into one unit and increasing manufacturing efficiency (Patel et al. 2013). HME can be simply defined as the process of forming a new material (extrudate) by enforcing a raw material or mixture (matrix and bioactive) with a rotating screw through a die under a set of temperature, pressure, rate of mixing, and feed rate, for the purpose of producing a stable product of uniform shape and density (Mamidwar et al. 2012). Ram extrusion (melt injection) is a process in which the raw materials are introduced in a heated cylindrical barrel with a rotating screw to melt. Afterward, a piston (a ram) pressurizes the molten mixtures through a die and transforms them into the desired shape. The resulting material has the consistency of a hard candy entrapping the active (Schultz et al. 1956, Swisher 1957, Castro et al. 2016). HME equipment consists of an extruder, auxiliary equipment for the extruder, downstream processing equipment, and other monitoring equipment used for performance and product quality evaluation. An extruder is typically composed of a feeding hopper, barrels, single or twin screws, and the die and screw-driving unit (Chokshi and Zia 2004) (Figure 5.4). For extrusion encapsulation very often, double screw extruders equipped with self-wiping screws are utilized, with length/diameter (L/D) ratio, typically between 20:1 and 40:1. The heat required to melt the material is generated

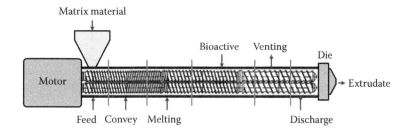

FIGURE 5.4 HME system for microencapsulation.

by friction as the material is sheared between the rotating screws and the wall of the barrel or by heating of the barrel. Downstream supplementary equipment serves for cooling, cutting, or collecting the products.

During the melt extrusion encapsulation process, a core material is dispersed in a molten shell formulation at a temperature of 85°C–140°C with or without the aid of a surfactant. In the case of flavor encapsulation, the melting temperatures should not exceed 140°C in order to avoid thermal degradation of the active compound. Typically the melting temperatures lie between 110°C and 140°C (Castro et al. 2016). Upon formation, the dispersion is extruded as filaments into a relatively cool environment that solidifies the extrudate. The environment can be either a gas phase or a suitable solvent. In the case of a gas phase, the cooled mass is simply broken up into particles and used, while in the case of a solvent, the solvent simultaneously cools and removes nonencapsulated or free core material from the filaments. Then the solidified product is dried and broken up to yield particles with the multinuclear structure. The particles are glass matrices loaded with dispersed core material (Yadav et al. 2015).

HME is a complex process; hence, the selection of the screw design parameters and the process operating conditions includes factors such as screw rotation speed, barrel temperature, feed flow rate, and component concentration (Figure 5.5). In addition, HME is a multivariable process characterized by a set of internal state and output variables, such as material filling ratio, pressure, material temperature, screw temperature, flow rate through the die, etc. (Singh and Mulvaney 1994, Grimard et al. 2016). HME has a number of advantages for encapsulation applications, such as (1) extruders are multifunctional systems, (2) permits incorporation of ingredients at different points, (3) encapsulates can be received in different shapes or sizes, (4) a small amount of water is needed to obtain the glassy matrices (no further drying), (5) payload can exceed 30%, and (6) favorable economics (Lakkis 2016) (Table 5.1).

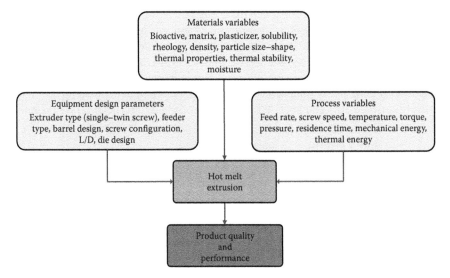

FIGURE 5.5   HME process variables.

The major disadvantage is the difficulty of accurately controlling the parameters of this complex set-up to ensure the good and constant quality of the final product (Zasypkin and Porzio 2004, Ubbink and Krüger 2006, Abbas et al. 2012, Castro et al. 2016) (Table 5.1). HME may be used to disperse bioactive compounds in a given matrix at the molecular level, thus forming solid solutions. In the case of poorly soluble compounds, it might increase the dissolution, absorption, and efficacy of bioactives (Kalepu and Nekkanti 2016). Active encapsulation compositions that are stable in the glassy state at ambient temperatures are prepared by melt extrusion of a ternary carrier blend comprising of (1) a food polymer, (2) a bioactive substance, and (3) a low molecular sugar or polyol. Such glassy matrices are useful for the encapsulation of encapsulates, in particular, flavors and medications (Zasypkin 2011).

Melt injection (ram extrusion) has been utilized to encapsulate lipid-based flavors. This process includes the preparation of a melt by boiling off sufficient water from high solids carbohydrate syrup, addition of flavorings along with an emulsifier, agitation under pressure in order to emulsify the oil in the melt, and injection of the mixture into a chilling, dehydrating solvent bath to obtain fine filaments (Zasypkin 2011). After solvent removal, the matrix is reduced in size and, in some cases, coated with an anti-caking agent before being packed. Alternatively, HME utilizes an extruder to form the carrier melt in a continuous process. The encapsulated flavor is either admixed or injected into the molten carbohydrate carrier. The encapsulating matrix is a carbohydrate, usually a mixture of sucrose and maltodextrin, which is dry fed and melted via a combination of shear and heat in the barrel. In this way, the crystalline structure is transformed into an amorphous phase. The bioactive is added through an opening in a cooled barrel situated toward the die end of the barrel (Figure 5.3) to avoid flashing off of low boiling point components (Lakkis 2016). The addition of the core material to the plasticized carrier matrix at a position near the die (e.g., a later stage of the screw-extrusion process) protects sensitive bioactives from the harsh extrusion conditions (Drusch and Mannino 2009, Sun-Waterhouse et al. 2011, Bakry et al. 2016). A rope of the amorphous mixture exits the die, which is then cooled quickly to form the solid glassy material. It should be noted that by incorporating plasticizers or high molecular weight polymers, a reduction or an increase of glass transition temperature (Tg), respectively, can be achieved, thus permitting the achievement almost any desired target glass transition temperature (Lakkis 2016).

So far a number of patents on melt extrusion appeared in literature, in which the matrix compositions were carefully defined to accommodate processing limitations of the extruder, as well as to generate a stable matrix in the glassy state characterized by a Tg higher than 40°C. The technique is used for the encapsulation of liquid flavors, essential oils, oleoresins, processed flavors, medications, pesticides, and vitamins (Zasypkin 2011). However, it should be noted that glass transition and consequently microcapsule stability are clearly related to matrix material properties and crystallization rates. Parker and Ring (1995) reported that in the glass transition region, small molecules are more mobile than might be expected from the high viscosity of the matrix. The mechanism of degradation of molecules entrapped in a glassy matrix is not fully understood but is speculated to be due to side chain flexibility or diffusion of small molecules (e.g., water and oxygen) through the glassy matrix (Bakry et al. 2016).

As it has been stated by Porzio (2008), a comparison between melt injection and melt extrusion demonstrates superior production efficiency and commercial scalability of the latter owing to its continuous process. Furthermore, melt injection limits the flavoring agents to materials with lower vapor pressure that can be admixed to the composition before melting. Flavorings that are in the form of aqueous extracts, water, or alcohol water solutions will result in a product with a Tg much below 25°C leading to plastic flow and loss of volatiles upon storage (Zasypkin 2011).

Various compounds such as flavorings, fish oil, enzymes, peptides, and bioactives have been successfully encapsulated, and their applications have shown great potential for food applications (Zuidam and Nedović 2010). Commonly used encapsulating materials in the food industry are carbohydrates, proteins, and lipids (Benshitrit et al. 2012). For example, diacetyl, vanilla extract, and beef flavors have been encapsulated using as matrix materials saccharide and maltodextrin. During this process, flavors were added into the premelted matrix materials prior to extrusion, and/or directly added into the molten encapsulation matrix materials already present in the extruder (Fulger and Popplewell 1999). Similarly, Porzio and Zasypkin (2009) examined additional matrix materials consisting of carbohydrates, cellulosics, proteins, gums, sugar, polyols, and mono- and disaccharides. Among the encapsulated flavors were butter, cinnamon, and lemonade flavors (Sobel et al. 2014). Further, it should be noted that extrusion is an especially useful technology for the encapsulation of lipophilic bioactives in an amorphous carbohydrate-based matrix (Madene et al. 2006). The potential of HME for delivery of lipophilic bioactives was demonstrated by studies on the encapsulation of vegetable oil, as model lipophilic carrier in a starch matrix (Van Lengerich 2001, Yılmaz et al. 2001). The main advantage of such a process is the reasonably long shelf life of the oxidation-labile bioactives due to strong impermeability of the carbohydrate matrices in the glassy state against oxygen (Gouin 2004, Lakkis 2016). However, from an economic point of view, the relatively limited load capacity in these systems, which is around 10%, makes it an unattractive encapsulation process (Gouin 2004, Emin 2013). Further, it has been demonstrated that adding the core material to the plasticized carrier matrix at a later stage of the screw-extrusion process protects sensitive polyunsaturated fatty acids (PUFA) from the harsh extrusion conditions. The oxidative stability of microencapsulated oils can be improved by admixing acidic antioxidants (e.g., citric acid, caffeic acid, ascorbic acid, or erythorbic acid) to the matrix to prevent contact of oxygen with the oil (Drusch and Mannino 2009, Sun-Waterhouse et al. 2011, Bakry et al. 2016). Extrusion encapsulation provides an effective method of protecting various bioactive compounds against evaporation and chemical reaction, as well as for controlling the delivery and preservation of stability of the bioactive compounds during processing and storage. In addition, it prevents the undesirable interactions with other food components and masking unpleasant tastes (e.g., bitterness) (Porzio 2008, Nedovic et al. 2011). This will play a significant role in the formulation and increase the efficacy of functional foods. Encapsulated bioactive food components offer to food processors the opportunity to improve the nutritional and health qualities of their food products. It should be noted that microencapsulation allows the protection of a wide range of food components, from small molecules and protein to probiotic bacteria (Alexe and Dima 2014).

5.5 MATERIALS USED FOR ENCAPSULATION

During the formulation of an active substance using HME, the required raw materials might include the active substance, a food grade polymer, plasticizers, surfactants, and antioxidants. A number of substances may be used in bioactive encapsulation. However, all these should comply with food regulations. The challenges in developing a commercially viable product depend on (1) selecting appropriate coating material, (2) selecting the most appropriate process to provide the desired morphology, stability, and release mechanism, and (3) economic feasibility of large-scale production (capital, operating, and other costs) (Daniel et al. 2015). Although it is difficult to number all criteria to select a proper material for encapsulation, the most important criteria for the selection of an encapsulation material include providing functionality to encapsulate and final product, restrictions for the coating material, concentration of encapsulates, type of release, stability requirements, and cost constraints. Generally, the materials should be food grade, biodegradable, and able to form a barrier between the internal phase and its surroundings. The majority of materials used for encapsulation in the food sector are biomolecules (Shahidi and Han 1993, Nedovic et al. 2011). In any case, a good knowledge of the physicochemical interactions occurring between bioactive compounds and the main constituents of foods such as polysaccharides, proteins, and lipids are required for effectiveness control of bioactives in food (Langourieux and Crouzet 1994, O'Neill 1996, Madene et al. 2006). Table 5.2 shows various materials used in extrusion encapsulation.

TABLE 5.2
Materials Used for Extrusion Encapsulation

Material	Description
Matrix	
Carbohydrates	Starch and modified starch, maltodextrins (DE < 20), cyclodextrins, glucose syrups (DE > 20), maltose, trehalose, sucrose, lactose, fructose, glucose, mannose, carboxymethyl cellulose, methyl cellulose, ethyl cellulose, nitrocellulose, acetyl cellulose, cellulose acetate-phthalate, cellulose acetate-butylate-phthalate
Plant extracts	Gums, galactomannans, pectins, etc.
Marine extracts	Alginate, carrageenans, agar
Proteins	Milk and whey proteins, gluten, gelatin derivatives, casein, soy proteins, pea proteins, cereal proteins, etc.
Lipids	Oils, fats, hardened oils, glycerides, fatty acids, fatty alcohols, phospholipids, waxes
Microbial and animal polysaccharides	Dextran, xanthan, gellan, chitosan
Polymers	Polyvinylpyrrolidone (PVP), polyvinyl acetate (PVAc), polyethylene glycol (PEG), poloxamer, polymethacrylates, etc.
Plasticizers	Water, glycerol, polyols, sugar alcohols, glycols, polyglycols, linear alcohols, glycerol, etc.
Surfactants	Lecithin, Tween 80, Tween 20, sodium lauryl sulfate
Antioxidants	Citric acid, caffeic acid, ascorbic acid, erythorbic acid, butylated hydroxyanisole

5.6 MICROENCAPSULATES IN FOOD PRODUCTS

In many cases food manufacturers often cannot simply add a bioactive compound to a food when formulating functional foods due to severe processing conditions, which can harm them. The careful design of the delivery system helps protect the sensitive bioactives from the environment and processing stresses encountered during food manufacturing. Microencapsulation technology can offer many benefits in developing functional foods containing bioactive agents. Such foods could be proven very beneficial to consumer's nutrition and health. This could be achieved by improving the physical and the chemical stability, bioavailability, and efficacy of various bioactive compounds through encapsulate incorporation into food formulation (Tolve et al. 2016). Moreover, microencapsulation can stabilize and protect bioactives from degradation during processing, in the final food product, and during intestinal transit until they are released at the desired site in the gastrointestinal tract to impart their targeted health effects (Augustin and Sanguansri 2015). Table 5.3 shows the extrusion encapsulation of various bioactives along with coating or matrix materials used.

Extrusion microencapsulation has been and is used almost exclusively for the encapsulation of volatile and unstable flavors in glassy carbohydrate matrices. The main advantages of this process are the very long shelf life to normally oxidation-sensitive flavor compounds, the taste and aroma differentiation, the masking of unpleasant tastes and odors, the stabilization of food ingredients, the increase of bioavailability, and to provide a controlled and sustained substance release (Zasypkin and Porzio 2004, Poshadri and Aparna 2010). It should be noted, for example, that the glassy matrix used in this process provides a perfect barrier to oxygen and could lead to shelf lives 5 times longer than those obtained by spray drying in the case of oils (Zasypkin and Porzio 2004). However, as it has been reported by Castro et al. (2016), the chemical diversity of encapsulated ingredients is a key issue, which should draw attention. Important is to emphasize that flavoring compounds might be unsaturated hydrocarbons, alcohols, ketones, aldehydes, esters, lactones, sulfur- and nitrogen-containing compounds, heterocycles, etc. Consequently, flavoring compounds are characterized by a broad range of physicochemical properties, and properties such as chemical groups, molecular weight, steric hindrance, vapor pressure, and relative solubility in oil and matrix phase are especially important for the encapsulation process as they affect and determine the interactions between ingredient and matrix (Goubet et al. 1998, Naknean and Meenune 2010). Hence, suitable interactions between bioactive and matrix material can lead to a reduction in the diffusion of flavor carbohydrate complex (Castro et al. 2016). In the case of sucrose/maltodextrin, the largest amounts of release occurred when the matrix was above its glass transition temperature, whether this was due to increased water content or elevated temperature (Gunning et al. 1999); smaller amounts of headspace release occurred when the water content of the encapsulated flavor system decreased from 3.5% w/w to 3.1% w/w. Jouquand et al. (2004), for maltodextrin matrix, reported opposite results, attributing the thermally enhanced retention to a change in the polarity of the matrix with increasing temperature. More recently, Heilig et al. (2016) reported that the equilibrium partition coefficient in aqueous, polysaccharide, and dairy matrices are

TABLE 5.3
Extrusion Encapsulation of Various Bioactives

Method	Encapsulant	Bioactive	Particle Size/Shape	Encapsulation Efficiency or Rate	References
Melt extrusion	Corn syrup/brominated vegetable oil/emargol	Orange oil			Swisher (1957)
Melt extrusion	Sucrose/corn syrup	Orange oil			Schultz et al. (1956)
Melt extrusion	High maltose corn syrup solids (maltose, dextrose, maltotriose, and higher saccharides)	ω-3 oils, β-carotene, fish oils	Oil droplets 2 to 20 μm		Saleeb and Arora (1999)
Melt extrusion	Maltodextrin	Diacetyl		35–50	Porzio and Popplewell (1999)
Melt extrusion	Protein/sucrose/maltodextrin/water	Cinnamic aldehyde			Porzio and Popplewell (1999)
Melt extrusion	Maltodextrin/corn syrup/methyl cellulose	Orange oil		8.3	Porzio and Popplewell (1999)
Melt extrusion	Maltodextrin DE-19/water/lecithin	Strawberry flavor			Benczedi and Bouquerand (2001)
Melt extrusion	Maltodextrin DE-10/miglyol/ lecithin	Ascorbic acid		18.9	Bouquerand (2007)
Melt extrusion	Maltodextrin/lecithin/miglyol	Ascorbic acid		96; 18.6	Chang et al. (2010)
Melt extrusion	Maltodextrin DE-19; DE-12; DE-6	Orange oil		8.3; 8.1; 7.9	Benczedi et al. (2011)
Melt extrusion	Corn starch/water	Eugenol		20.5	Kollengode and Hanna (1997)
Melt extrusion	Potato starch/capsule/glycerol/water	Orange oil			Quellet et al. (2001)
Melt extrusion	Native maize starch	Medium-chain triglyceride			Emin and Schuchmann (2013)
Melt extrusion	Whey protein/sucrose/maltodextrin/ water	Cinnamic aldehyde			Black et al. (1998)
Melt extrusion	Sucrose/maltodextrins	Cherry, pepper mint flavors		10; 7.4	Gunning et al. (1999)

(Continued)

TABLE 5.3 (*Continued*)
Extrusion Encapsulation of Various Bioactives

Method	Encapsulant	Bioactive	Particle Size/Shape	Encapsulation Efficiency or Rate	References
Melt extrusion	Cap100/EmCap12639/lactose; EmCap12634/Hi-Cap100/lactose	Lemon flavor			Zasypkin and Porzio (2004)
Melt extrusion	Native corn starch/β-cyclodextrin	d-Limonene			Yuliani et al. (2006)
Melt extrusion	Modified starch/maltodextrin/lecithin	Vitamin E		93	Chang et al. (2014)
Melt extrusion	Maltodextrin DE-12 and Maltodextrin DE-17/sucrose	Orange terpenes and tocopherol		67	Tackenberg et al. (2015)
Co-extrusion	Sodium alginate solution, 1.5%	Olive oil and caffeic acid	NM	60.6	Sun-Waterhouse et al. (2011)
Extrusion	Sodium alginate, 0.5, 1, 2, 4, or 8 w/v%	Clove, thyme, and cinnamon oil	NM	90–94	Soliman et al. (2013)
Extrusion	Sodium alginate	Vitamin C		74%	Thangaraj and Seethalakshmi (2014)
Extrusion	Amidated pectin	Bilberry pomace extract			Baum et al. (2014)
Extrusion	Calcium alginate	Astaxanthin	Beads, 686–1044 μm	26–98	Lin et al. (2016)
Electrostatic Extrusion	Alginate	Ethyl vanillin	Uniformly sized spheres, ~450 μm		Manojlovic et al. (2008)
Electrostatic extrusion	Alginate–chitosan + ascorbic acid	Raspberry leaf, hawthorn, ground ivy, yarrow, nettle, and olive leaf extracts	Microbeads, 781–1786 μm	80–89%	Belščak-Cvitanović et al. (2011)
Electrostatic extrusion	Alginate	Yeast	Microbeads 50–600 μm		Manojlovic et al. (2006)
Extrusion	Denatured whey protein isolates	*Lb. rhamnosus*	Beads (~3 mm)		Ainsley Reid et al. (2005)

(*Continued*)

TABLE 5.3 (*Continued*)
Extrusion Encapsulation of Various Bioactives

Method	Encapsulant	Bioactive	Particle Size/Shape	Encapsulation Efficiency or Rate	References
Extrusion	Whey proteins	*Saccharomyces cerevisiae*	Beads 2.6 mm		Hébrard et al. (2006)
Extrusion	Chitosan, poly-L-lysine, sodium alginate	*Lactobacillus cs 547, Bifidobacterium bifidum ATCC 1994 y Lactobacillus casei 01, Lactococcus lactis ssp. cremoris*	Beads, 1.89 mm		Krasaekoopt et al. (2004)
Extrusion	1.8% alginate + 1% Hi-Maize starch	*Lactobacillus acidophilus, Bifidobacterium lactis*			Kailasapathy (2006)
Extrusion	Alginate	*Bifidobacterium longum*	1.03–2.62 mm		Chen et al. (2006)
Extrusion	Alginate	*Lactobacillus reuteri*	2000–3000 µm		Muthukumarasamy et al. (2006)
Extrusion	Alginate	*Lactobacillus acidophilus*	200–300 µm		Özer et al. (2009)
Extrusion	Denatured whey protein isolates/alginate	*Saccharomyces boulardii*			Hébrard et al. (2010)
Extrusion	Denatured whey protein isolates hydrolyzed whey protein	*Lb. rhamnosus GG*	Irregular-shaped gel particles immobilized cells, phase-separated from the surrounding protein matrix		Doherty et al. (2010)
Extrusion	Sodium alginate, amidated low-methoxyl pectin, and blends	*Lactobacillus casei*	Beads, 0.71–0.97 mm	54–79	Sandoval-Castilla et al. (2010)

(Continued)

TABLE 5.3 (Continued)
Extrusion Encapsulation of Various Bioactives

Method	Encapsulant	Bioactive	Particle Size/Shape	Encapsulation Efficiency or Rate	References
Extrusion	Denatured whey protein isolates	Lb. rhamnosus GG	Beads (~200 μm), high strength		Doherty et al. (2011)
Extrusion	Denatured whey protein isolates/pectin	Lb. rhamnosus	Irregular microcapsules (185 ± 20 μm)		Gerez et al. (2012)
Extrusion	Pectin/Whey proteins coat	Lb. acidophilus	Regular particles		Gebara et al. (2013)
Extrusion	Alginate–whey protein	Lactobacillus delbrueckii subsp. bulgaricus	Microspheres, 1.0–3.0 mm		Chen et al. (2014)
Extrusion	Alginate, potato starch and type 4 resistant starch (RS4)	L. acidophilus DDS1-10	1.22–2.61 mm		Mo (2015)

difficult to relate and compare due to different experimental conditions, such as the aroma compound selection, matrix composition, and equilibration temperature.

The extrusion encapsulation of the flavoring materials process starts by forming a low moisture (5%–10%) carbohydrate melt (110°C–130°C), the addition of an emulsifier into the melt, and then of the flavoring material. The product of such a process contains 8%–20% flavor load and is exceptionally stable to deterioration by oxidation (Risch 1988). It should be noted that a number of patents that apply melt extrusion appeared in the past, demanding the matrix compositions to be carefully defined to accommodate processing limitations of the extruder, as well as to generate a stable matrix in the glassy state characterized by a glass transition temperature of greater than 40°C. These are referred to liquid flavor, essential oils, and processed flavor encapsulation (Swisher 1957, 1962, Garwood et al. 1995, Black et al. 1998, Saleeb and Arora 1999, Porzio and Popplewell 2002, Porzio and Zasypkin 2009, Zasypkin 2011, Fuisz 2013, Zasypkin et al. 2014).

Garwood et al. (1995) reported a method of encapsulating volatile aroma compounds in which an inert gas (CO_2) is dissolved in an aromatized edible liquid, and the aromatized, gasified liquid is co-extruded with a molten carbohydrate material having a glass transition temperature of between 20°C and 80°C. In such a way, a continuous stream is formed in which the outer shell of the carbohydrate material surrounds the inner core of the aromatized, gasified liquid. This stream is extruded into a pressure chamber with a pressure higher than the internal pressure of the inert gas in the core, and the pressure is maintained until carbohydrate shell cooling to a temperature lower than the glass transition. In this way, the shell of capsules hardens surrounding the inner core of gasified, aromatized liquid. This method has been applied for the preparation of coffee aroma capsules using a 100% coffee derived material (Garwood et al. 1995).

Using a HME process encapsulation partially crystalline carbohydrate extrudates containing maximally 6.0% orange terpenes and 9.2% α-tocopherol were obtained. The decreased encapsulated orange terpene content in contrast to α-tocopherol was due to evaporation of the active substance at die's exit (Tackenberg et al. 2015). During storage, water sorption, crystallization, and a partly collapse of the cylindrical extrudates occurred, but the active substance content remained constant.

Dolçà et al. (2015) prepared microcapsules of rosemary essential oil by co-extrusion technique using alginate as wall material and calcium chloride as cross linker. They found that rosemary oil has pesticidal properties, and its microencapsulation allows knowing that these properties remain inside the microcapsules.

Microencapsulation is a technology, which can be used for the protection, stabilization, and the slow release of carotenoids, anthocyanins, and chlorophylls. For this purpose, a number of different techniques and wall materials could be used that result to overcome instability, solubility, and handling problems of natural food colorants (Özkan and Bilek 2014). However, it should be emphasized that encapsulation by spray drying has been the primary method used to encapsulate anthocyanins (Robert and Fredes 2015), though anthocyanins from berry fruits have been encapsulated by the oil dispersed phase, double emulsion (w/o/w), extrusion, emulsification/heat gelation, microgel synthesis, freeze drying, supercritical CO_2, and ionic gelification techniques (Oidtmann et al. 2012).

In the past, it was hypothesized that HME would assist in improving the solubility and bioavailability of polyphenols. Polyphenols exist in and can be extracted from the seeds, roots, fruits, and leaves of certain plants. These include items such as (1) phenolic acids, which include polymeric structures, such as hydrolyzable tannins, lignans, stilbenes, and flavonoids, and (2) flavonoids, which include flavonols (e.g., quercetin, kaempferol, flavones, isoflavones, flavanones, anthocyanidin pigments, flavanols, and catechin monomers and proanthocyanidin polymers) (Duvuri 2011). Many encapsulation methods are described in the literature, among which some have been successfully applied to plant polyphenols and can be classified into physical, physicochemical, chemical methods, and other connected stabilization methods (Munin and Edwards-Lévy 2011). A patent by Huynh and Cormier (2013) for dry particulate encapsulation includes: (1) mixing of a labile active agent in a water-insoluble matrix and water, to form a melt with said encapsulate and said matrix; (2) extrusion of the melt to obtain an extrudate and cooling; (3) drying of the extrudate to a moisture content of about 5% to about 10% by weight; and finally (4) dry milling of grinding of the dried extrudate to a particle size of about 150 µm or less than 150 µm. The matrix material comprises of about 70% or greater than 70% proteins by weight. The labile active agent has improved stability to light, heat, oxygen-related changes, or any combination thereof, within this composition as a result of the extrusion process. According to inventors, the coloring agents that may be used might be carotenoids, porphyrins, and flavonoids.

Duvuri (2011) employed HME with a novel polymeric solubilizer, Soluplus®, for the encapsulation of resveratrol and quercetin. It was observed that there was an intimate interaction between resveratrol and Soluplus® and quercetin and Soluplus® during melt extrusion that might have produced certain changes at the molecular level in the polyphenols. Furthermore, it was shown that solubility in water and miscibility of the HME mixtures were improved. In addition, the HME mixtures remained uniform with respect to resveratrol and quercetin being used in different concentrations along with Soluplus®, and particle size affected dissolution and release of bioactive.

da Rosa et al. (2014), using phenolic extracts of blackberry, reported that encapsulation efficiency depended on the phenolic compound and the encapsulated coating used. For example, gallic acid and epicatechin were predominantly in microcapsules coated with β-cyclodextrin and xanthan. Further, the controlled release of phenolic extract capsules was influenced by coating, solvent, and pH.

Different methods of microencapsulation have been used to encapsulate marine, vegetable, and essential oils; these include emulsification, spray drying, coaxial electrospray system, freeze drying, coacervation, in situ polymerization, melt extrusion, supercritical fluid technology, and fluidized bed coating. Spray drying and coacervation are the most commonly used techniques for the microencapsulation of oils (Bakry et al. 2016). Microencapsulation can enhance the oxidative stability, thermostability, shelf life, and biological activity of oils. In the past few years, extrusion techniques have been used to encapsulate some vegetable and essential oils, including olive, clove, thyme, and cinnamon oils, for the food industries (Sun-Waterhouse et al. 2011, Soliman et al. 2013), using almost exclusively a carbohydrate matrix. Clove, thyme, and cinnamon oils were microencapsulated through the extrusion technique to reduce the evaporation rate, thereby increasing their antifungal activity (Soliman et al. 2013, Bakry et al. 2016). HME has a potential for the delivery of lipophilic bioactives and oils (Van Lengerich 2001, Yılmaz et al. 2001, Van Lengerich et al. 2010) using starch matrix as a model lipophilic carrier. Sun-Waterhouse et al. (2012) reported that the combined use of co-extrusion encapsulation and fortification with an antioxidant is an effective strategy for improving the oxidative stability of avocado oil, suppressing hydrolytic rancidity.

Extrusion process can be used for the encapsulation of fish oils. Fish oils can be encapsulated into a mixture or dough by using an extruder with one or more screws in a continuous process (Van Lengerich et al. 2007). This process utilizes temperatures below 30°C and pressures in the range of 500–5000 kPa. Moreover, emulsions using protein, gum, or modified starch as an emulsifier can be prepared and mixed into 45%–75% by weight of matrix materials (starch, flours, proteins, gums, etc.) along with relatively high amounts of plasticizer and 0.5%–4% by weight of an acidic antioxidant using a twin screw extruder with a barrel temperature between 5°C and 10°C (Kaur et al. 2015). The oxidative stability of microencapsulated oils can be improved by admixing acidic antioxidants to the matrix to prevent contact of oxygen with the oil (Drusch and Mannino 2009, Sun-Waterhouse et al. 2011, Bakry et al. 2016).

Encapsulation can provide higher shelf life and stability of vitamin C. Thangaraj and Seethalakshmi (2014) developed microcapsules of vitamin C by an extrusion process, using sodium alginate used as a coating material with an average efficiency rate of 74%. Extrusion microencapsulation is a promising technique to ensure the stability of ascorbic acid. Extrusion at relatively low temperatures has been used for the encapsulation of heat labile vitamins. In a process, Abbas et al. (2012) utilized starch/oligosaccharides as the wall-forming plasticized mass at relatively reduced temperatures (60°C–80°C) prior to admixing with the sensitive vitamin to avoid thermal degradation. The key point in this process was to ensure the stability of the core material by decreasing the melt temperature at the later stage of extrusion, as well as the residence time. Tsuei et al. (1996) invented a patent for vitamin C encapsulation in which the preferred solid matrix-forming materials were carnauba wax, yellow beeswax, white beeswax, paraffin, and linear or branched polyethylenes. More recently, Bouquerand (2007) also published a patent for the encapsulation of vitamin C by extrusion using a glassy matrix of carbohydrate material containing vitamin C, mono- and disaccharides, and solid of vitamin C. The invention permits glassy matrix to be formulated in various ways; for example, it can contain maltodextrin 10 DE, lecithin, and medium-chain triglycerides. The mono- and disaccharide content of the matrix material should not be above 10% by weight, relative to the weight of carbohydrate material. The main objective of the invention is to provide glassy vitamin C products with a sufficiently high glass transition temperature (Tg) to warrant stability at room temperature. Hence, the Tg of the vitamin C powder should be above 40°C (between 40°C and 100°C) and preferably between 40°C and 60°C (Bouquerand 2007).

Microencapsulation has been used to provide protection for probiotics all through food processing and marketing until they reach the target site in the gastrointestinal tract (Abd El-Salam and El-Shibiny 2015). Extrusion technology is very useful in the encapsulation of probiotics. For this purpose, a number of different technologies are available.

The selection of the best method is related to different aspects as desired size, acceptable dispersion size, production scale, and the maximum shear that the probiotic cells can support (Chávarri et al. 2012). Extrusion is among these methods and could be divided in two groups, dropwise and jet breakage, and the limit between them is established according to the minimum jet speed (Chávarri et al. 2012). As it has been stated by de Vos et al. (2010), it should be relatively gentle, excluding deleterious solvents, while it can be done under both aerobic and anaerobic conditions. Extrusion microencapsulation of probiotics can be based on extruding a mixture of concentrated viable probiotic cells and whey proteins or whey proteins-polysaccharide solution through a nozzle to form droplets that fall and harden in a solution (Abd El-Salam and El-Shibiny 2015). In addition, extrusion can be performed in two steps in which probiotics are first entrapped in polysaccharide beads followed by coating the obtained beads with whey proteins. The size of the formed capsules depends on the extrusion conditions (Anal and Singh 2007). Shi et al. (2013) using extrusion entrapped the probiotic *Lactobacillus bulgaricus* in alginate-milk microspheres. The efficiency of encapsulation was reported to be nearly 100% and the viability of the entrapped microorganism was almost unchanged in simulated gastric conditions. Jiang et al. (2016) noted that microencapsulation of *L. acidophilus* NCFM using polymerized whey protein as wall material can be used in the development of fermented dairy products with better survivability of probiotic organism. Seyedain-Ardabili et al. (2016) reported the production of symbiotic bread by using microencapsulation and can enhance the viability and thermal resistance of probiotic bacteria, and therefore significantly improve their survival in bread and other bakery products. They used alginate and starch beads with and without chitosan coating, and found that viable microorganisms survived after baking, while breads met the criteria for probiotic products. Having this in mind, melt extrusion might be proven a suitable technique for the probiotic encapsulation after further studying of process conditions and matrix composition.

During food production, a major requirement for the products is to have excellent or very good organoleptic characteristics, for example, without any unpleasant taste, odor, and flavor. However, due to material origin, processing, and storage, various products can naturally have an unpleasant taste (e.g., bitterness), change their attributes, and/or develop unpleasant taste, odor, and aroma. Microencapsulation has been considered as an effective method to mask the unpleasant taste of certain ingredients. More recently, it was examined as a technique to delay lipid oxidation of PUFA in various food products (Gouin 2004, Kaur et al. 2015). So, fish oils, especially rich in ω-3 fatty acids, can be encapsulated to prevent off-flavor through (1) minimization of contact between oxygen and fish oil, (2) prevention of contact between metal ions and fish oil, (3) prevention of direct exposure to light, and (4) trapping off-flavors (Garg et al. 2006, Beindorff and Zuidam 2010, Kaur et al. 2015). In fish oil encapsulation, a variety of materials such as gelatin, whey protein and cyclodextrin, ethyl cellulose, zein, and chitosan has been used (Sobel et al. 2014). Among the methods used, encapsulation by spray drying finds more application.

Fish oils can be encapsulated by melt injection using a starch matrix along with antioxidants, sugars, emulsifiers, and water at a temperature above 100°C. The extrudate is collected in an organic solvent bath (Valentinotti et al. 2006). In this way, microencapsulated oil withstands oxidation. However, it should be stored under low water activity conditions (Beindorff and Zuidam 2010). Fish oils can also be encapsulated using an extruder in a continuous process. The process is performed at relatively low temperatures (<30°C) and pressures in the range of 500–5000 kPa process (Van Lengerich et al. 2004, 2007, 2010). The process is performed in a twin screw extruder and involves the use of a plasticizer in relatively high amounts. Flours, starch, proteins, etc., can be used as matrix materials with relatively high amounts of plasticizer and a small amount of an acidic antioxidant. The process includes an emulsification stage using a protein and a gum, and/or modified starch as an emulsifier. Diguet et al. (2005) in an invention propose the fish oil to be added in and encapsulated with another technique such as spray drying, before extrusion.

HME can be proven an advantageous technique, compared to the other available conventional techniques, for taste masking. HME can achieve taste masking of bitter compounds via various mechanisms such as the formation of solid dispersions and intermolecular interactions, and this has led to its widespread use in formulation pharmaceuticals (Maniruzzaman et al. 2014).

Caffeine, having a bitter taste, is usually delivered in these food products in an encapsulated form to reduce the perception of bitterness. Caffeine encapsulated with mono- and diglycerides, polymeric materials, lipids, and oils can be used in several food products, among which are breakfast bars, icings, and other food (Bohannon 2008, Sobel et al. 2014). In addition, calcium polysaccharide gel-coated pellets with caffeine, theophylline, and theobromine were prepared by an extrusion–spheronization method (Sriamornsak and Kennedy 2007). Beindorff and Zuidam (2015) found that one or more bitter tasting ingredients may be effectively delivered and their flavor masked by an encapsulation using a specific type of fat augmented with an emulsifier having a specific set of characteristics. The encapsulating layer comprises of a fat or fat blend having a melting point between 10°C to 55°C, and an emulsifier having a hydrophilic/lipophilic balance of at most 13. Microencapsulates have a diameter in the range of 5–750 μm and at most 5% water by weight. Further, techniques used for polyphenolics encapsulation also contribute to mask the bitter taste of these compounds. Encapsulation using extrusion technology has been successfully applied for increasing the stability of polyphenols and preserving thus their bioactivity (Belščak-Cvitanović et al. 2011, López-Córdoba et al. 2014).

Aceval Arriola et al. (2016) encapsulated aqueous leaf extract of *Stevia rebaudiana* Bertoni using sodium alginate. An evaluation of the total phenolic content and antioxidant stability also performed. High encapsulation efficiency values were achieved for the wet (69.8%) and for the lyophilized (97.7%) beads. Both the wet and the lyophilized beads showed stability in total phenolic content and preserved antioxidant potential throughout 30 days of storage at 4°C. They found that encapsulation of stevia extracts in alginate beads is a promising technique for food supplementation with natural antioxidants. Cherukuri and Siris (1999) invented a method and apparatus for forming an encapsulated product matrix. The apparatus for forming an encapsulated product matrix is comprised of at least one dry spray device for upwardly ejecting a product additive into a feed-stream of substantially solid particles, and of at least one extruding device for upwardly ejecting a product encapsulant into a feed-stream of solid particles. The encapsulating substances can act as a taste mask to improve the organoleptic qualities of a food product.

5.7 CONCLUSION

Although a lot of data on microencapsulation processes are available, data and information on microencapsulate use in real food matrices are not so extensive. Hence, there is a need to examine the stability and behavior of microencapsulated bioactives in real foods. Of course, well-documented applications in beverages, bakery, and dairy exist, including products such as bread, pasta, cheese, and yogurt. It should be noted that in some cases, legal or regulatory issues exist and need to be solved.

Extrusion and especially HME might be proven as an encapsulation technology of the future because it permits the development of glassy extrudates with a variety of formulations. In addition, HME has the possibility to encapsulate more than one bioactive compound or ingredient into one matrix using either a single or multiple steps. This will permit the development of new or novel or functional food products. In this direction, it should be emphasized that process parameters, as well as compatibility and suitability of matrix materials for this process, and the mechanisms of interactions between bioactives and various matrices should be further explored as well.

REFERENCES

Abbas, S., C. Da Wei, K. Hayat, and Z. Xiaoming. 2012. Ascorbic acid: Microencapsulation techniques and trends—A review. *Food Reviews International* 28(4):343–374. doi: 10.1080/87559129.2011.635390.

Aceval Arriola, N.D., P.M. de Medeiros, E.S. Prudencio, C.M.O. Müller, and R.D. de Mello Castanho Amboni. 2016. Encapsulation of aqueous leaf extract of *Stevia rebaudiana* Bertoni with sodium alginate and its impact on phenolic content. *Food Bioscience* 13:32–40. doi: http://dx.doi.org/10.1016/j.fbio.2015.12.001.

Ainsley Reid, A., J.C. Vuillemard, M. Britten, Y. Arcand, E. Farnworth, and C.P. Champagne. 2005. Microentrapment of probiotic bacteria in a Ca^{2+}-induced whey protein gel and effects on their viability in a dynamic gastro-intestinal model. *Journal of Microencapsulation* 22(6):603–619. doi: 10.1080/02652040500162840.

Alexe, P., and C. Dima. 2014. Microencapsulation in food products. *AgroLife Scientific Journal* 3(1):9–14.

Anal, A.K., and H. Singh. 2007. Recent advances in microencapsulation of probiotics for industrial applications and targeted delivery. *Trends in Food Science & Technology* 18(5):240–251. doi: http://dx.doi.org/10.1016/j.tifs.2007.01.004.

Augustin, M.A., and L. Sanguansri. 2015. Challenges and solutions to incorporation of nutraceuticals in foods. *Annual Review of Food Science and Technology* 6(1):463–477. doi: 10.1146/annurev-food-022814-015507.

Bakry, A.M., S. Abbas, B. Ali, H. Majeed, M.Y. Abouelwafa, A. Mousa, and L. Liang. 2016. Microencapsulation of oils: A comprehensive review of benefits, techniques, and applications. *Comprehensive Reviews in Food Science and Food Safety* 15(1):143–182. doi: 10.1111/1541-4337.12179.

Baum, M., M. Schantz, S. Leick, S. Berg, M. Betz, K. Frank, H. Rehage et al. 2014. Is the antioxidative effectiveness of a bilberry extract influenced by encapsulation? *Journal of the Science of Food and Agriculture* 94(11):2301–2307. doi: 10.1002/jsfa.6558.

Beindorff, C.M., and N.J. Zuidam. 2015. Encapsulated food composition. Google Patent, WO 2015032816 A1.

Beindorff, C.M., and N.J. Zuidam. 2010. Microencapsulation of fish oil. In *Encapsulation Technologies for Active Food Ingredients and Food Processing*, edited by N.J. Zuidam and V. Nedovic, pp. 161–185. New York: Springer.

Belščak-Cvitanović, A., R. Stojanović, V. Manojlović, D. Komes, I. Juranović Cindrić, V. Nedović, and B. Bugarski. 2011. Encapsulation of polyphenolic antioxidants from medicinal plant extracts in alginate–chitosan system enhanced with ascorbic acid by electrostatic extrusion. *Food Research International* 44(4):1094–1101. doi: http://dx.doi.org/10.1016/j.foodres.2011.03.030.

Benczedi, D., and P.E. Bouquerand. 2001. Process for the preparation of granules for the controlled release of volatile compounds. Google Patent, US 20010036503 A1.

Benczedi, D., P.E. Bouquerand, and E. Steinboeck. 2011. Process for the preparation of extruded delivery systems. Google Patent, US 8017060 B2.

Benshitrit, R.C., C.S. Levi, S.L. Tal, E. Shimoni, and U. Lesmes. 2012. Development of oral food-grade delivery systems: Current knowledge and future challenges. *Food & Function* 3(1):10–21. doi: 10.1039/c1fo10068h.

Betoret, E., N. Betoret, D. Vidal, and P. Fito. 2011. Functional foods development: Trends and technologies. *Trends in Food Science & Technology* 22(9):498–508. doi: http://dx.doi.org/10.1016/j.tifs.2011.05.004.

Bhaskaran, S., and P.K. Lakshmi. 2010. Extrusion spheronization—A review. *International Journal of PharmTech Research* 2(4):2429–2433.

Bisharat, G.I., A.E. Lazou, N.M. Panagiotou, M.K. Krokida, and Z.B. Maroulis. 2015. Antioxidant potential and quality characteristics of vegetable-enriched corn-based extruded snacks. *Journal of Food Science and Technology* 52(7):3986–4000. doi: 10.1007/s13197-014-1519-z.

Black, M., L.M. Popplewell, and M.A. Porzio. 1998. Controlled release encapsulation compositions. Google Patent, US 5756136 A.

Bohannon, R. 2008. Food products having caffeine incorporated therein. Google Patent, US 20080152763 A1.

Bordoloi, R., and S. Ganguly. 2014. Extrusion technique in food processing and a review on its various technological parameters. *Indian Journal of Scientific Research and Technology* 2(1):1–3.

Bouquerand, P.E. 2007. Extruded glassy vitamin C particles. Google Patent, EP 1836902 A1.

Bouvier, J.-M., and O.H. Campanella (eds.). 2014. Extrusion equipment. *Extrusion Processing Technology*, pp. 13–51. Chichester, West Sussex, UK: John Wiley & Sons, Ltd.

Brennan, C., M. Brennan, E. Derbyshire, and B.K. Tiwari. 2011. Effects of extrusion on the polyphenols, vitamins and antioxidant activity of foods. *Trends in Food Science & Technology* 22(10):570–575. doi: 10.1016/j.tifs.2011.05.007.

Bugarski, B.M., B. Obradovic, V.A. Nedovic, and M.F.A. Goosen. 2005. Electrostatic droplet generation technique for cell immobilization. In *Finely Dispersed Particles*, edited by A.M. Spasic and J.-P. Hsu, pp. 869–886. Boca Raton, FL: CRC Press.

Bugarski, B.M., B. Obradovic, V.A. Nedovic, and D. Poncelet. 2004. Immobilization of cells and enzymes using electrostatic droplet generation. In *Fundamentals of Cell Immobilisation Biotechnology*, edited by V. Nedović and R. Willaert, pp. 277–294. Dordrecht, the Netherlands: Springer.

Burgain, J., C. Gaiani, M. Linder, and J. Scher. 2011. Encapsulation of probiotic living cells: From laboratory scale to industrial applications. *Journal of Food Engineering* 104(4):467–483. doi: http://dx.doi.org/10.1016/j.jfoodeng.2010.12.031.

Camire, M.E. 2011. Nutritional changes during extrusion cooking. In *Advances in Food Extrusion Technology*, edited by M. Maskan and A. Altan, pp. 87–102. Boca Raton, FL: CRC Press.

Castro, N., V. Durrieu, C. Raynaud, A. Rouilly, L. Rigal, and C. Quellet. 2016. Melt extrusion encapsulation of flavors: A review. *Polymer Reviews* 56(1):137–186. doi: 10.1080/15583724.2015.1091776.

Chang, D.W., X.M. Zhang, and J.M. Kim. 2014. Encapsulation of vitamin E in glassy carbohydrates by extrusion. *Advanced Materials Research* 842:95–99.

Chang, D., S. Abbas, K. Hayat, S. Xia, X. Zhang, M. Xie, and J.M. Kim. 2010. Original article: Encapsulation of ascorbic acid in amorphous maltodextrin employing extrusion as affected by matrix/core ratio and water content. *International Journal of Food Science and Technology* 45(9):1895–1901. doi: 10.1111/j.1365-2621.2010.02348.x.

Chávarri, M., I. Marañón, and M.C. Villarán. 2012. Encapsulation technology to protect probiotic bacteria. In *Probiotics*, edited by E. Rigobelo, pp. 501–550. InTech.

Chen, L., G.E. Remondetto, and M. Subirade. 2006. Food protein-based materials as nutraceutical delivery systems. *Trends in Food Science & Technology* 17(5):272–283. doi: http://dx.doi.org/10.1016/j.tifs.2005.12.011.

Chen, M.-Y., W. Zheng, Q.-Y. Dong, Z.-H. Li, L.-E. Shi, and Z.-X. Tang. 2014. Activity of encapsulated *Lactobacillus bulgaricus* in alginate-whey protein microspheres. *Brazilian Archives of Biology and Technology* 57:736–741.

Cherukuri, S.R., and S. Siris. 1999. Method and apparatus for forming an encapsulated product matrix. Google Patent, WO 1999061145 A1.

Chokshi, R., and H. Zia. 2004. Hot-melt extrusion technique: A review. *Iranian Journal of Pharmaceutical Research* 3(1):3–16.

da Rosa, C.G., C.D. Borges, R.C. Zambiazi, J.K. Rutz, S.R. da Luz, F.D. Krumreich, E.V. Benvenutti, and M.R. Nunes. 2014. Encapsulation of the phenolic compounds of the blackberry (*Rubus fruticosus*). *LWT—Food Science and Technology* 58(2):527–533. doi: http://dx.doi.org/10.1016/j.lwt.2014.03.042.

Daniel, M., S. Kushwaha, and Shakti. 2015. Microencapsulating food ingredients. *International Journal of Scientific and ResearchPublications (IJSRP)* 5(4), 2015 edition.

Darrington, H. 1987. A long-running cereal. *Food Manufacturing* 3:47–48.

de Vos, P., M.M. Faas, M. Spasojevic, and J. Sikkema. 2010. Encapsulation for preservation of functionality and targeted delivery of bioactive food components. *International Dairy Journal* 20(4):292–302. doi: http://dx.doi.org/10.1016/j.idairyj.2009.11.008.

Desai, K.G.H., and H. Jin Park. 2005. Recent developments in microencapsulation of food ingredients. *Drying Technology* 23(7):1361–1394. doi: 10.1081/drt-200063478.

Dhanasekharan, K.M., and J.L. Kokini. 2003. Design and scaling of wheat dough extrusion by numerical simulation of flow and heat transfer. *Journal of Food Engineering* 60(4):421–430. doi: http://dx.doi.org/10.1016/S0260-8774(03)00065-7.

Diguet, S., K. Feltes, N. Kleemann, B. Leuenberger, and J. Ulm. 2005. Extrusion-stable polyunsaturated fatty-acid compositions for food products. Google Patent, WO 2005089569 A1.

Doherty, S.B., V.L. Gee, R.P. Ross, C. Stanton, G.F. Fitzgerald, and A. Brodkorb. 2010. Efficacy of whey protein gel networks as potential viability-enhancing scaffolds for cell immobilization of *Lactobacillus rhamnosus* GG. *Journal of Microbiological Methods* 80(3):231–241. doi: http://dx.doi.org/10.1016/j.mimet.2009.12.009.

Doherty, S.B., V.L. Gee, R.P. Ross, C. Stanton, G.F. Fitzgerald, and A. Brodkorb. 2011. Development and characterisation of whey protein micro-beads as potential matrices for probiotic protection. *Food Hydrocolloids* 25(6):1604–1617. doi: http://dx.doi.org/10.1016/j.foodhyd.2010.12.012.

Dolçà, C., M. Ferrándiz, L. Capablanca, E. Franco, E. Mira, F. López, and D. García. 2015. Microencapsulation of rosemary essential oil by co-extrusion/gelling using alginate as a wall material. *Journal of Encapsulation and Adsorption Sciences* 5:121–130. doi: 10.4236/jeas.2015.53010.

Đorđević, V., B. Balanč, A. Belščak-Cvitanović, S. Lević, K. Trifković, A. Kalušević, I. Kostić, D. Komes, B. Bugarski, and V. Nedović. 2015. Trends in encapsulation technologies for delivery of food bioactive compounds. *Food Engineering Reviews* 7(4):452–490. doi: 10.1007/s12393-014-9106-7.

Drusch, S., and S. Mannino. 2009. Patent-based review on industrial approaches for the microencapsulation of oils rich in polyunsaturated fatty acids. *Trends in Food Science & Technology* 20(6–7):237–244. doi: http://dx.doi.org/10.1016/j.tifs.2009.03.007.

Duvuri, A. 2011. The formulation of naturally occuring polyphenolic nutraceutical agents using hot-melt extrusion. Open Access Master's Theses, University of Rhode Island, Kingston, RI, p. 88.

Abd El-Salam, M.H., and S. El-Shibiny. 2015. Preparation and properties of milk proteins-based encapsulated probiotics: A review. *Dairy Science & Technology* 95(4):393–412. doi: 10.1007/s13594-015-0223-8.

Emin, M.A., and H.P. Schuchmann. 2013. Droplet breakup and coalescence in a twin-screw extrusion processing of starch based matrix. *Journal of Food Engineering* 116(1): 118–129. doi: http://dx.doi.org/10.1016/j.jfoodeng.2012.12.010.

Emin, M.A. 2013. *Dispersive Mixing of Oil in Plasticized Starch by Extrusion Processing to Design Functional Foods.* München, Germany: Verlag Dr. Hut.

Fuisz, R.C. 2013. Extrudable and extruded compositions for delivery of bioactive agents, method of making same and method of using same. Google Patent, US 8613285 B2.

Fulger, C.V., and L.M. Popplewell. 1999. Flavor encapsulation. Google Patent, US 5958502 A.

Garg, M.L., L.G. Wood, H. Singh, and P.J. Moughan. 2006. Means of delivering recommended levels of long chain n-3 polyunsaturated fatty acids in human diets. *Journal of Food Science* 71(5):R66–R71. doi: 10.1111/j.1750-3841.2006.00033.x.

Garwood, R.E., Z.I. Mandralis, and S.A. Westfall. 1995. Encapsulation of volatile aroma compounds. Google Patent, US 5399368 A.

Gebara, C., K.S. Chaves, M.C.E. Ribeiro, F.N. Souza, C.R.F. Grosso, and M.L. Gigante. 2013. Viability of *Lactobacillus acidophilus* La5 in pectin–whey protein microparticles during exposure to simulated gastrointestinal conditions. *Food Research International* 51(2):872–878. doi: http://dx.doi.org/10.1016/j.foodres.2013.02.008.

Gerez, C.L., G. Font de Valdez, M.L. Gigante, and C.R.F. Grosso. 2012. Whey protein coating bead improves the survival of the probiotic *Lactobacillus rhamnosus* CRL 1505 to low pH. *Letters in Applied Microbiology* 54(6):552–556. doi: 10.1111/j.1472-765X.2012.03247.x.

Gharsallaoui, A., G. Roudaut, O. Chambin, A. Voilley, and R. Saurel. 2007. Applications of spray-drying in microencapsulation of food ingredients: An overview. *Food Research International* 40(9):1107–1121. doi: http://dx.doi.org/10.1016/j.foodres.2007.07.004.

Gibbs, B.F., S. Kermasha, I. Alli, and C.N. Mulligan. 1999. Encapsulation in the food industry: A review. *International Journal of Food Sciences and Nutrition* 50(3):213–224. doi: 10.1080/096374899101256.

Goubet, I., J.L. Le Quere, and A.J. Voilley. 1998. Retention of aroma compounds by carbohydrates: Influence of their physicochemical characteristics and of their physical state. A review. *Journal of Agricultural and Food Chemistry* 46(5):1981–1990. doi: 10.1021/jf970709y.

Gouin, S. 2004. Microencapsulation: Industrial appraisal of existing technologies and trends. *Trends in Food Science & Technology* 15(7–8):330–347. doi: http://dx.doi.org/10.1016/j.tifs.2003.10.005.

Grimard, J., L. Dewasme, and A.V. Wouwer. 2016. A review of dynamic models of hot-melt extrusion. *Processes* 4(2):19.

Gunning, Y.M., P.A. Gunning, E. Kate Kemsley, R. Parker, S.G. Ring, R.H. Wilson, and A. Blake. 1999. Factors affecting the release of flavor encapsulated in carbohydrate matrixes. *Journal of Agricultural and Food Chemistry* 47(12):5198–5205. doi: 10.1021/jf990039r.

Harper, J.M. 1989. Food extruders and their applications. In *Extrusion Cooking*, edited by C. Mercier, P. Linko, and J.M. Harper, pp. 1–15. St. Paul, MN: American Association of Cereal Chemists.

Harper, J.M., and J.P. Clark. 1979. Food extrusion. *Critical Reviews in Food Science and Nutrition* 11(2):155–215. doi: 10.1080/10408397909527262.

Hébrard, G., S. Blanquet, E. Beyssac, G. Remondetto, M. Subirade, and M. Alric. 2006. Use of whey protein beads as a new carrier system for recombinant yeasts in human digestive tract. *Journal of Biotechnology* 127(1):151–160. doi: http://dx.doi.org/10.1016/j.jbiotec.2006.06.012.

Hébrard, G., V. Hoffart, E. Beyssac, J.-M. Cardot, M. Alric, and M. Subirade. 2010. Coated whey protein/alginate microparticles as oral controlled delivery systems for probiotic yeast. *Journal of Microencapsulation* 27(4):292–302. doi: 10.3109/02652040903134529.

Heilig, A., A. Sonne, P. Schieberle, and J. Hinrichs. 2016. Determination of aroma compound partition coefficients in aqueous, polysaccharide, and dairy matrices using the phase ratio variation method: A review and modeling approach. *Journal of Agricultural and Food Chemistry* 64(22):4450–4470. doi: 10.1021/acs.jafc.6b01482.

Huynh, K.U., and F. Cormier. 2013. Natural water-insoluble encapsulation compositions and processes for preparing same. Google Patent, CA 2591772 C.

Jagtap, P.S., S.S. Jain, N. Dand, K.R. Jadhav, and V.J. Kadam. 2012. Hot melt extrusion technology, approach of solubility enhancement: A brief review. *Der Pharmacia Lettre* 4(1):42–53.

Jiang, Y., Z. Zheng, T. Zhang, G. Hendricks, and M. Guo. 2016. Microencapsulation of *Lactobacillus acidophilus* NCFM using polymerized whey proteins as wall material. *International Journal of Food Sciences and Nutrition* 67(6):670–677. doi: 10.1080/09637486.2016.1194810.

Jouquand, C., V. Ducruet, and P. Giampaoli. 2004. Partition coefficients of aroma compounds in polysaccharide solutions by the phase ratio variation method. *Food Chemistry* 85(3):467–474. doi: http://dx.doi.org/10.1016/j.foodchem.2003.07.023.

Kailasapathy, K. 2006. Survival of free and encapsulated probiotic bacteria and their effect on the sensory properties of yoghurt. *LWT—Food Science and Technology* 39(10):1221–1227. doi: http://dx.doi.org/10.1016/j.lwt.2005.07.013.

Kalepu, S., and V. Nekkanti. 2016. Improved delivery of poorly soluble compounds using nanoparticle technology: A review. *Drug Delivery and Translational Research* 6(3):319–332. doi: 10.1007/s13346-016-0283-1.

Kaur, M., S. Basu, and U.S. Shivhare. 2015. Omega-3 fatty acids: Nutritional aspects, sources, and encapsulation strategies for food fortification. *Direct Research Journal of Health and Pharmacology (DRJHP)* 3(1):12–31.

Kleinebudde, P. 2013. Solid lipid extrusion. In *Melt Extrusion: Materials, Technology and Drug Product Design*, edited by A. Michael Repka, N. Langley, and J. DiNunzio, pp. 299–328. New York: Springer.

Kollengode, A.N.R., and M.A. Hanna. 1997. Cyclodextrin complexed flavors retention in extruded starches. *Journal of Food Science* 62(5):1057–1060. doi: 10.1111/j.1365-2621.1997.tb15037.x.

Kostov, G., V. Shopska, R. Denkova, M. Ivanova, T. Balabanova, and R. Vlaseva. 2016. Encapsulation of plant and animal oils used in dairy industry: A review. *Acta Universitatis Cibiniensis. Series E: Food Technology* 20(1):21–40.

Krasaekoopt, W., B. Bhandari, and H. Deeth. 2004. The influence of coating materials on some properties of alginate beads and survivability of microencapsulated probiotic bacteria. *International Dairy Journal* 14(8):737–743. doi: http://dx.doi.org/10.1016/j.idairyj.2004.01.004.

Kumar, P., K. Sandeep, and S. Alavi. 2010. Extrusion of foods. In *Mathematical Modeling of Food Processing*, edited by M.M. Farid, pp. 795–827. Boca Raton, FL: CRC Press.

Lakkis, J.M. 2016. Introduction. In *Encapsulation and Controlled Release Technologies in Food Systems*, pp. 1–15. Chichester, West Sussex, UK: John Wiley & Sons, Ltd.

Langourieux, S., and J. Crouzet. 1994. Study of aroma compounds-polysaccharides interactions by dynamic exponential dilution. *LWT—Food Science and Technology* 27(6):544–549. doi: http://dx.doi.org/10.1006/fstl.1994.1107.

Lazou, A.E., P.A. Michailidis, S. Thymi, M.K. Krokida, and G.I. Bisharat. 2007. Structural properties of corn-legume based extrudates as a function of processing conditions and raw material characteristics. *International Journal of Food Properties* 10(4):721–738. doi: 10.1080/10942910601154305.

Lazou, A., M. Krokida, and C. Tzia. 2010. Sensory properties and acceptibility of corn and lentil extruded puffs. *Journal of Sensory Studies* 25(6):838–860. doi: 10.1111/j.1745-459X.2010.00308.x.

Lazou, A., and M. Krokida. 2010. Structural and textural characterization of corn–lentil extruded snacks. *Journal of Food Engineering* 100(3):392–408. doi: http://dx.doi.org/10.1016/j.jfoodeng.2010.04.024.

Li, Y.O., and L.L. Diosady. 2011. Microencapsulation: Applications in micronutrient fortification through "engineered" staple foods. *Bioencapsulation Innovations* 2011:15–18, March/April.

Lin, S.-F., Y.-C. Chen, R.-N. Chen, L.-C. Chen, H.-O. Ho, Y.-H. Tsung, M.-T. Sheu, and D.-Z. Liu. 2016. Improving the stability of astaxanthin by microencapsulation in calcium alginate beads. *PLoS ONE* 11(4):e0153685. doi: 10.1371/journal.pone.0153685.

Liu, G., X. Miao, W. Fan, R. Crawford, and Y. Xiao. 2010. Porous PLGA microspheres effectively loaded with BSA protein by electrospraying combined with phase separation in liquid nitrogen. *Journal of Biomimetics, Biomaterials and Tissue Engineering* 6:1–18.

López-Córdoba, A., L. Deladino, and M. Martino. 2014. Release of yerba mate antioxidants from corn starch–alginate capsules as affected by structure. *Carbohydrate Polymers* 99:150–157. doi: http://dx.doi.org/10.1016/j.carbpol.2013.08.026.

Low, K.G., and S.F. Lim. 2014. Study on electrostatic extrusion method for synthesizing calcium alginate encapsulated iron oxide. *Journal of Applied Science & Process Engineering* 1(1):9–27.

Madene, A., M. Jacquot, J. Scher, and S. Desobry. 2006. Flavour encapsulation and controlled release—A review. *International Journal of Food Science and Technology* 41(1):1–21. doi: 10.1111/j.1365-2621.2005.00980.x.

Mamidwar, S., S. Hodge, V. Deshmukh, and V. Borkar. 2012. Hot-melt extusion. *International Journal of Pharmaceutical Sciences Review & Research* 15(1):105–112.

Maniruzzaman, M., J.S. Boateng, B.Z. Chowdhry, M.J. Snowden, and D. Douroumis. 2014. A review on the taste masking of bitter APIs: Hot-melt extrusion (HME) evaluation. *Drug Development and Industrial Pharmacy* 40(2):145–156. doi: 10.3109/03639045.2013.804833.

Maniruzzaman, M., J.S. Boateng, M.J. Snowden, and D. Douroumis. 2012. A review of hot-melt extrusion: Process technology to pharmaceutical products. *ISRN Pharmaceutics* 2012:9. doi: 10.5402/2012/436763.

Manojlovic, V., J. Djonlagic, B. Obradovic, V. Nedovic, and B. Bugarski. 2006. Immobilization of cells by electrostatic droplet generation: A model system for potential application in medicine. *International Journal of Nanomedicine* 1(2):163–171.

Manojlovic, V., N. Rajic, J. Djonlagic, B. Obradovic, V. Nedovic, and B. Bugarski. 2008. Application of electrostatic extrusion—Flavour encapsulation and controlled release. *Sensors (Basel, Switzerland)* 8(3):1488–1496.

Mans, J. 1982. Extruders. *Processed Prepared Foods* 151(11):60–63.

Mo, L. 2015. Viability of *Lactobacillus acidophilus* DDS 1-10 encapsulated with an alginate-starch matrix. Dissertations and Theses in Food Science and Technology. Paper 58. University of Nebraska–Lincoln, Lincoln, NE.

Mollet, B., and C. Lacroix. 2007. Where biology and technology meet for better nutrition and health. *Current Opinion in Biotechnology* 18(2):154–155. doi: http://dx.doi.org/10.1016/j.copbio.2007.03.003.

Munin, A., and F. Edwards-Lévy. 2011. Encapsulation of natural polyphenolic compounds: A review. *Pharmaceutics* 3(4):793–829. doi: 10.3390/pharmaceutics3040793.

Muthukumarasamy, P., P. Allan-Wojtas, and R.A. Holley. 2006. Stability of *Lactobacillus reuteri* in different types of microcapsules. *Journal of Food Science* 71(1):M20–M24. doi: 10.1111/j.1365-2621.2006.tb12395.x.

Naknean, P., and M. Meenune. 2010. Factors affecting retention and release of flavour compounds in food carbohydrates. *International Food Research Journal* 17:23–34.

Navale, S.A., S.B. Swami, and N.J. Thakor. 2015. Extrusion cooking technology for foods: A review. *Journal of Ready to Eat Food* 2(3):66–80.

Nedovic, V., A. Kalusevic, V. Manojlovic, S. Levic, and B. Bugarski. 2011. 11th International Congress on Engineering and Food (ICEF11): An overview of encapsulation technologies for food applications. *Procedia Food Science* 1:1806–1815. doi: http://dx.doi.org/10.1016/j.profoo.2011.09.265.

Nikmaram, N., M.H. Kamani, and R. Ghalavand. 2015. The effects of extrusion cooking on antinutritional factors, chemical propertiesand contaminating microorganisms of food. *International Journal of Farming and Allied Sciences* 4(4):352–354.

O'Neill, T.E. 1996. Flavor binding by food proteins: An overview. In *Flavor-Food Interactions*, edited by R.J. McGorrin and J.V. Leland, pp. 59–74. Washington, DC: American Chemical Society.

Oidtmann, J., M. Schantz, K. Mäder, M. Baum, S. Berg, M. Betz, U. Kulozik et al. 2012. Preparation and comparative release characteristics of three anthocyanin encapsulation systems. *Journal of Agricultural and Food Chemistry* 60(3):844–851. doi: 10.1021/jf2047515.

Oxley, J.D. 2012. Chapter 6—Coextrusion for food ingredients and nutraceutical encapsulation: Principles and technology. In *Encapsulation Technologies and Delivery Systems for Food Ingredients and Nutraceuticals*, pp. 131–150. Philadelphia, PA: Woodhead Publishing.

Oxley, J. 2014. Chapter 4—Overview of microencapsulation process technologies. In *Microencapsulation in the Food Industry*, edited by A.G. Gaonkar, N. Vasisht, A.R. Khare, and R. Sobel, pp. 35–46. San Diego, CA: Academic Press.

Özer, B., H.A. Kirmaci, E. Şenel, M. Atamer, and A. Hayaloğlu. 2009. Improving the viability of *Bifidobacterium bifidum* BB-12 and *Lactobacillus acidophilus* LA-5 in white-brined cheese by microencapsulation. *International Dairy Journal* 19(1):22–29. doi: http://dx.doi.org/10.1016/j.idairyj.2008.07.001.

Özkan, G., and S.E. Bilek. 2014. Microencapsulation of natural food colourants. *International Journal of Nutrition and Food Sciences* 3(3):145–156. doi: 10.11648/j.ijnfs.20140303.13.

Parker, R., and S.G. Ring. 1995. Diffusion in maltose-water mixtures at temperatures close to the glass transition. *Carbohydrate Research* 273(2):147–155. doi: http://dx.doi.org/10.1016/0008-6215(95)00120-I.

Patel, A., D. Sahu, A. Dashora, and R. Garg. 2013. A review of hot melt extrusion technique. *International Journal of Innovative Research in Science, Engineering and Technology* 2(6):2194–2198.

Patil, H., R.V. Tiwari, and M.A. Repka. 2016. Hot-melt extrusion: From theory to application in pharmaceutical formulation. *AAPS PharmSciTech* 17(1):20–42. doi: 10.1208/s12249-015-0360-7.

Poncelet, D., A. Picot, and S. El Mafadi. 2011. Encapsulation: An essential technology for functional food applications. *Innovations in Food Technology* February 2011: 32–34.

Porzio, M.A., and L.M. Popplewell. 1999. Encapsulation compositions. Google Patent, US 5897897 A.

Porzio, M.A., and L.M. Popplewell. 2002. Encapsulation compositions. Google Patent, US 6416799 B1.

Porzio, M.A., and D. Zasypkin. 2009. Encapsulation compositions and processes for preparing the same. Google Patent, US 7488503 B1.

Porzio, M. 2008. Melt extrusion and melt injection. *Perfumer & Flavorist* 33:48–53.

Poshadri, A., and K. Aparna. 2010. Microencapsulation technology: A review. *Journal of Research ANGRAU* 38(1):86–102.

Quellet, C., M. Taschi, and J.B. Ubbink. 2001. Encapsulated liquid. Google Patent, EP 1116515 A2.

Ramachandra Rao, H.G., and M.L. Thejaswini. 2015. Extrusion technology: A novel method of food processing. *International Journal of Innovative Science, Engineering & Technology* 2(4):358–369.

Rathore, S., P.M. Desai, C.V. Liew, L.W. Chan, and P.W. Sia Heng. 2013. Microencapsulation of microbial cells. *Journal of Food Engineering* 116(2):369–381. doi: http://dx.doi.org/10.1016/j.jfoodeng.2012.12.022.

Reineccius, G.A. 1989. Flavor encapsulation. *Food Reviews International* 5(2):147–176. doi: 10.1080/87559128909540848.

Riaz, M.N. 2000. *Extruders in Food Applications*. Boca Raton, FL: CRC Press.

Risch, S.J. 1988. Encapsulation of flavors by extrusion. In *Flavor Encapsulation*, pp. 103–109. Washington, DC: American Chemical Society.

Robert, P., and C. Fredes. 2015. The encapsulation of anthocyanins from berry-type fruits. *Trends in Foods Molecules* 20(4):5875.

Ruiz-Gutiérrez, M., C. Amaya-Guerra, A. Quintero-Ramos, E. Pérez-Carrillo, T. Ruiz-Anchondo, J. Báez-González, and C. Meléndez- Pizarro. 2015. Effect of extrusion cooking on bioactive compounds in encapsulated red cactus pear powder. *Molecules* 20(5):8875.

Saleeb, F.Z., and V.K. Arora. 1999. Method of preparing glass stabilized material. Google Patent, US 5972395 A.

Sandoval-Castilla, O., C. Lobato-Calleros, H.S. García-Galindo, J. Alvarez-Ramírez, and E.J. Vernon-Carter. 2010. Textural properties of alginate–pectin beads and survivability of entrapped *Lb. casei* in simulated gastrointestinal conditions and in yoghurt. *Food Research International* 43(1):111–117. doi: http://dx.doi.org/10.1016/j.foodres.2009.09.010.

Schultz, T.H., K.P. Dimick, and B. Makower. 1956. Incorporation of natural fruit flavors into fruit juice powders. I. Locking of citrus oils in sucrose and dextrose. *Food Technology* 10(1):57.

Seyedain-Ardabili, M., A. Sharifan, and B.G. Tarzi. 2016. The production of synbiotic bread by microencapsulation. *Food Technology and Biotechnology* 54(1):103–107. doi: 10.17113/ftb.54.01.16.4234.

Shahidi, F., and X.-Q. Han. 1993. Encapsulation of food ingredients. *Critical Reviews in Food Science and Nutrition* 33(6):501–547. doi: 10.1080/10408399309527645.

Shahidi, F., and R.B. Pegg. 2007. Encapsulation, stabilization, and controlled release of food ingredients and bioactives. In *Handbook of Food Preservation*, 2nd edition, edited by M. Shafiur Rahman, pp. 509–568. Boca Raton, FL: CRC Press.

Shi, L.-E., Z.-H. Li, D.-T. Li, M. Xu, H.-Y. Chen, Z.-L. Zhang, and Z.-X. Tang. 2013. Encapsulation of probiotic *Lactobacillus bulgaricus* in alginate–milk microspheres and evaluation of the survival in simulated gastrointestinal conditions. *Journal of Food Engineering* 117(1):99–104. doi: http://dx.doi.org/10.1016/j.jfoodeng.2013.02.012.

Singh, B., and S.J. Mulvaney. 1994. Modeling and process control of twin-screw cooking food extruders. *Journal of Food Engineering* 23(4):403–428. doi: http://dx.doi.org/10.1016/0260-8774(94)90102-3.

Singh, S., S. Gamlath, and L. Wakeling. 2007. Nutritional aspects of food extrusion: A review. *International Journal of Food Science and Technology* 42(8):916–929. doi: 10.1111/j.1365-2621.2006.01309.x.

Singhal, S., V.K. Lohar, and V. Arora. 2011. Hot melt extrusion technique. *WebmedCentral Pharmaceutical Sciences* 2(1):WMC001459

Sobel, R., M. Gundlach, and C.-P. Su. 2014. Chapter 33—Novel concepts and challenges of flavor microencapsulation and taste modification. In *Microencapsulation in the Food Industry*, edited by A.G. Gaonkar, N. Vasisht, A.R. Khare, and R. Sobel, pp. 421–442. San Diego, CA: Academic Press.

Soliman, E., A. El-Moghazy, M. El-Din, and M. Massoud. 2013. Microencapsulation of essential oils within alginate: Formulation and in vitro evaluation of antifungal activity. *Journal of Encapsulation and Adsorption Sciences* 3(1):48–55. doi: 10.4236/jeas.2013.31006

Sriamornsak, P., and R.A. Kennedy. 2007. Effect of drug solubility on release behavior of calcium polysaccharide gel-coated pellets. *European Journal of Pharmaceutical Sciences* 32(3):231–239. doi: http://dx.doi.org/10.1016/j.ejps.2007.08.001.

Steel, C.J., M.G.V. Leoro, M. Schmiele, R.E. Ferreira, and Y.K. Chang. 2012. Thermoplastic extrusion in food processing. In *Thermoplastic Elastomers*, edited by A. El-Sonbati, pp. 265–290. InTech.

Stojanovic, R., A. Belscak-Cvitanovic, V. Manojlovic, D. Komes, V. Nedovic, and B. Bugarski. 2012. Encapsulation of thyme (*Thymus serpyllum* L.) aqueous extract in calcium alginate beads. *Journal of the Science of Food and Agriculture* 92(3):685–696. doi: 10.1002/jsfa.4632.

Sun-Waterhouse, D., J. Zhou, G.M. Miskelly, R. Wibisono, and S.S. Wadhwa. 2011. Stability of encapsulated olive oil in the presence of caffeic acid. *Food Chemistry* 126(3):1049–1056. doi: http://dx.doi.org/10.1016/j.foodchem.2010.11.124.

Sun-Waterhouse, D., L. Penin-Peyta, S.S. Wadhwa, and G.I.N. Waterhouse. 2012. Storage stability of phenolic-fortified avocado oil encapsulated using different polymer formulations and co-extrusion technology. *Food and Bioprocess Technology* 5(8):3090–3102. doi: 10.1007/s11947-011-0591-x.

Swisher, H.E. 1957. Solid flavoring composition and method of preparing the same. Google Patent, US 2809895 A.

Swisher, H.E. 1962. Solid essential oil flavoring composition and process for preparing the same. Google Patent, US 3041180 A.

Tackenberg, M.W., R. Krauss, A. Marmann, M. Thommes, H.P. Schuchmann, and P. Kleinebudde. 2015. Encapsulation of liquids using a counter rotating twin screw extruder. *European Journal of Pharmaceutics and Biopharmaceutics* 89:9–17. doi: http://dx.doi.org/10.1016/j.ejpb.2014.11.017.

Thangaraj, S., and M. Seethalakshmi. 2014. Microencapsulation of vitamin C through extrusion process. *International Journal of Advanced Research in Biological Sciences* 1(7):16–21.

Tolve, R., F. Galgano, M.C. Caruso, F.L. Tchuenbou-Magaia, N. Condelli, F. Favati, and Z. Zhang. 2016. Encapsulation of health-promoting ingredients: Applications in foodstuffs. *International Journal of Food Sciences and Nutrition* 67(8):1–31. doi: 10.1080/09637486.2016.1205552.

Tsuei, A.C., L. Kogl, and D.B. Pendergrass. 1996. Method of encapsulation and microcapsules produced thereby. Google Patent, US 5589194 A.

Ubbink, J., and J. Krüger. 2006. Physical approaches for the delivery of active ingredients in foods. *Trends in Food Science & Technology* 17(5):244–254. doi: http://dx.doi.org/10.1016/j.tifs.2006.01.007.

Valentinotti, S., L. Armanet, and J. Porret. 2006. Encapsulated polyunsaturated fatty acids. Google Patent, US 20060134180 A1.

Van Lengerich, B., G. Walther, and B. Van Auken. 2007. Encapsulation of readily oxidizable components. Google Patent, US 20070098853 A1.

Van Lengerich, B.H. 2001. Embedding and encapsulation of controlled release particles. Google Patent, US 6190591 B1.

Van Lengerich, B.H., L. Leung, S.C. Robie, J. Lakkis, Y. Kang, and T.M. Jarl. 2004. Encapsulation of sensitive components using pre-emulsification. Google Patent, WO 2004009054 A3.

Van Lengerich, B.H., G. Walther, and B. van Auken. 2010. Encapsulation of readily oxidizable components. Google Patent, US 7803413 B2.

Viswanath, D., and V. Krishna. 2015. Nutraceuticals: Beneficence par excellence. *EC Nutrition* 1(3):137–139.

Wang, L., and T. Bohn. 2012. Health-promoting food ingredients and functional food processing. In *Nutrition, Well-Being and Health*, edited by J. Bouayed, pp. 201–224. Rijeka, Croatia: InTech.

Yadav, V., A. Sharma, and S.K. Singh. 2015. Microencapsulation techniques applicable to food flavours research and development: A comprehensive review. *International Journal of Food and Nutritional Sciences* 4(3):119–124.

Yeo, L.Y., Z. Gagnon, and H.-C. Chang. 2005. AC electrospray biomaterials synthesis. *Biomaterials* 26(31):6122–6128. doi: http://dx.doi.org/10.1016/j.biomaterials.2005.03.033.

Yılmaz, G., R.O.J. Jongboom, H. Feil, and W.E. Hennink. 2001. Encapsulation of sunflower oil in starch matrices via extrusion: Effect of the interfacial properties and processing conditions on the formation of dispersed phase morphologies. *Carbohydrate Polymers* 45(4):403–410. doi: http://dx.doi.org/10.1016/S0144-8617(00)00264-2.

Yuliani, S., P.J. Torley, B. D'Arcy, T. Nicholson, and B. Bhandari. 2006. Extrusion of mixtures of starch and d-limonene encapsulated with β-cyclodextrin: Flavour retention and physical properties. *Food Research International* 39(3):318–331. doi: http://dx.doi.org/10.1016/j.foodres.2005.08.005.

Zasypkin, D. 2011. Melt extrusion encapsulation of flavors and other encapsulates in a carrier containing spices and herbs. Google Patent, US 20110256199 A1.

Zasypkin, D., S. Paranjpe, M. Reick, and S. Johnson. 2014. Encapsulation compositions comprising of spices, herbs, fruit, and vegetable powders. Google Patent, WO 2014146092 A1.

Zasypkin, D., and M. Porzio. 2004. Glass encapsulation of flavours with chemically modified starch blends. *Journal of Microencapsulation* 21(4):385–397. doi: 10.1080/02652040410001695924.

Zuidam, N.J., and V.A. Nedović. 2010. Introduction. In *Encapsulation Technologies for Active Food Ingredients and Food Processing*, edited by N.J. Zuidam and V. Nedovic, pp. 1–2. New York: Springer.

Zuidam, N.J., and E. Shimoni. 2010. Overview of microencapsulates for use in food products or processes and methods to make them. In *Encapsulation Technologies for Active Food Ingredients and Food Processing*, edited by N.J. Zuidam and V. Nedovic, pp. 3–29. New York: Springer.

6 Osmotic Dehydration and Combined Processes for Tissue Modification and Selective Ingredient Impregnation and Encapsulation

E. Dermesonlouoglou, P. Taoukis, and M. Giannakourou

CONTENTS

6.1 INTRODUCTION

Osmotic dehydration (OD), also known as osmotic drying or dewatering-impregnation soaking, involves immersing a solid food in a hypertonic water solution to induce loss of water from the food into the solution and solute transfer from the solution into the food (Figure 6.1). Typically it is applied to plant origin materials, such as fruits and vegetables that consist of tissues or organizations of cells with different characteristics and complexity. Water and solute mass transfer strongly depend on the

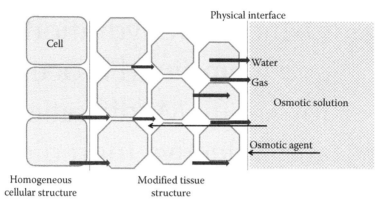

FIGURE 6.1 Mass transfer phenomena during osmotic dehydration. (Adapted by Shi, J. and Xue, J.S., Application and development of osmotic dehydration technology in food processing, in: Ratti, C. (ed.), *Advances in Food Dehydration*, CRC Press, Boca Raton, FL, 2009.)

properties of the osmotic solution (including type and concentration of osmo-active solute and osmotic gradient), working pressure, and the structure of the solid food being processed. To optimize the process and food product quality, it is important to understand the parameters affecting mass transfer during OD.

OD has been reported as a feasible treatment for incorporating physiologically active compounds, such as minerals, vitamins, and antioxidants into plant tissues without destroying the initial food matrix, especially as a pretreatment to further processing. Through this selective incorporation into the food system, it is possible, to a certain extent, to change the nutritional and functional properties, achieving a specific formulation of the product while maintaining to a great extent its integrity. The combined application of OD with novel hybrid technologies such as high hydrostatic pressure, ultrasound, pulsed electric field, and γ-gamma irradiation can increase the value of the final product, with regard to microbial stability, nutritional, functional, and sensory characteristics, while at the same time intensifying the process of mass transfer. New aspects also include the use of alternative osmotic solutes in the osmotic solution, such as alternative carbohydrates, emulsions, and the combined use of OD and edible coating.

In this chapter, the basics of OD including the mechanism of the process, the parameters affecting the process, and common applications as well as novel approaches will be presented. Although OD as an impregnation/incorporation method has been widely studied, the application of this mild thermal treatment as an encapsulation method is newly introduced, but seems rather promising and warrants further investigation.

6.2 MODELING OF MASS TRANSFER

Extensive work has been done in developing models to predict the mass transfer kinetics of OD at atmospheric pressure. Water loss (WL) and solid gain (SG), which can be calculated by the following equations, respectively, characterize the OD process. The WL and SG during OD depend on several parameters (detailed in Section 6.3). The selection of the values of the process parameters is performed considering the final application and characteristics of the final product:

$$WL = \frac{\left(M_0 - m_0\right) - \left(M - m\right)}{m_0} \left(\text{g of water/g of initial dry matter}\right) \qquad (6.1)$$

$$SG = \frac{m - m_0}{m_0} \left(\text{g of total solids/g of initial dry matter}\right) \qquad (6.2)$$

Empirical, semiempirical (Azuara's, Peleg's, Page's, Magee's, Webull's distribution model and the penetration model), mechanistic (such as Toupin et al.'s, Marcotte et al.'s, the hydrodynamic model, Spiazzi and Mascheroni's, Segui et al.'s—not presented), and phenomenological approaches (Crank's model) have been proposed (Assis et al. 2015) (Table 6.1).

Empirical and semiempirical models correlate the process parameters with WL and SG, without taking into account the underlying simultaneous transport phenomena of water and solutes, and include multivariable regressions, response-surface analysis, and mass balances (Azuara et al., 1992; Kaymak-Ertekin and Sultanoglu, 2000; Sereno et al., 2001; Mercali et al., 2012; Herman-Lara et al., 2013). They do not take into account the process complexity and their applicability is specific because they are only capable of representing data at conditions similar to those on which such models were developed (Ochoa-Martínez and Ayala-Aponte, 2007). They can be applied to nonclassical geometries and some can also predict the equilibrium values. However, knowledge on microscopic histological and cellular features, such as the types of tissue, the geometric properties of the cell, the cellular water potential, the mechanical properties of the cell wall, and the presence of intercellular spaces is important to understand structure–property relationships and then to estimate the rate of mass transfer of water and solutes in food (Aguilera et al., 2000; Mebatsion et al., 2008; Mercali et al., 2012). Mechanistic models have been developed taking into account this kind of structure and/or macroscopic and microscopic approaches. Phenomenological models can determine the diffusion coefficient but they require the equilibrium values and are only suitable for classical geometries. Although they give a good description of the mass transfer mechanism, the diffusion approach has a number of assumptions (such as equilibrium values, classical geometries, no external resistance, no shrinkage of the sample, no effective diffusivity changes with temperature), and the effective diffusion coefficients that are calculated strongly depend on the experimental conditions and the raw material characteristics. Among these, the most used model comprises a group of analytical solutions proposed by Crank (1975), based on Fick's second law to represent the diffusional mechanism in OD performed at atmospheric pressure (Assis et al., 2015). As such, an example is presented on the use of Crank's analytical solution for infinite slab geometry.

TABLE 6.1

Models Used in the Osmotic Process of Fruits, Vegetables, Meat, and Fish

Model		Applications
Azuara's	$WL = \dfrac{s_1 * t * WL_\infty}{1 + s_1 * t}$ $SG = \dfrac{s_1 * t * SG_\infty}{1 + s_1 * t}$	Apple, apricot, banana, beetroot, carrot, chayote, cherry tomato, elephant foot yam, mackerel, eggplant, green bean, guava, kiwifruit, mango, nectarine, papaya, pineapple, pomegranate arils, pumpkin, chicken, beef meat, goat meat, shark
Peleg's	$WL = \dfrac{t}{k_1^w + k_2^w * t}$ $SG = \dfrac{t}{k_1^s + k_2^s * t}$	Apple, cherry tomato, elephant foot yam, guava, kiwifruit, papaya, pear, pomegranate arils, potato, yacon, yam, catfish, chicken, beef meat, goat meat
Page's	$\dfrac{WL}{WL_\infty} = 1 - \exp\left(-A_w * t^{B_w}\right)$ $\dfrac{SG}{SG_\infty} = 1 - \exp\left(-A_s * t^{B_s}\right)$	Sardine sheet, apple, banana, carrot, cherry tomato, guava, kiwifruit, mango, pear, pomegranate arils, strawberry
Penetration model	$WL = k^w * t^{1/2}$ $SG = k^s * t^{1/2}$	Banana, carrot, pomegranate arils, yam
Magee's	$k = A * C_o^x * T^y$	Apple, banana, elephant foot yam
Weibull	$\dfrac{WL}{WL_\infty} = \exp\left(-\left(\dfrac{t}{\alpha_w}\right)^{\beta_w}\right)$ $\dfrac{SG}{SG_\infty} = \exp\left(-\left(\dfrac{t}{\alpha_s}\right)^{\beta_s}\right)$	Apple, mango, chicken, chub mackerel, goat meat, sardine sheet, strawberry
Crank's	*Infinite plate* $\dfrac{WL}{WL_\infty}$ or $\dfrac{SG}{SG_\infty} = 1 - \dfrac{8}{\pi^2} \sum_{n=0}^{n=\infty} \dfrac{1}{(2n+1)^2} \exp\left[-\left(n+\dfrac{1}{2}\right)^2 \pi^2 F_{ow}\right]$ *Infinite sphere* $\dfrac{WL}{WL_\infty}$ or $\dfrac{SG}{SG_\infty} = 1 - \dfrac{6}{\pi^2} \sum_{n=0}^{n=\infty} \dfrac{1}{n^2} \exp\left[-n^2 \pi^2 D_e \dfrac{1}{r^2}\right]$ *Infinite cylinder* $\dfrac{WL}{WL_\infty}$ or $\dfrac{SG}{SG_\infty} = 1 - 4\sum_{n=0}^{n=\infty} \dfrac{1}{\alpha_n^2} \exp\left[-\alpha^2 \dfrac{D_e t}{r_C^2}\right]$	Apple, banana, beetroot, coconut, carrot, cherry tomato, chub mackerel, eggplant, green bean, jenipapo, kiwifruit, litchi, mango, melon, nectarine, papaya, pineapple, potato, pumpkin, yacon, radish, chicken, beef, goat meat, shark

6.2.1 Example of Calculations for Osmotic Dehydration

Pumpkin (*Cucurbita maxima*, var. Long Island Cheese) fruit samples, cut into 4 × 3 × 1 cm slabs were osmotically dehydrated in OD sugar solutions of different concentrations (65, 70, and 75°Brix), under different process temperatures (75, 85, and 95°C) for time duration up to 200 min (Lazou et al., 2016). Necessary calculations include the following steps:

1. Mass transfer parameters and changes are estimated by using WL (Equation 6.1) and SG (Equation 6.2), where M_0 is the initial mass of fresh material before the osmotic treatment, M is the mass of pumpkin samples after time t of osmotic treatment, m is the dry mass of pumpkin after time t of osmotic treatment, and m_0 is the dry mass of fresh material. The corresponding figures (Figure 6.2) illustrate in a straightforward way the mass transport into and out of the cellular tissue, for the case of a concentration of 70°Brix of the OD solution, for all temperatures studied.

2. For simplification purposes, pumpkin slabs are considered as having an infinite slab shape of thickness $2L$, initially at uniform moisture and solute concentrations C_{w0} and C_{s0}, respectively. At $t = 0$, the material is immersed in a sugar solution with constant concentration and temperature. It is assumed that the solution temperature and concentration remain constant during OD. This condition is well established by selecting a high solution-to-solid mass ratio (approximately 10/1). Also, constant equilibrium moisture and solute concentrations at the surface (negligible external resistance to mass transfer) of the material are considered. In these conditions, the unsteady-state one-dimensional mass transfer in the solid material can be described by the following general equation (6.3) based on Fick's law of diffusion:

$$\frac{\partial C}{\partial t} = D_{eff} \frac{\partial^2 C}{\partial x^2} \tag{6.3}$$

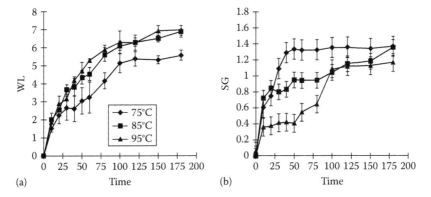

(a)   Time   (b)   Time

FIGURE 6.2 WL (a) and SG (b) *versus* time for the OD of pumpkin slabs with a 70°Brix sucrose solution (error bars represent the standard deviation of three replicates).

where D_{eff} is the effective moisture diffusivity (m²/min). The following boundary conditions are assumed (6.4):

$$C(x,0) = C_0 \quad \text{at } t = 0, \quad \frac{\partial C}{\partial x} = 0 \quad \text{at } x = 0, \quad C(L,t) = C_e \quad \text{at } x = L \quad (6.4)$$

where
  x is the spatial coordinate
  $C = C(x, t)$
  subscripts 0 and e refer to initial and equilibrium states, respectively

Average dimensionless concentrations (space-mean concentrations), $\varphi(t)$, could be obtained by applying the method of separation of variables to Equation 6.5, to yield concentration distribution, $C(x, t)$, and then taking the spatially average, arriving at (Crank, 1975; Souraki et al., 2014):

$$\varphi(t) = \frac{\overline{C}(t) - C_e}{C_0 - C_e} = \frac{8}{\pi^2} \sum_{n=0}^{\infty} \frac{1}{(2n+1)^2} \exp\left[-\frac{\pi^2 (2n+1)^2 D_{eff}}{4L^2} t \right] \quad (6.5)$$

in which C is the mass of water (w) or solute (s) per volume of the osmo-dehydrated tissue. C_0 is the initial content per volume, C_e is the equilibrium content per volume, D_{eff} is the effective moisture diffusivity (m²/min), L is the half thickness (drying from both sides) of pumpkin slabs (m) ($L = 0.05 \pm 0.010$ mm), and t is the osmotic treatment time (min). For long drying times, $n = 1$, the series of Equation 6.5 can be simplified by using just the first term and the following mathematical expression can be obtained:

$$\ln(\varphi) = \ln\left(\frac{8}{\pi^2} \right) - \left(\frac{\pi^2 D_{eff}}{4L^2} \right) t \quad (6.6)$$

Therefore, for water, Equation 6.6 can be written as follows:

$$\ln \frac{(C_{wt} - C_{w\infty})}{(C_{w0} - C_{w\infty})} = \ln\left(\frac{8}{\pi^2} \right) - \left(\frac{\pi^2 D_{effw}}{4L^2} \right) t \quad (6.7)$$

and, for solute, correspondingly:

$$\ln \frac{(C_{st} - C_{s\infty})}{(C_{s0} - C_{s\infty})} = \ln\left(\frac{8}{\pi^2} \right) - \left(\frac{\pi^2 D_{effs}}{4L^2} \right) t \quad (6.8)$$

where

C_{wt} and C_{st} are the water and solute concentration at time t

C_{w0} and C_{s0} are the initial water and solute concentration

$C_{w\infty}$ and $C_{s\infty}$ are the equilibrium water and solute concentration, which can be determined by experiments at a long time length

Without considering samples' shrinkage during the osmotic procedure, the volume of the material and thus L in the right hand side of Equations 6.7 and 6.8 is constant (Rodriguez et al., 2012). Also, concentrations of the left hand side of Equations 6.7 and 6.8 can be replaced by the following drying data, for water (Equation 6.9) and the solute (Equation 6.10), respectively:

$$\frac{\left(C_{wt} - C_{w\infty}\right)}{\left(C_{w0} - C_{w\infty}\right)} = \frac{\dfrac{M - m}{V_t} - \dfrac{M_\infty - m_\infty}{V_\infty}}{\dfrac{M_0 - m_0}{V_0} - \dfrac{M_\infty - m_\infty}{V_\infty}} \tag{6.9}$$

where M_∞ and m_∞ are the sample mass and the dry mass, respectively, at the equilibrium phase:

$$\frac{\left(C_{st} - C_{s\infty}\right)}{\left(C_{s0} - C_{s\infty}\right)} = \frac{\dfrac{m}{V_t} - \dfrac{m_\infty}{V_\infty}}{\dfrac{m_0}{V_0} - \dfrac{m_\infty}{V_\infty}} \tag{6.10}$$

Assuming $V_0 = V_t = V_\infty$, concentrations in the left hand side part of Equations 6.9 and 6.10 are replaced by mass data, already measured during the osmotic procedure. Since values of moisture and solids content are experimentally determined, the first part of Equations 6.9 and 6.10 is calculated. Therefore, diffusivities (both for water and solute) are typically determined by plotting experimental data (first part of Equations 6.9 and 6.10) versus drying time t. It is a linear plot with a slope of $(\pi^2 D_{eff(w\ or\ s)})/(4L^2)$ easily determined by linear regression (Alibas, 2014).

For our example concerning OD of pumpkin slabs, the described methodology resulted in the following values for diffusivities (Table 6.2).

The experimental values of the effective diffusivity (Table 6.2) were in the range 10^{-9} to 10^{-10} (m^2/s) and are in general agreement with other similar works in recent literature (Souraki et al., 2012, 2013, 2014; Porciuncula et al., 2013; Simpson et al., 2015, etc.). The temperature of the osmotic solution had a significant influence mostly on water-effective diffusion coefficients, confirming the well-recognized assumption that diffusion processes, as well as other similar mechanisms, are temperature-dependent phenomena. The dependence of the diffusion coefficients on temperature is accordingly usually described by the Arrhenius equation.

TABLE 6.2

Diffusion Coefficients of Water and Solute (Sucrose) Estimated Using Fick's Law for OD of Pumpkin Slabs Using a 70°Brix Sucrose Solution

T_{osm} (°C)	D_{effw} (m²/s)[a] (10⁹)	D_{effs} (m²/s)[a] (10⁹)
75	2.97 ± 0.24	0.28 ± 0.10
85	2.55 ± 0.22	0.37 ± 0.10
95	2.83 ± 0.60	0.98 ± 0.59

[a] Mean of four replicates ± standard deviation.

6.3 PARAMETERS AFFECTING THE OSMOTIC DEHYDRATION PROCESS

The effect of parameters on mass transfer and product quality has been discussed by Torreggiani (1993), Rastogi and Raghavarao (1994, 1997), Rastogi et al. (2000, 2002), Fito et al. (2001a,b), Lewicki and Lenart (2006), Khan (2012), and Ahmed et al. (2016).

Raw material characteristics include variability in the variety and maturity of plant products affecting tissue structure (total porosity as well as the size and shape distribution of pores and their interactions with external liquid), tissue compactness, initial insoluble and soluble solids content, intercellular spaces, enzymatic activity, and size and geometry of sample (Fito et al., 1996; Panagiotou et al., 1998; Yan et al., 2008; Bekele and Ramaswamy, 2010). The osmotic process is also affected by the physicochemical properties of the solutes (molecular weight, ionic state, solute solubility in water, cell membrane permeability, as well as sensory properties) and cost (Bekele and Ramaswamy, 2010).

The immersion time and temperature affect the mass transfer and the product quality. Studies on the optimization of duration of osmosis process indicated that mass exchange took place at the maximum rate within the first two hours of the osmotic treatment increasing with temperature (Gaspartero et al., 2003; Mokhtarian et al., 2014). The immersion of fruits in osmotic solutions for a relatively long time causes leaching of water-soluble components such as minerals, vitamins, and organic acids (Yadav and Singh, 2014). To avoid losses of such compounds, these have been added to osmotic solutions to counteract the leaching effect of OD.

High temperatures of the osmotic solution lead to the decrease of the viscosity of the osmotic solution and thus the external resistance to mass transfer rate; it enhances the removal of water and uptake of solids (Phisut, 2012). Although the mass transfer rate increases with temperature, it is limited up to 60°C as higher temperature destroys the cell membranes (mostly reported for plant tissue). The pH of the solution can also affect the osmotic process. Acidification increases the rate of water removal by changes in tissue properties and consequential changes in the texture of fruits and vegetables. As the concentration of solution increases, the osmotic solution is more viscous and therefore the solutes exhibit slower penetration (Oladejo et al., 2013); thus the mass transfer resistance in the solution adjacent to the surface of samples increases. On the other hand, the higher the solution concentration, higher is the equilibrium level. Agitation is one of the key factors (Rastogi et al., 2002) to enhance mass transfer and to prevent the formation of a dilute solution film around the samples, and agitation or stirring process can be applied during OD (Panagiotou et al., 1998; Rastogi et al., 2002; Akbarian et al., 2014). However, agitation may be difficult and may cause damage to the products. Solution-to-sample ratio should be chosen appropriately so that the driving force for the removal of moisture is adequate till the end of the process. The driving force decreases when osmotic solutions become dilute. The commonly proposed and applied solution-to-product ratios are 3:1 to 5:1 (Beaudry et al., 2003; Ruiz-Lopez et al., 2010; Jain et al., 2011; Gupta et al., 2012).

6.4 OSMOTIC DEHYDRATION EFFECT ON QUALITY AND NUTRITIONAL PROPERTIES

OD is conducted at mild temperatures (30–50°C) avoiding damages related to thermal processes. OD aims to reduce the water content with minimal damage to product quality such as improved texture, color, appearance, and loss of volatile flavoring constituents. Oxidative browning of fruits and vegetables is prevented by the use of highly concentrated sugar/salt solution. Osmo-dehydrated products present improved organoleptic characteristics such as sweeter taste compared to conventionally dried products. The final products are very pleasant for direct consumption due to their better physicochemical properties and nutritional profile (Tortoe, 2010). Structural damage during subsequent processes such as drying or freezing is significantly reduced (Maestrelli et al., 2001). The uptake of osmotic solutes and moisture reduction by OD exert a cryoprotectant effect on the texture and color of several frozen fruits (Chiralt et al., 2001). Osmotically pretreated frozen products have longer shelf life and retained color, flavor, and texture after thawing (Maestrelli et al., 2001; Dermesonlouoglou et al., 2007a,c, 2016). OD can improve quality retention by properly selecting solutes and by maintaining an equilibrated ratio of impregnation and water removal (Tortoe, 2010). OD can be applied with the aim of modifying the composition of food by partial water removal and impregnation (Torreggiani, 1993). Through this selective incorporation into the food system, it is possible, to a certain extent, to change the nutritional and functional properties, achieving a specific formulation of the product without significantly modifying its integrity. Although a large number of studies on OD have been published in the last years, few studies focus on the improvement of nutritional value. Most recent studies are presented in Table 6.3.

TABLE 6.3

Representative Research Studies on Osmotic Dehydration of Fruits and Vegetables

Raw Material	Osmotic Agents/Osmotic Preprocessing Conditions	Parameter	Conclusions	Reference
Apple, banana, potato, model food	Sucrose (50% w/w), sodium chloride (10% w/w), grape seed extract as the source of phenolics (0.63% w/w) Air drying (55°C)	Total phenolics, Antiradical scavenging capacity	During OT, total phenolic content and antiradical scavenging capacity were significantly increased. The extent of grape phenolic impregnation was controlled by food structure and the kind of osmo-active solute: plant tissue showed a lower grape phenolic infusion than that of the model food. OT, as a pretreatment, protected against grape phenolic degradation during further convective air drying.	Rózek et al. (2010a,b)
Apple	Sucrose 65°Brix Stirring 25°C; 2 or 8 h	Polyphenol oxidase activity	The low PPO activity found in the edible parenchyma of osmotically dehydrated apples is attributed to penetration by the osmotic agent and flooding of the intercellular spaces, which produces low moisture content and a limited O_2 concentration in the immediate environment of the enzyme. OD prevents enzyme–substrate interaction. Thus, the low PPO activity would reduce browning of the product.	Quiles et al. (2005)
Banana	Blanching pretreatment: Ascorbic acid 0.5% (w/w) and citric acid 1% (w/w) at 70°C for 90s Sucrose 45, 55, 65% (w/w). Stirring at 80rpm Food: solution 1:11 (w/w). 30, 40, 50°C; 60 or 180 min	Phenolic compounds, tannins, antioxidant activity	The use of high concentrations of osmotic solution favored a higher retention of antioxidant capacity. The solute incorporation in the banana tissue had a protector, mainly those which are responsible for antioxidant capacity, such as tannins and phenolic acids.	Almeida et al. (2014)

(Continued)

TABLE 6.3 (*Continued*)
Representative Research Studies on Osmotic Dehydration of Fruits and Vegetables

Raw Material	Osmotic Agents/Osmotic Preprocessing Conditions	Parameter	Conclusions	Reference
Coconut	Sucrose (0–50% w/v), 0.13 g/100 mL curcumin powder/1:1 propylene glycol:polysorbate; 1–5 h Food: solution 1:5 (w/w) Ultrasound (35 kHz, amplitude 100%, 1 min on and 15 min off)	Curcuminoids	High performance liquid chromatography (HPLC) analysis indicated that all curcuminoids were infused into the matrix. Application of ultrasound resulted in enhancement of moisture, solid as well as curcuminoid mass transfer.	Bellary et al. (2011)
Indian gooseberry	FOS solution (70°Brix); food: solution 1:5 (w/v); 30, 40, 50°C	Total phenolics, total flavonoids, antioxidant activity	The antioxidant activity and phenolic as well as flavonoid contents of the FOS osmotic dehydrated samples were found to be significantly higher as compared to fresh fruits in both in vitro and cell line assays.	Nambiar et al. (2016)
Mango	Sucrose (45°Brix) with and without 1% w/w calcium chloride or 1% w/w ascorbic acid; 25°C, 15 h Food: solution 1:10 (w/w)	Vitamin C, β-carotene	Addition of calcium in the OD solution significantly improved vitamin C retention. The retention of all-*trans*-β-carotene was significantly lower in all OD treated mango samples dried at 50°C but remained unchanged in OD-treated mango samples with calcium or vitamin C dried at 70°C. Moreover, osmotic dehydration with and without additives reduced the ratio of 13-*cis*-β-carotene to all-*trans*-β-carotene.	Guiamba et al. (2016)

(Continued)

TABLE 6.3 (*Continued*)

Representative Research Studies on Osmotic Dehydration of Fruits and Vegetables

Raw Material	Osmotic Agents/Osmotic Preprocessing Conditions	Parameter	Conclusions	Reference
Sweet cherry, mango, blueberry	Cherry and blueberry: Sucrose 25% (w/w), FOS and soy lecithin 1% (w/w), wine polyphenols 1% (w/w), vitamin C 0.25% (w/w), mango flavor 0.1% (w/w) Food: solution 1:1 (w/v) Air drying; 45°C; 6–7 h	Total polyphenols, mass spectral analyses of phenolic component, sugar and lecithin analysis of infused fruits, antioxidant activity	During infusion, loss of anthocyanins from blueberry and cherry fruits was observed, but this did not reduce the visual appeal of infused fruits. In conjunction with FOS solution, fruits could be infused with soy lecithin (phospholipids) and excised mango assimilated significant amounts of phospholipids. Structurally, FOS solution–infused cherry fruits showed preservation of tissue structure similar to that in a fresh fruit. DPPH radical scavenging activity of fresh and infused fruit extracts did not differ significantly suggesting that the antioxidant activity of infused fruits is not impaired by the process.	Jacob and Gopinadhan (2012)
Sour cherry (frozen and/ or thawed)	Apple concentrate (40°Brix); 40°C; 180 min Food: solution 1:1 (w/v)	Total polyphenols, antioxidant activity	OD presented a negative effect on total phenolic content and antioxidant capacity in sour cherry fruits. However, OD was an important pretreatment in dehydration, since it reduced the water activity in over 50% and, in this case, on following treatments of convective drying and vacuum microwave drying, the time, the temperature, and the microwave power required to complete the dehydration were lower.	Nowicka et al. (2015)
Tomato	Sucrose 55% and NaCl 10% (w/w), sucrose 65% (w/w); Agitation 0–1000 rpm; 20–40°C; 6 h	Carotenoids	The carotenoid content was higher for the processed samples as compared to the fresh ones, due to weight loss and consequent pigment concentration. OT can be considered as an efficient process with regard to tomato quality, since it favors water removal and from the tissue, without affecting its carotenoid content.	Tonon et al. (2007)

(Continued)

TABLE 6.3 (*Continued*)

Representative Research Studies on Osmotic Dehydration of Fruits and Vegetables

Raw Material	Osmotic Agents/Osmotic Preprocessing Conditions	Parameter	Conclusions	Reference
Tomato, cherry	Sucrose (55°Brix) and NaCl (20% w/w), sucrose (27.5°Brix) and NaCl (10% w/w), 30, 40, 50°C, 24 h	Lycopene, β-carotene	An increase in lycopene and b-carotene was observed in samples osmotically dehydrated at moderate temperatures (30 and 40°C) with the solutions that include sucrose on its composition. Microscopic observations revealed a direct relationship between the integrity of the cellular matrix and the preservation or even synthesis of lycopene and β-carotene.	Heredia et al. (2009)
Tomato, cherry	Sucrose 55% (w/w); 30°C, 120 min; ternary solution of sucrose 27.5% and NaCl 10% (w/w); 40°C, 60 min	Lycopene	OT limited the isomerization during the later stage of drying, whereas both the loss of total lycopene and the *trans-cis* isomerization, mainly to the 13-*cis* form, were favored by an increase in temperature and the microwave power.	Heredia et al. (2010)
	Sucrose 55% (w/w); 30°C, 120 min; Sucrose 42.1% and 5% NaCl (w/w); 30°C, 210 min		When OT was applied, a 10% rate of isomerization was registered, although it was the nonosmotically dehydrated samples that underwent the greatest degree of isomerization in the subsequent drying stage.	
	Combination of OD, convective drying and microwave (0, 1, 3 W/g) assisted air drying (40, 55, 80°C)		Regarding MW, samples dehydrated at 3 W/g, and esp. at 80°C, showed a greater level of isomerization but a lower percentage of residual total lycopene.	

Effect on quality and nutritional properties.

6.5 DEVELOPMENTS IN OSMOTIC DEHYDRATION

6.5.1 ALTERNATIVE AGENTS IN THE OSMOTIC SOLUTIONS

6.5.1.1 Alternative Carbohydrates as Osmotic Agents

When addressing the issue of osmotic agents mostly used in impregnation procedures, the solute chosen must have high water solubility, low cost, and positive effect on the sensory properties and must ensure the stability of the final product (Brochier et al., 2015). Sugar and/or common salts are the most popular solutes, both for fruit and vegetable tissues, as well as in the fish industry. There is research activity for appropriate substitutes of sucrose that can be used for OD and yield a product with similar or better quality characteristics, namely, structural, textural, and sensory properties. In this context, the use of alternative osmotic agents and their impact on the quality and functionality of the final products is discussed in this section.

The type of osmotic agent used has a strong effect on the mass transfer during osmosis because of the difference in molecular weight and in the ability to bind water (Sritongtae et al., 2011). The rate of WL increases with increasing concentrations (osmolarity) of the osmotic solution, but the molecular size of the solute also has an effect on the rate of both water loss and solid gain. Low molecular weight agents result in higher osmotic pressure gradients and are easier to penetrate into the cell of fruit and vegetables as compared to high molecular weight osmotic agents (Tortoe, 2010). Penetration studies showed that the rate of solute penetration is directly related to the solution concentration and inversely related to the size of the sugar molecule (Panagiotou et al., 1998; Giraldo et al., 2003). The partial replacement of sucrose with polyhydric alcohols or invert sugar has the potential to improve the texture and quality of the dried fruit and especially to maintain its moistness. Sorbitol and glycerol (common humectants) promote mass transfer phenomena (water loss and solid gain) compared to conventional sucrose solutions. Sorbitol has low caloric content (2.4 kcal/g), making it consistent with the objective of weight control, is noncariogenic, and reduces water activity. Sorbitol is slowly absorbed and, consequently, the rise in blood glucose and the insulin response related to the ingestion of glucose is significantly reduced. It exerts a protective effect on color, on the ascorbic acid content of the dehydrated product, and on the fermentative process of fruits (Rizzolo et al., 2007). According to Johnson et al. (2013), sorbitol is also prebiotic, with proven health-promoting properties, for example, reduction of serum cholesterol. Researchers used sorbitol in the dehydration of apricots (Toğrul and İspir, 2007), strawberries (Rizzolo et al., 2007), yacon (de Mendonça et al., 2016), pineapple (Paul et al., 2014; Vilela et al., 2016), and green pepper (Ozdemir et al., 2008). Maltitol is a sugar polyol, about 90% as sweet as sugar, noncariogenic, and has half the calories of sugar. Maltitol is absorbed in amounts of 50–75% and thus provides about 30–50% lower energy than sucrose. Based on its superior functional properties, maltitol has been proposed as an alternative osmotic agent for yacon slices (Mendonça et al., 2017), for the osmotic pretreatment of frozen green peas (Giannakourou and Taoukis, 2003), or for candied fruits (Vilela et al., 2016).

Glycerol is used as a plasticizer in foods to improve their texture and is classified in Codex Alimentarius as a humectant/thickener and has antimicrobial properties (Moreira et al., 2007; Codex Alimentarius, 2012). Its caloric value is 1.8×10^7 J/kg (4.3 kcal/g) and is a by-product of biodiesel production. Scientific works of OD with glycerol include chestnuts (Moreira et al., 2007), Andes berries, tamarillos (Osorio et al., 2007), carambola slices (Barman and Badwaik, 2017), papaya (Thalerngnawachart and Duangmal, 2016), yacon (Brochier et al., 2014, 2015), mandarin (Therdthai et al., 2011), cantaloupe (Sritongtae et al., 2011), and litchi (Mahayothee et al., 2009), where glycerol showed the best potential for browning retardation. Maltodextrin is a nonsweet nutritive saccharide polymer that consists of D-glucose units linked primarily by [alpha]-1–4 bonds, has a dextrose equivalent of more than 20, and is classified as a GRAS ingredient (Nurhadi et al., 2016). Its relative sweetness, when compared to sucrose, is approximately 20% and some of their important functional properties include bulking, gelling, crystallization prevention, promotion of dispersibility, freezing control, and binding (Chronakis, 1998). It is also used as an osmotic agent. A series of published works have studied the use of maltodextrin in the osmotic solution, such as tomatoes and watermelon (Dermesonlouoglou et al., 2007a,b,c, 2008a), gilthead sea bream fillets (Tsironi and Taoukis, 2010, 2014), and pineapple (Corrêa et al., 2016). Another carbohydrate that can be used as an effective humectant is trehalose. Its application has been studied in a variety of food tissues. Vicente et al. (2012) studied the impact of OD with glucose and trehalose on the micro- and ultrastructure, the rheological properties, and the state of water of apple tissues, showing that the type of osmotic solute has a strong effect on tissue rheological behavior. Aktas et al. (2013) investigated the application of different OD solutions, based on sucrose or trehalose to decrease the drying time, and decrease the loss of quality after drying and during the storage of apples. Their results showed the significant decrease of the drying time, as well as the effective protection provided against nonenzymatic browning, as well as an adequate retention of quality attributes when trehalose was used. Bearing in mind that sucrose is not suitable for diabetic patients, alternative osmotic solutes need to be explored considering the needs of this special group of consumers (Nambiar et al., 2016). **Fructooligosaccharides** (FOS) is one such compound, being nondigestible oligosaccharides, which can be used as alternative sweeteners, dietary fibers, or prebiotics. Angilelli et al. (2015) studied and modeled mass transfer during osmotic treatment with FOS solutions into melon tissue. Nambiar et al. (2016) investigated the infusion with fructooligosaccharides in Indian gooseberry and assessed the superior antioxidant properties of the modified tissue and Rubio-Arraez et al. (2015) published a comparative kinetic study of the OD of lemon slices using novel sweeteners, such as oligofructose, tagatose, and an aqueous extract of stevia leaves. A series of published works have studied the use of oligofructose in the osmotic solution in a pretreatment prior to freezing for producing improved dehydro-frozen products, such as frozen green peas (Giannakourou and Taoukis, 2003), tomatoes, (Dermesonlouoglou et al., 2007c), cucumber (Dermesonlouoglou et al., 2008b), strawberry (Dermesonlouoglou et al., 2016), and watermelon (Dermesonlouoglou et al., 2007a).

Recently, the use of multicomponent solutions has gained increasing interest. The mixture of salt and sugar in a ternary solution has been reported as an advantageous method for OD in various applications, leading to higher water loss/solute update without excessively over-sweetening or over-salting the product (Rodrigues and Fernandes, 2007; Heredia and Andrés, 2008; Aminzadeh et al., 2012; Mercali et al., 2012; Vasconcelos et al., 2012; Agustinelli et al., 2014). Studies on the use of ternary solutions to better control the main mass fluxes (water loss vs. solid gain) are among the main research interests in the field of OD aiming at the optimization of the procedure, depending on the final use of the modified tissue. In this context, depending on the main purpose of the OD (dehydration, salting, candying, etc.), not only the concentration but also the type of the osmotic solute plays a significant role.

6.5.2 NOVEL HYBRID METHODS OF OSMOTIC DEHYDRATION: APPLICATION OF NOVEL TREATMENTS COMBINED WITH OD

OD is a rather slow process, allowing for the mass transfer to occur between food tissue and OD solution. In this context, one of the main reasons for applying an additional step, combined with OD is to facilitate and enhance mass exchange by further cell permeabilization. Numerous procedures such as vacuum or pulsed vacuum, pulsed electric fields, ultrasound, etc., are therefore applied before or at the same time with OD in order to accelerate and maximize mass transfer phenomena.

Vacuum osmotic dehydration or vacuum impregnation (VI) and pulsed vacuum osmotic dehydration (PVOD) has been the subject of numerous studies and is currently applied by the food industry. Andres et al. (2001), Zhao and Xie (2004), and Fito et al. (2001a,b) have described in depth the principles, mechanisms, benefits, and effects of applying vacuum combined to OD process. According to Andres et al. (2001), VI of a porous food with an external solution leads to a rapid transfer of liquids, promoting fast compositional changes by the action of the hydrodynamic mechanism (HDM). HDM is a fast mass transfer process that occurs when porous structures are immersed in a liquid phase and is responsible for the VI processes of porous products when a vacuum step is imposed in a solid-liquid system followed by a successive step at atmospheric pressure. Numerous studies have been published the last 10 years, demonstrating the effectiveness of the VI (or PVOD) technique on several tissues. An interesting and very promising aspect of VI is the modification of plant tissue composition and structure for several industrial uses (e.g., minimal processing, freezing pretreatment, functional foods design, etc.) (Perez-Cabrera et al., 2011). VI has been applied as an effective technique to incorporate compounds of added value in plant tissues, as presented in Betoret et al. (2003), who proposed a method, based on VI in order to impregnate probiotics into apple tissue. Fito et al. (2001) explored the feasibility of VI treatments in the development of products fortified with calcium and iron salts from fresh fruits and vegetables. Applications have been reported for apple (Schulze et al., 2012; Wang et al., 2015; Neri et al., 2016), pepper (Derossi et al., 2010), figs (Şahin and Öztürk, 2016), mango (Lin et al., 2016), strawberry (Moreno et al., 2012, 2016a), pineapple (de Lima et al., 2016), truffles (Derossi et al., 2015), gilthead sea bream (Andrés-Bello et al., 2015), and potato (Erihemu et al., 2014).

The pulsed electric field process has been recently proposed for OD enhancement due to the electroporation (electropermeabilization) phenomena, which depends on electrically induced perforation of the cell membrane. It can appear as a formation (or growth of existing) of small plasmolemma pores. Recent publications have studied this mass transfer enhancement and a rapid osmo-dehydration process in a variety of food matrices (strawberries [Taiwo et al., 2003], apples, blueberries [Moreno et al., 2016b], kiwifruit [Dermesonlouoglou et al., 2016], and red peppers [Ade-Omowaye et al., 2003a,b]). OD is improved by ultrasound in part by the cavitation phenomena and mass transfer enhancement (Chemat et al., 2011) as demonstrated in strawberry (Garcia-Noguera et al., 2010), quince (Noshad et al., 2012), kiwifruit (Nowacka et al., 2014), and broccoli (Xin et al., 2014). The ultrasonic waves can cause a rapid series of alternative compressions and expansions in materials and the creation of microscopic channels leading to an increase of sugar loss and water diffusivity (Barman and Badwaik, 2017). Ultrasonic OD technology uses lower solution temperatures to obtain higher water loss and solute gain rates. Due to the lower temperatures during dehydration and the shorter treatment times, food quality attributes such as flavor, color, and nutritional value are better retained (Chemat et al., 2011). An important modification accomplished by γ-irradiation is the increase of the permeability of the plant tissue, by causing severe disintegration of the interior structure, leading to a significant increase of the mass transfer rates during OD process (Ahmed et al., 2016) applied in apple, potato, and carrot (Wang and Chao, 2002, 2003a,b; Nayak et al., 2006; Rastogi et al., 2006).

OD reduces water activity (a_w) of the food product, extending significantly its stability, but microorganisms growth is not hindered at the levels of a_w obtained. That is why OD treatment is often combined with other techniques such as high hydrostatic pressure (HHP) (Ciurzyńska et al., 2016). Application of high pressure (HP) has been reported to accelerate the diffusion of components into the food and alter the diffusion coefficients (Rastogi and Niranjan, 1998); Rastogi et al., 2007). This combined technique, often referred to as "high pressure impregnation (HPI)," (Pérez-Won et al., 2016) is rather novel and could lead to cell dehydration while gaining in soluble solids until equilibrium in which net transport phenomena is negligible. A lot of studies have been recently published focusing on the application of this combined technique, obtained in two successive steps, first an HHP step and then an OD process, focusing on the mass transfer increase during OD by pretreatment of food tissues at high pressures (Rastogi et al., 2002). Another aim of the combined methodology is to alter functional, nutritional, and sensory attributes of the final product, leading to a more stable food with new rheological and textural properties. Applications include strawberries (Taiwo et al., 2003; Núñez-Mancilla et al., 2014), banana slices (Verma et al., 2014), potato slices (Rastogi et al., 2000), and cherry tomatoes (Dermesonlouoglou et al., 2008a).

Finally, another significant advantage when applying an additional method is the decrease of energy consumption, such as in the case of osmotic treatment before microwave-vacuum drying (MWODS) (Erle and Schubert, 2001). Various combinations of osmotic and microwave-based dehydration techniques have been proposed in recent literature. Most commonly, conventional osmotic treatment is initially performed and then the products are finish dried using a microwave–convective or microwave–vacuum treatment, with the scope of replacing hot air drying. This final step is deemed necessary to reduce the water activity to a shelf-stable level, since foods, immediately after OD process, are considered as intermediate moisture foods, still perishable. Another possible configuration is the simultaneous application of microwave and OD process, in order to enhance the rate of moisture diffusion from the sample aiming at reducing process time. This combined methodology was initially developed as an immersion-based configuration, where apple cylinders were submerged in a continuously circulating bath of osmotic solution and microwaves were applied to the syrup and product. In a further stage, this process evolved into a spray-based set-up (Wray and Ramaswamy, 2016), often called MWODS, where the samples were continuously coated in a thin layer of osmotic solution, an arrangement that allowed for more microwave power to be immediately directed onto the sample itself, resulting in higher moisture loss and lower solid uptake. MWODS was applied in several tissues such as banana (Pereira et al., 2007), pineapple (Corrêa et al., 2011), potato cubes (Sutar et al., 2012), cranberries (Wray and Ramaswamy, 2015b), cherry tomatoes (Heredia et al., 2007), mushrooms (Torringa et al., 2001), apples (Azarpazhooh and Ramaswamy, 2012), and whole cranberries (Wray and Ramaswamy, 2015a, 2016).

The process of applying an OD step prior to freezing, often termed osmodehydrofreezing, produces fruit and vegetable products of extended shelf life with good quality characteristics, especially in texture and structure (Dermesonlouoglou et al., 2016). Since the moisture content of OD-pretreated samples is decreased, the available water for freezing is less and quality deterioration is minimized after thawing (Tregunno and Goff, 1996). It lowers energy requirements and distribution/packaging costs. James et al. (2014) reviewed different dehydration pretreatments and freezing methods and current commercial production and developments focusing on the effects of dehydrofreezing compared to conventional freezing (sour cherries [Nowicka et al., 2015], mango [Rincon and Kerr, 2010; Zhao et al., 2014], pineapple [Ramallo and Mascheroni, 2010], strawberries [Blanda et al., 2009], carrots [Ando et al., 2012, 2016], and kiwi [Talens et al., 2002]).

6.5.3 INCORPORATION BY COATINGS AND EMULSIONS

In the recent years, the application of appropriate coatings on the surface of raw material prior to OD has been proposed to minimize the undesirable structural changes induced by cell alteration due to the deformation and breakage of cellular elements. It is possible to reduce solute incorporation by coating the tissue with an edible covering with high affinity for water, allowing it to flow out of the fruit, and low affinity for the solute, reducing its incorporation. Coating of the food that will subsequently be submitted to OD with an artificial barrier on the surface may efficiently hinder the penetration of solute inside the food, not affecting much the rate of water removal (Kotovicz et al., 2014). Relevant works studying this novel technique involve Khin et al. (2006, 2007), Jalaee et al. (2011), Azam et al. (2013), and Rodriguez et al. (2016). Akbarian et al. (2015) pretreated quince slices by pectin-carboxymethyl-cellulose-based edible coating and OD at the same time, by incorporating the coating solutions into the OD solutions. Application of edible coatings on osmo-dehydrated tissues (after OD process is completed) modifies end product's properties. For example, Mohebbi et al. (2011) applied OD in the solution of NaCl and methyl cellulose was used for gum coating. Silva et al. (2015) studied the influence of two different edible coatings on air drying kinetics and characteristics of osmo-dehydrated pineapple slices. Osmo-treated pineapple slices were coated with pectin, and a mix of whey protein isolate + locust bean gum + glycerol and were then hot-air-dried. Results confirmed the protection provided by edible coatings to important quality attributes of the final dried product.

The use of emulsions in the OD solution was only recently considered as a means to incorporate microcapsules into the intercellular spaces of vegetative tissue. Salazar-López et al. (2015) analyzed the simultaneous application of microencapsulation (ME) and OD in the development of an innovative process. In this context, osmo-dehydrated pineapple impregnated with microcapsules (OPIM) was obtained when fresh pineapple rings were osmo-dehydrated in emulsions prepared with water, sugar, gum Arabic, and piquin pepper oleoresin (EP) (Figure 6.3). The simultaneous application of OD with emulsions and ME made it possible to incorporate microcapsules into the intercellular spaces of pineapple tissue. When the proposed technique was applied to prepare pineapple with functional properties, well-defined intercellular spaces and microcapsules were observed. Structural changes after OD and freeze drying were observed by scanning electron microscope (SEM) and shown in Figure 6.4a through c. During the OD process, the cells changed their shape and reduced their size due to liquid loss. Fresh pineapple was characterized by cells with a well-defined shape (Figure 6.4a). Conversely, the cells in the freeze-dried OPIM tissue were irregularly shaped and a moderate collapse and tissue disruption were observed with respect to fresh pineapple (Figure 6.4b). Nevertheless, when the proposed technique was applied to prepare pineapple with functional properties, well-defined microcapsules were observed. SEM images showed the presence of

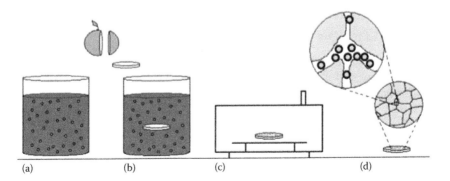

FIGURE 6.3 Scheme of the impregnation-dehydration process: (a) emulsion preparation, (b) OD in the emulsion, (c) drying, and (d) final product impregnated with microcapsules. (Adapted by Salazar-López, E.I. et al., *Food Bioprocess Technol.*, 8(8), 1699, 2015.)

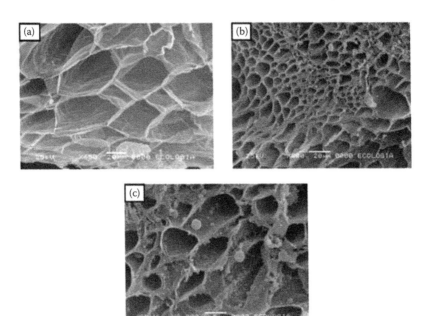

FIGURE 6.4 SEM images of pineapple products. Fresh pineapple (a), freeze-dried OPIM samples (pineapple samples were osmo-dehydrated during 120 min) (b, c). (Adapted by Salazar-López, E.I. et al., *Food Bioprocess Technol.*, 8(8), 1699, 2015.)

microcapsules of 1–5 μm into the intercellular spaces of osmo-dehydrated pine-apple tissue (Figure 6.4c).

The successful ME obtained seems really promising and, as a consequence this flexible technique can also be implemented in other fruits and vegetables. Its versatility offers the possibility to design healthy and functional products combining flavors and substances of added nutritional value in an effective way.

6.6 CONCLUSIONS

Osmotic dehydration is a method traditionally used for a variety of food matrices for the partial removal of moisture and the uptake of solute, resulting in alteration of the composition of the original product. Temperature, concentration of the solution, type of osmotic compound, variety, geometry, and other properties of the raw material are among the most important process parameters and the optimization of their values is crucial for achieving the best final products. Application of treatments such as pulsed electric field, high pressure, gamma irradiation, vacuum, microwave, centrifugal force, or ultrasound in conjunction with OD procedure can accelerate the mass transfer and drying rates by the modification of the cell membranes. Through these combined techniques, apart from the enhancement of the OD process, the nutritional, functional, and sensory properties of the final product can be significantly improved. Among the different combinations, the simultaneous application of ME and OD, although only recently proposed in recent literature, seems promising as an innovative process to incorporate microcapsules into the intercellular spaces of fruit tissues. The most important advantage of this combined process is its versatility that allows for designing tailor made, healthy products with desired flavors and functional substances. In this context, further studies concerning the influence of OD and emulsion parameters on product stability and quality are necessary to sustain the benefits of this novel methodology.

REFERENCES

Ade-Omowaye, B. I. O., K. A. Taiwo, N. M. Eshtiaghi, A. Angersbach, and D. Knorr. 2003a. Comparative evaluation of the effects of pulsed electric field and freezing on cell membrane permeabilisation and mass transfer during dehydration of red bell peppers. *Innovative Food Science and Emerging Technologies* 4 (2):177–188.

Ade-Omowaye, B. I. O., P. Talens, A. Angersbach, and D. Knorr. 2003b. Kinetics of osmotic dehydration of red bell peppers as influenced by pulsed electric field pretreatment. *Food Research International* 36 (5):475–483.

Aguilera, J. M., D. W. Stanley, and K. W. Baker. 2000. New dimensions in microstructure of food products. *Trends in Food Science and Technology* 11:3–9.

Agustinelli, S. P., D. Menchón, D. Agüería, P. Sanzano, and M. I. Yeannes. 2014. Osmotic dehydration dynamic of common carp (*Cyprinus carpio*) fillets using binary and ternary solutions. *Journal of Aquatic Food Product Technology* 23 (2):115–128.

Ahmed, I., I. M. Qazi, and S. Jamal. 2016. Developments in osmotic dehydration technique for the preservation of fruits and vegetables. *Innovative Food Science & Emerging Technologies* 34:29–43.

Akbarian, M., N. Ghasemkhani, and M. Fatemeh. 2014. Osmotic dehydration of fruits in food industrial: A review. *International Journal of Biosciences* 4 (1):42–57.

Akbarian, M., B. Ghanbarzadeh, M. Sowti, and J. Dehghannya. 2015. Effects of pectin-CMC-based coating and osmotic dehydration pretreatments on microstructure and texture of the hot-air dried quince slices. *Journal of Food Processing and Preservation* 39 (3):260–269.

Aktas, T., P. Ulger, F. Daglioglu, and F. Hasturk. 2013. Changes of nutritional and physical quality characteristics during storage of osmotic pretreated apple before hot air drying and sensory evaluation. *Journal of Food Quality* 36 (6):411–425.

Alibas, I. 2014. Mathematical modeling of microwave dried celery leaves and determination of the effective moisture diffusivities and activation energy. *Food Science and Technology (Campinas)* 34:394–401.

Almeida, J. A. R., L. P. Mussi, D. B. Oliveira, and N. P. Pereira. 2014. Effect of temperature and sucrose concentration on the retention of polyphenol compounds and antioxidant activity of osmotically dehydrated bananas. *Journal of Food Processing and Preservation* 38:1–9.

Alzamora, S. M., D. Salvatori, M. S. A. Tapia, A. Lopez-Malo, J. Welti-Chanes, and P. Fito. 2005. Novel functional foods from vegetable matrices impregnated with biologically active compounds. *Journal of Food Engineering* 67 (1–2):205–214.

Aminzadeh, R., J. Sargolzaei, and M. Abarzani. 2012. Preserving melons by osmotic dehydration in a ternary system followed by air-drying. *Food and Bioprocess Technology* 5 (4):1305–1316.

Ando, H., K. Kajiwara, S. Oshita, and T. Suzuki. 2012. The effect of osmotic dehydrofreezing on the role of the cell membrane in carrot texture softening after freeze-thawing. *Journal of Food Engineering* 108 (3):473–479.

Ando, Y., Y. Maeda, K. Mizutani, N. Wakatsuki, S. Hagiwara, and H. Nabetani. 2016. Effect of air-dehydration pretreatment before freezing on the electrical impedance characteristics and texture of carrots. *Journal of Food Engineering* 169:114–121.

Andres, A., D. Salvatori, A. Albors, A. Chiralt, and P. Fito. 2001. Vacuum impregnation viability of some fruits and vegetables. In P. Fito, A. Chiralt, J. M. Barat, W. E. L. Spiess, D. Besnilian (eds.), *Osmotic Dehydration and Vacuum Impregnation*. Lancaster, U.K.: Technomic Publishing Co.

Andrés-Bello, A., C. De Jesús, P. García-Segovia, M. J. Pagán-Moreno, and J. Martínez-Monzó. 2015. Vacuum impregnation as a tool to introduce biopreservatives in gilthead sea bream fillets (*Sparus aurata*). *LWT—Food Science and Technology* 60 (2, Part 1):758–765.

Angilelli, K. G., J. R. Orives, H. C. da Silva, R. L. Coppo, I. Moreira, and D. Borsato. 2015. Multicomponent diffusion during osmotic dehydration process in melon pieces: Influence of film coefficient. *Journal of Food Processing and Preservation* 39 (4):329–337.

Assis, F. R., R. M. S. C. Morais, and A. M. M. B. Morais. 2015. Mass transfer in osmotic dehydration of food products: Comparison between mathematical models. *Food Engineering Reviews* 8 (2):116–133.

Azam, M., M. A. Haq, and A. Hasnain. 2013. Osmotic dehydration of mango cubes: Effect of novel gluten-based coating. *Drying Technology* 31 (1):120–127.

Azarpazhooh, E. and H. S. Ramaswamy. 2012. Evaluation of factors influencing microwave osmotic dehydration of apples under continuous flow medium spray (MWODS) conditions. *Food and Bioprocess Technology* 5 (4):1265–1277.

Azuara, E., C. I. Beristain, and H. S. Garcia. 1992. Development of a mathematical model to predict kinetics of osmotic dehydration. *Journal of Food Science and Technology* 29:239–242.

Barman, N. and L. S. Badwaik. 2017. Effect of ultrasound and centrifugal force on carambola (*Averrhoa carambola* L.) slices during osmotic dehydration. *Ultrasonics Sonochemistry* 34:37–44.

Beaudry, C., G. S. V. Raghvan, and T. J. Rennie. 2003. Micro wave finish drying of osmotically dehydrated cranberries. *Drying Technology* 21:1797–1810.

Bekele, Y. and H. Ramaswamy 2010. Going beyond conventional osmotic dehydration for quality advantage and energy savings. *Ethiopian Journal of Applied Sciences and Technology* 1 (1):1–15.

Bellary, A. N., H. B. Sowbhagya, and N. K. Rastogi. 2011. Osmotic dehydration assisted impregnation of curcuminoids in coconut slices. *Journal of Food Engineering* 105 (3):453–459.

Betoret, N., L. Puente, M. J. Díaz, M. J. Pagán, M. J. García, M. L. Gras, J. Martínez-Monzó, and P. Fito. 2003. Development of probiotic-enriched dried fruits by vacuum impregnation. *Journal of Food Engineering* 56 (2–3):273–277.

Blanda, G., L. Cerretani, A. Cardinali, S. Barbieri, A. Bendini, and G. Lercker. 2009. Osmotic dehydrofreezing of strawberries: Polyphenolic content, volatile profile and consumer acceptance. *LWT—Food Science and Technology* 42 (1):30–36.

Brochier, B., L. D. F. Marczak, and C. P. Z. Noreña. 2014. Osmotic dehydration of yacon using glycerol and sorbitol as solutes: Water effective diffusivity evaluation. *Food and Bioprocess Technology* 8 (3):623–636.

Brochier, B., L. D. F. Marczak, and C. P. Z. Noreña. 2015. Use of different kinds of solutes alternative to sucrose in osmotic dehydration of yacon. *Brazilian Archives of Biology and Technology* 58 (1):34–40.

Chemat, F., H. Zille, and M. Kamran Khan. 2011. Applications of ultrasound in food technology: Processing, preservation and extraction. *Ultrasonics Sonochemistry* 18 (4):813–835.

Chiralt, A., N. Martínez-Navarrete, J. Martínez-Monzó, P. Talens, G. Moraga, A. Ayala, and P. Fito. 2001. Changes in mechanical properties throughout osmotic processes: Cryoprotectant effect. *Journal of Food Engineering* 49 (2–3):129–135.

Chronakis, I. S. 1998. On the molecular characteristics, compositional properties, and structural functional mechanisms of maltodextrins: A review. *Critical Reviews in Food Science and Nutrition* 38 (7):599–637.

Ciurzyńska, A., H. Kowalska, K. Czajkowska, and A. Lenart. 2016. Osmotic dehydration in production of sustainable and healthy food. *Trends in Food Science and Technology* 50:186–192.

Codex Alimentarius. 2012. Codex general standard for food additives (GSFA, Codex STAN 192-1995), Available in http://www.fao.org/fao-who-codexalimentarius/standards/gsfa/en/. Accessed September 15, 2016.

Corrêa, J. L. G., S. R. S. Dev, Y. Gariepy, and G. S. V. Raghavan. 2011. Drying of pineapple by microwave-vacuum with osmotic pretreatment. *Drying Technology* 29 (13):1556–1561.

Corrêa, J. L. G., D. B. Ernesto, and K. S. de Mendonça. 2016. Pulsed vacuum osmotic dehydration of tomatoes: Sodium incorporation reduction and kinetics modeling. *LWT—Food Science and Technology* 71:17–24.

Crank, J. 1975. *The Mathematics of Diffusion*. Oxford, U.K.: Clarendon Press.

de Lima, M. M., G. Tribuzi, J. A. R. de Souza, I. G. de Souza, J. B. Laurindo, and B. A. Mattar Carciofi. 2016. Vacuum impregnation and drying of calcium-fortified pineapple snacks. *LWT—Food Science and Technology* 72:501–509.

de Mendonça, K. S., J. L. Gomes Corrêa, J. R. de Jesus Junqueira, M. C . de Angelis Pereira, and M. Barbosa Vilela. 2016. Optimization of osmotic dehydration of yacon slices. *Drying Technology* 34 (4):386–394.

Dermesonlouoglou, E., M. Giannakourou, and P. Taoukis. 2007a. Kinetic modelling of the quality degradation of frozen watermelon tissue: Effect of the osmotic dehydration as a pre-treatment. *International Journal of Food Science and Technology* 42 (7):790–798.

Dermesonlouoglou, E. K., S. Boulekou, and P. S. Taoukis. 2008a. Mass transfer kinetics during osmotic dehydration of cherry tomatoes pre-treated by high hydrostatic pressure. *Acta Horticulturae* 802 (14):127–133.

Dermesonlouoglou, E. K., M. Giannakourou, and P. S. Taoukis. 2016. Kinetic study of the effect of the osmotic dehydration pre-treatment with alternative osmotic solutes to the shelf life of frozen strawberry. *Food and Bioproducts Processing* 99:212–221.

Dermesonlouoglou, E. K., M. C. Giannakourou, and P. Taoukis. 2007b. Stability of dehydrofrozen tomatoes pretreated with alternative osmotic solutes. *Journal of Food Engineering* 78 (1):272–280.

Dermesonlouoglou, E. K., M. C. Giannakourou, and P. S. Taoukis. 2007c. Kinetic modelling of the degradation of quality of osmo-dehydrofrozen tomatoes during storage. *Food Chemistry* 103 (3):985–993.

Dermesonlouoglou, E. K., S. Pourgouri, and P. S. Taoukis. 2008b. Kinetic study of the effect of the osmotic dehydration pre-treatment to the shelf life of frozen cucumber. *Innovative Food Science & Emerging Technologies* 9 (4):542–549.

Derossi, A., T. De Pilli, and C. Severini. 2010. Reduction in the pH of vegetables by vacuum impregnation: A study on pepper. *Journal of Food Engineering* 99 (1):9–15.

Derossi, A., A. Iliceto, T. De Pilli, and C. Severini. 2015. Application of vacuum impregnation with anti-freezing proteins to improve the quality of truffles. *Journal of Food Science and Technology* 52 (11):7200–7208.

Erihemu, K. Hironaka, Y. O., and H. Koaze. 2014. Iron enrichment of whole potato tuber by vacuum impregnation. *LWT—Food Science and Technology* 59 (1):504–509.

Erle, U., and H. Schubert. 2001. Combined osmotic and microwave-vacuum dehydration of apples and strawberries. In P. Fito, A. Chiralt, J. M. Barat, W. E. L. Spiess, D. Besnilian (eds.), *Osmotic Dehydration and Vacuum Impregnation*. Lancaster, U.K.: Technomic Publishing Co.

Fito, P., A. Andrés, A. Chiralt, and P. Pardo. 1996. Coupling of hydrodynamic mechanism and deformation relaxation phenomena during vacuum treatments in solid porous food–liquid systems. *Journal of Food Engineering* 27:229–240.

Fito, P., A. Chiralt, J. M. Barat, A. Andrés, J. Martínez-Monzó, and N. Martínez-Navarrete. 2001a. Vacuum impregnation for development of new dehydrated products. *Journal of Food Engineering* 49 (4):297–302.

Fito, P., A. Chiralt, N. Betoret, M. Gras, M. Cháfer, J. Martínez-Monzó, A. Andrés, and D. Vidal. 2001b. Vacuum impregnation and osmotic dehydration in matrix engineering: Application in functional fresh food development. *Journal of Food Engineering* 49 (2–3):175–183.

Garcia-Noguera, J., F. I. P. Oliveira, M. I. Gallão, C. L. Weller, S. Rodrigues, and F. A. N. Fernandes. 2010. Ultrasound-assisted osmotic dehydration of strawberries: Effect of pretreatment time and ultrasonic frequency. *Drying Technology* 28 (2):294–303.

Gaspartero, O. C. P., P. D. L. Silva, and E. Gertrudes. 2003. Study of conservation of banana by osmotic dehydration and drying in a conventional dryer. *Journal of Chemical Engineering* 3:25–29.

Giannakourou, M. C. and P. S. Taoukis. 2003. Stability of dehydrofrozen green peas pretreated with nonconventional osmotic agents. *Journal of Food Science* 68 (6):2002–2010.

Giraldo, G., P. Talens, P. Fito, and A. Chiralt. 2003. Influence of sucrose solution concentration on kinetics and yield during osmotic dehydration of mango. *Journal of Food Engineering* 58:33–43.

Guiamba, I., L. Ahrné, M. A. M. Khan, and U. Svanberg. 2016. Retention of -carotene and vitamin C in dried mango osmotically pretreated with osmotic solutions containing calcium or ascorbic acid. *Food and Bioproducts Processing* 98:320–326.

Gupta, S., A. K. Jaiswal, and A.-G. Nissreen. 2012. Statistical optimization of blanching time and temperature of Irish York Cabbage using desirability function. *Journal of Food Processing and Preservation* 36 (5):412–422.

Heredia, A. and A. Andrés. 2008. Mathematical equations to predict mass fluxes and compositional changes during osmotic dehydration of cherry tomato halves. *Drying Technology* 26 (7):873–883.

Heredia, A., C. Barrera, and A. Andrés. 2007. Drying of cherry tomato by a combination of different dehydration techniques. Comparison of kinetics and other related properties. *Journal of Food Engineering* 80 (1):111–118.

Heredia, A., I. Peinado, C. Barrera, and A. Grau. 2009. Influence of process variables on colour changes, carotenoids retention and cellular tissue alteration of cherry tomato during osmotic dehydration. *Journal of Food Composition and Analysis* 22:285–294.

Heredia, A., I. Peinado, E. Rosa, and A. Andres. 2010. Effect of osmotic pre-treatment and microwave heating on lycopene degradation and isomerization in cherry tomato. *Food Chemistry* 123:92–98.

Herman-Lara, E., C. E. Martinez-Sanchez, H. Pacheco-Angulo, R. Carmona-Garcia, H. Ruiz-Espinosa, and H. Ruiz-Lopez. 2013. Mass transfer modeling of equilibrium and dynamic periods during osmotic dehydration of radish in NaCl solutions. *Food and Bioproducts Processing* 91:216–224.

Jacob, J. K. and P. Gopinadhan. 2012. Infusion of fruits with nutraceuticals and health regulatory components for enhanced functionality. *Food Research International* 45:93–102.

Jain, S. K., R. C. Verma, L. K. Murdia, H. K. Jain and G. P. Sharma. 2011. Optimization of process parameters for osmotic dehydration of papaya cubes. *Journal of Food Science and Technology* 48 (2):211–217.

Jalaee, F., A. Fazeli, H. Fatemian, and H. Tavakolipour. 2011. Mass transfer coefficient and the characteristics of coated apples in osmotic dehydrating. *Food and Bioproducts Processing* 89 (4):367–374.

James, C., G. Purnell, and S. J. James. 2014. A critical review of dehydrofreezing of fruits and vegetables. *Food and Bioprocess Technology* 7 (5):1219–1234.

Johnson, C. R., D. Thavarajah, G. F. Combs Jr., and P. Thavarajah. 2013. Lentil (*Lens culinaris* L.): A prebiotic-rich whole food legume. *Food Research International* 51 (1):107–113.

Kaymak-Ertekin, F. and M. Sultanoglu. 2000. Modelling of mass transfer during osmotic dehydration of apples. *Journal of Food Engineering* 46:243–250.

Khan, M. R. 2012. Osmotic dehydration technique for fruits preservation—A review. *Pakistan Journal of Food Sciences* 22 (2):71–85.

Khin, M. M., W. Zhou, and C. O. Perera. 2006. A study of the mass transfer in osmotic dehydration of coated potato cubes. *Journal of Food Engineering* 77 (1):84–95.

Khin, M. M., W. Zhou, and S. Y. Yeo. 2007. Mass transfer in the osmotic dehydration of coated apple cubes by using maltodextrin as the coating material and their textural properties. *Journal of Food Engineering* 81 (3):514–522.

Kotovicz, V., L. S. N. Ellendersen, M. M. Clarindo, and M. L. Masson. 2014. Influence of process conditions on the kinetics of the osmotic dehydration of yacon (*Polymnia sonchifolia*) in fructose solution. *Journal of Food Processing and Preservation* 38 (3):1385–1397.

Lazou, A. E., M. G. Giannakourou, T. I. Lafka, and E. S. Lazos. 2016. Kinetic study of the osmotic pretreatment and quality evaluation of traditional greek candied pumpkin. *Journal of Food and Nutrition Sciences* 2016 (1):28–36.

Lewicki, P. P. and A. Lenart. 2006. Osmotic dehydration of fruits and vegetables. In A. S. Mujumdar (ed.), *Handbook of Industrial Drying*, 3rd edn., pp. 665–688. Boca Raton, FL: Taylor & Francis Group.

Lin, X., C. Luo, and Y. Chen. 2016. Effects of vacuum impregnation with sucrose solution on mango tissue. *Journal of Food Science* 81 (6):E1412–E1418.

Maestrelli, A., R. Lo Scalzo, D. Lupi, G. Bertolo, and D. Torreggiani. 2001. Partial removal of water before freezing: Cultivar and pre-treatments as quality factors of frozen muskmelon (*Cucumis melo*, cv reticulatus Naud). *Journal of Food Engineering* 49 (2–3):255–260.

Mahayothee, B., P. Udomkun, M. Nagle, M. Haewsungcharoen, S. Janjai, and J. Mueller. 2009. Effects of pretreatments on colour alterations of litchi during drying and storage. *European Food Research and Technology* 229 (2):329–337.

Mebatsion, H. K., P. Verboven, Q. T. Ho, B. E. Verlinden, and B. M. Nicolai. 2008. Modelling fruit (micro) structures, why and how? *Trends in Food Science and Technology* 19:59–66.

Mendonça, K., J. Correa, J. Junqueira, M. Angelis-Pereira, and M. Cirillo. 2017. Mass transfer kinetics of the osmotic dehydration of yacon slices with polyols. *Journal of Food Processing and Preservation* 41(1):e12983.

Mercali, G. D., L. D. Ferreira Marczak, I. C. Tessaro, and C. Pelayo Zapata Norena. 2012. Osmotic dehydration of bananas (*Musa sapientum*, shum.) in ternary aqueouw solutions of sucrose and sodium chloride. *Journal of Food Process Engineering* 35 (1):149–165.

Mohebbi, M., M. Fathi, and F. Shahidi. 2011. Genetic algorithm–artificial neural network modeling of moisture and oil content of pretreated fried mushroom. *Food and Bioprocess Technology* 4 (4):603–609.

Mokhtarian, M., M. H. Majd, F. Koushki, H. Bakhshabadi, A. D. Garmakhany, and S. Rashidzadeh. 2014. Optimisation of pumpkin mass transfer kinetic during osmotic dehydration using artificial neural network and response surface methodology modelling. *Quality Assurance and Safety Crops and Foods* 6:201–214.

Moreira, R., F. Chenlo, M. D. Torres, and G. Vázquez. 2007. Effect of stirring in the osmotic dehydration of chestnut using glycerol solutions. *LWT—Food Science and Technology* 40 (9):1507–1514.

Moreno, J., C. Espinoza, R. Simpson, G. Petzold, H. Nuñez, and M. P. Gianelli. 2016a. Application of ohmic heating/vacuum impregnation treatments and air drying to develop an apple snack enriched in folic acid. *Innovative Food Science & Emerging Technologies* 33:381–386.

Moreno, J., M. Gonzales, P. Zúñiga, G. Petzold, K. Mella, and O. Muñoz. 2016b. Ohmic heating and pulsed vacuum effect on dehydration processes and polyphenol component retention of osmodehydrated blueberries (cv. *Tifblue*). *Innovative Food Science & Emerging Technologies* 36:112–119.

Moreno, J., R. Simpson, N. Pizarro, K. Parada, N. Pinilla, J. E. Reyes, and S. Almonacid. 2012. Effect of ohmic heating and vacuum impregnation on the quality and microbial stability of osmotically dehydrated strawberries (cv. *Camarosa*). *Journal of Food Engineering* 110 (2):310–316.

Nambiar, S. S., A. Basu, N. P. Shetty, N. K. Rastogi, and S. G. Prapulla. 2016. Infusion of fructooligosaccharide in Indian gooseberry (*Emblica officinalis*) fruit using osmotic treatment and its effect on the antioxidant activity of the fruit. *Journal of Food Engineering* 190:139–146.

Nayak, C. A., K. Suguna, and N. K. Rastogi. 2006. Combined effect of gamma-irradiation and osmotic treatment on mass transfer during rehydration of carrots. *Journal of Food Engineering* 74 (1):134–142.

Neri, L., L. Di Biase, G. Sacchetti, C. Di Mattia, V. Santarelli, D. Mastrocola, and P. Pittia. 2016. Use of vacuum impregnation for the production of high quality fresh-like apple products. *Journal of Food Engineering* 179:98–108.

Noshad, M., M. Mohebbi, F. Shahidi, and S. A. Mortazavi. 2012. Effect of osmosis and ultrasound pretreatment on the moisture adsorption isotherms of quince. *Food and Bioproducts Processing* 90 (2):266–274.

Nowacka, M., U. Tylewicz, L. Laghi, M. Dalla Rosa, and D. Witrowa-Rajchert. 2014. Effect of ultrasound treatment on the water state in kiwifruit during osmotic dehydration. *Food Chemistry* 144:18–25.

Nowicka, P., A. Wojdyło, K. Lech, and A. Figiel. 2015. Influence of osmodehydration pretreatment and combined drying method on the bioactive potential of sour cherry fruits. *Food and Bioprocess Technology* 8 (4):824–836.

Núñez-Mancilla, Y., A. Vega-Gálvez, M. Pérez-Won, L. Zura, P. García-Segovia, and K. Di Scala. 2014. Effect of osmotic dehydration under high hydrostatic pressure on microstructure, functional properties and bioactive compounds of strawberry (*Fragaria vesca*). *Food and Bioprocess Technology* 7 (2):516–524.

Nurhadi, B., Y. H. Roos, and V. Maidannyk. 2016. Physical properties of maltodextrin DE 10: Water sorption, water plasticization and enthalpy relaxation. *Journal of Food Engineering* 174:68–74.

Ochoa-Martínez, C. I. and A. A. Ayala-Aponte. 2007. Prediction of mass transfer kinetics during osmotic dehydration of apples using neural networks. *LWT—Food Science and Technology* 40 (4):638–645.

Oladejo, D., B. I. O. Ade-Omowaye, and A. Abioye. 2013. Experimental study on kinetics, modeling and optimisation of osmotic dehydration of mango (*Mangifera indica* L.). *The International Journal of Engineering and Science* 2 (4):1–8.

Osorio, C., M. S. Franco, M. P. Castaño, M. L. González-Miret, F. J. Heredia, and A. L. Morales. 2007. Colour and flavour changes during osmotic dehydration of fruits. *Innovative Food Science & Emerging Technologies* 8 (3):353–359.

Ozdemir, M., B. F. Ozen, L. L. Dock, and J. D. Floros. 2008. Optimization of osmotic dehydration of diced green peppers by response surface methodology. *LWT—Food Science and Technology* 41 (10):2044–2050.

Panagiotou, N. M., V. T. Karathanos, and Z. B. Maroulis. 1998. Mass transfer modelling of the osmotic dehydration of some fruits. *International Journal of Food Science and Technology* 33 (3):267–284.

Patel, S. and A. Goyal. 2011. Functional oligosaccharides: Production, properties and applications. *World Journal of Microbiology and Biotechnology* 27 (5):1119–1128.

Paul, P. K., S. K. Ghosh, D. K. Singh, and N. Bhowmick. 2014. Quality of osmotically pre-treated and vacuum dried pineapple cubes on storage as influenced by type of solutes and packaging materials. *Journal of Food Science and Technology* 51 (8):1561–1567.

Pereira, N. R., A. Marsaioli Jr., and L. M. Ahrné. 2007. Effect of microwave power, air velocity and temperature on the final drying of osmotically dehydrated bananas. *Journal of Food Engineering* 81 (1):79–87.

Perez-Cabrera, L., M. Chafer, A. Chiralt, and C. Gonzalez-Martinez. 2011. Effectiveness of antibrowning agents applied by vacuum impregnation on minimally processed pear. *LWT—Food Science and Technology* 44 (10):2273–2280.

Pérez-Won, M., R. Lemus-Mondaca, G. Tabilo-Munizaga, S. Pizarro, S. Noma, N. Igura, and M. Shimoda. 2016. Modelling of red abalone (*Haliotis rufescens*) slices drying process: Effect of osmotic dehydration under high pressure as a pretreatment. *Innovative Food Science & Emerging Technologies* 34:127–134.

Phisut, N. 2012. Factors affecting mass transfer during osmotic dehydration of fruits. *International Food Research Journal* 19 (1):7–18.

Porciuncula, B. D. A., M. F. Zotarelli, B. A. M. Carciofi, and J. B. Laurindo. 2013. Determining the effective diffusion coefficient of water in banana (*Prata variety*) during osmotic dehydration and its use in predictive models. *Journal of Food Engineering* 119:490–496.

Quiles, A., I. Hernando, I. Pérez-Munuera, V. Larrea, E. Llorca, and M. A. Lluch. 2005. Polyphenoloxidase (PPO) activity and osmotic dehydration in Granny Smith apple. *Journal of Science and Food Agricultural* 85:1017–1020.

Rahman, M. S., S. S. Sablani, and M. A. Al-Ibrahim. 2001. Osmotic dehydration of potato: Equilibrium kinetics. *Drying Technology* 19 (6):1163–1176.

Ramallo, L. A. and R. H. Mascheroni. 2010. Dehydrofreezing of pineapple. *Journal of Food Engineering* 99 (3):269–275.

Rastogi, N. K., A. Angersbach, and D. Knorr. 2000. Synergistic effect of high hydrostatic pressure pretreatment and osmotic stress on mass transfer during osmotic dehydration. *Journal of Food Engineering* 45 (1):25–31.

Rastogi, N. K. and K. Niranjan. 1998. Enhanced mass transfer during osmotic dehydration of high pressure treated pineapple. *Journal of Food Science* 63 (3):508–511.

Rastogi, N. K. and K. S. M. S. Raghavarao. 1994. Effect of temperature and concentration on osmotic dehydration of coconut. *Lebensmittel Wissenschaft und Technologie* 27:564–567.

Rastogi, N. K., K. S. M. S. Raghavarao, V. M. Balasubramaniam, K. Niranjan, and D. Knorr. 2007. Opportunities and challenges in high pressure processing of foods. *Critical Reviews in Food Science and Nutrition* 47 (1):69–112.

Rastogi, N. K., K. S. M. S. Raghavarao, and K. Niranjan. 1997. Mass transfer during osmotic dehydration of banana: Fickian diffusion in cylindrical configuration. *Journal of Food Engineering* 31:423–432.

Rastogi, N. K., K. S. M. S. Raghavarao, K. Niranjan, and D. Knorr. 2002. Recent developments in osmotic dehydration: Methods to enhance mass transfer. *Trends in Food Science and Technology* 13 (2):48–59.

Rastogi, N. K., K. Suguna, C. A. Nayak, and K. S. M. S. Raghavarao. 2006. Combined effect of γ-irradiation and osmotic pretreatment on mass transfer during dehydration. *Journal of Food Engineering* 77 (4):1059–1063.

Rincon, A. and W. L. Kerr. 2010. Influence of osmotic dehydration, ripeness and frozen storage on physicochemical properties of mango. *Journal of Food Processing and Preservation* 34 (5):887–903.

Rizzolo, A., F. Gerli, C. Prinzivalli, S. Buratti, and D. Torreggiani. 2007. Headspace volatile compounds during osmotic dehydration of strawberries (cv *Camarosa*): Influence of osmotic solution composition and processing time. *LWT—Food Science and Technology* 40 (3):529–535.

Rodrigues, S. and F. A. N. Fernandes. 2007. Dehydration of melons in a ternary system followed by air-drying. *Journal of Food Engineering* 80 (2):678–687.

Rodriguez, A., M. A. García, and L. A. Campañone. 2016. Experimental study of the application of edible coatings in pumpkin sticks submitted to osmotic dehydration. *Drying Technology* 34 (6):635–644.

Rodriguez, M. M., J. R. Arballo, L. A. Campanone, M. B. Cocconi, A. M. Pagano, and R. H. Mascheroni. 2012. Osmotic dehydration of nectarines: Influence of the operating conditions and determination of the effective diffusion coefficients. *Food Bioprocess Technology* 6:2708–2720.

Rózek, A., I. Achaerandio, C. Guell, F. Lopez, and M. Ferrando. 2010a. Use of commercial grape phenolic extract to supplement solid foodstuff. *LWT—Food Science and Technology* 43:623–631.

Rózek, A., V. J. Garcia-Perez, F. Lopez, C. Guell, and M. Ferrando. 2010b. Infusion of grape phenolics into fruits and vegetables by osmotic treatment: Phenolic stability during air drying. *Journal of Food Engineering* 99:142–150.

Rubio-Arraez, S., J. V. Capella, M. D. Ortolá, and M. L. Castelló. 2015. Modelling osmotic dehydration of lemon slices using new sweeteners. *International Journal of Food Science and Technology* 50 (9):2046–2051.

Ruiz-Lopez, I. I., I. R. Huerta-Mora, M. A. Vivar-Vera, C. E. Martınez Sanchez, and E. Herman-Lara. 2010. Effect of osmotic dehydration on air-drying characteristics of chayote. *Drying Technology* 28:1201–1212.

Şahin, U. and H. K. Öztürk. 2016. Effects of pulsed vacuum osmotic dehydration (PVOD) on drying kinetics of figs (*Ficus carica* L.). *Innovative Food Science & Emerging Technologies* 36:104–111.

Salazar-López, E. I., M. Jiménez, R. Salazar, and E. Azuara. 2015. Incorporation of microcapsules in pineapple intercellular tissue using osmotic dehydration and microencapsulation method. *Food and Bioprocess Technology* 8 (8):1699–1706.

Schulze, B., S. Peth, E. M. Hubbermann, and K. Schwarz. 2012. The influence of vacuum impregnation on the fortification of apple parenchyma with quercetin derivatives in combination with pore structures x-ray analysis. *Journal of Food Engineering* 109 (3):380–387.

Sereno, A. M., R. Moreira, and E. Martinez. 2001. Mass transfer coefficients during osmotic dehydration of apple in single and combined aqueous solutions of sugar and salt. *Journal of Food Engineering* 47:43–49.

Shi, J. and J. S. Xue. 2009. Application and development of osmotic dehydration technology in food processing. In C. Ratti (ed.), *Advances in Food Dehydration*. Boca Raton, FL: CRC Press.

Silva, K. S., C. C. Garcia, L. R. Amado, and M. A. Mauro. 2015. Effects of edible coatings on convective drying and characteristics of the dried pineapple. *Food and Bioprocess Technology* 8 (7):1465–1475.

Simpson, R., C. Ramírez, V. Birchmeier, A. Almonacid, J. Moreno, H. Nuñez, and A. Jaques. 2015. Diffusion mechanisms during the osmotic dehydration of Granny Smith apples subjected to a moderate electric field. *Journal of Food Engineering* 166:204–211.

Siramard, S. and S. Charoenrein. 2014. Effect of ripening stage and infusion with calcium lactate and sucrose on the quality and microstructure of frozen mango. *International Journal of Food Science and Technology* 49 (9):2136–2141.

Souraki, B. A., A. Ghaffari, and Y. Bayat. 2012. Mathematical modeling of moisture and solute diffusion in the cylindrical green bean during osmotic dehydration in salt solution. *Food Bioproducts Processing* 90:64–67.

Souraki, B. A., M. Ghavami, and H. Tondro. 2014. Correction of moisture and sucrose effective diffusivities for shrinkage during osmotic dehydration of apple in sucrose solution. *Food Bioproducts and Processing* 92:1–8.

Souraki, B. A., H. Tondro, and M. Guavami. 2013. Modeling of mass transfer during osmotic dehydration of apple using an enhanced lumped model. *Drying Technology* 31:595–604.

Sritongtae, B., T. Mahawanich, and K. Duangmal. 2011. Drying of osmosed cantaloupe: Effect of polyols on drying and water mobility. *Drying Technology* 29 (5):527–535.

Sutar, P. P., G. V. S. Raghavan, Y. Gariepy, S. Prasad, and A. Trivedi. 2012. Optimization of osmotic dehydration of potato cubes under pulsed microwave vacuum environment in ternary solution. *Drying Technology* 30 (13):1449–1456.

Taiwo, K. A., M. N. Eshtiaghi, B. I. O. Ade-Omowaye, and D. Knorr. 2003. Osmotic dehydration of strawberry halves: Influence of osmotic agents and pretreatment methods on mass transfer and product characteristics. *International Journal of Food Science and Technology* 38 (6):693–707.

Talens, P., N. Martínez-Navarrete, P. Fito, and A. Chiralt. 2002. Changes in optical and mechanical properties during osmodehydrofreezing of kiwi fruit. *Innovative Food Science & Emerging Technologies* 3 (2):191–199.

Thalerngnawachart, S. and K. Duangmal. 2016. Influence of humectants on the drying kinetics, water mobility, and moisture sorption isotherm of osmosed air-dried papaya. *Drying Technology* 34 (5):574–583.

Therdthai, N., W. Zhou, and K. Pattanapa. 2011. Microwave vacuum drying of osmotically dehydrated mandarin cv. (*Sai-Namphaung*). *International Journal of Food Science and Technology* 46 (11):2401–2407.

Toğrul, I. T. and A. İspir. 2007. Effect on effective diffusion coefficients and investigation of shrinkage during osmotic dehydration of apricot. *Energy Conversion and Management* 48 (10):2611–2621.

Tonon, R. V., A. F. Baroni, and M. D. Hubinger. 2007. Osmotic dehydration of tomato in ternary solutions: Influence of process variables on mass transfer kinetics and an evaluation of the retention of carotenoids. *Journal of Food Engineering* 82:509–517.

Torreggiani, D. 1993. Osmotic dehydration in fruits and vegetable processing. *Food Research International* 26:59–68.

Torringa, E., E. Esveld, I. Scheewe, R. van den Berg, and P. Bartels. 2001. Osmotic dehydration as a pre-treatment before combined microwave-hot-air drying of mushrooms. *Journal of Food Engineering* 49 (2–3):185–191.

Tortoe, C. 2010. A review of osmodehydration for food industry. *African Journal of Food Science* 4 (6):303–324.

Tregunno, N. B. and H. D. Goff. 1996. Osmodehydrofreezing of apples: Structural and textural effects. *Food Research International* 29 (5–6):471–479.

Tsironi, T. N. and P. S. Taoukis. 2010. Modeling microbial spoilage and quality of gilthead seabream fillets: Combined effect of osmotic pretreatment, modified atmosphere packaging, and nisin on shelf life. *Journal of Food Science* 75 (4):M243–M251.

Tsironi, T. N. and P. S. Taoukis. 2014. Effect of processing parameters on water activity and shelf life of osmotically dehydrated fish filets. *Journal of Food Engineering* 123:188–192.

Vasconcelos, J. I. L. A., S. A. C. Andrade, M. I. S. Maciel, N. B. Guerra, and M. A. S. Vasconcelos. 2012. Osmotic dehydration of the Indian fig (*Opuntia ficus indica*) with binary and ternary solutions. *International Journal of Food Science and Technology* 47 (11):2359–2365.

Verma, D., N. Kaushik, and P. Srinivasa Rao. 2014. Application of high hydrostatic pressure as a pretreatment for osmotic dehydration of banana slices (*Musa cavendishii*) finish-dried by dehumidified air drying. *Food and Bioprocess Technology* 7 (5):1281–1297.

Vicente, S., A. B. Nieto, K. Hodara, M. A. Castro, and S. M. Alzamora. 2012. Changes in structure, rheology, and water mobility of apple tissue induced by osmotic dehydration with glucose or trehalose. *Food and Bioprocess Technology* 5 (8):3075–3089.

Vilela, A., C. Sobreira, A. S. Abraão, A. M. Lemos, and F. M. Nunes. 2016. Texture quality of candied fruits as influenced by osmotic dehydration agents. *Journal of Texture Studies* 47 (3):239–252.

Wang, J. and Y. Chao. 2002. Drying characteristics of irradiated apple slices. *Journal of Food Engineering* 52 (1):83–88.

Wang, J. and Y. Chao. 2003a. Effect of 60Co irradiation on drying characteristics of apple. *Journal of Food Engineering* 56 (4):347–351.

Wang, J. and Y. Chao. 2003b. Effect of gamma irradiation on quality of dried potato. *Radiation Physics and Chemistry* 66 (4):293–297.

Wang, Z., T. Wei, and M. Zhang. 2015. Effects of vacuum and normal pressure impregnation on water loss and solid gain of apple (*Malus pumila* Mill). *Journal of Food Processing and Preservation* 39 (6):1045–1050.

Wray, D. and H. S. Ramaswamy. 2015a. Development of a microwave–vacuum-based dehydration technique for fresh and microwave–osmotic (MWODS) pretreated whole cranberries (*Vaccinium macrocarpon*). *Drying Technology* 33 (7):796–807.

Wray, D. and H. S. Ramaswamy. 2015b. Microwave-osmotic/microwave-vacuum drying of whole cranberries: Comparison with other methods. *Journal of Food Science* 80 (12):E2792–E2802.

Wray, D. and H. S. Ramaswamy. 2016. Recycling of osmotic solutions in microwave–osmotic dehydration: Product quality and potential for creation of a novel product. *Journal of the Science of Food and Agriculture* 96 (10):3515–3523.

Xin, Y., M. Zhang, and B. Adhikari. 2014. Freezing characteristics and storage stability of broccoli (*Brassica oleracea* L. var. *botrytis* L.) under osmodehydrofreezing and ultrasound-assisted osmodehydrofreezing treatments. *Food and Bioprocess Technology* 7 (6):1736–1744.

Yadav, A. K. and S. V. Singh. 2014. Osmotic dehydration of fruits and vegetables: A review. *Journal of Food Science and Technology-Mysore* 51 (9):1654–1673.

Yan, Z., F. A. R. Oliviera, and M. J. Sousa-Gallagher. 2008. Shrinkage and porosity of banana, pineapple and mango slices during air-drying. *Journal of Food Engineering* 84 (3):430–440.

Yusof, N. L., A. G. Rasmusson, and F. Gómez Galindo. 2016. Reduction of the nitrate content in baby spinach leaves by vacuum impregnation with sucrose. *Food and Bioprocess Technology* 9 (8):1358–1366.

Zhao, J.-H., R. Hu, H.-W. Xiao, Y. Yang, F. Liu, Z.-L. Gan, and Y.-Y. Ni. 2014. Osmotic dehydration pretreatment for improving the quality attributes of frozen mango: Effects of different osmotic solutes and concentrations on the samples. *International Journal of Food Science and Technology* 49 (4):960–968.

Zhao, Y. and J. Xie. 2004. Practical applications of vacuum impregnation in fruit and vegetable processing. *Trends in Food Science and Technology* 15 (9):434–451.

7 Other Technologies for Encapsulation (Air Suspension Coating, Pan Coating, and Vacuum Drying)

Panagiota Eleni and Magdalini K. Krokida

CONTENTS

7.1   INTRODUCTION

Encapsulation is a rapidly expanding technology that offers a wide range of potential applications in several areas, including pharmaceutical and food industries (Ezhilarasi et al. 2013). It can be defined as the process of incorporating two substances, entrapping one within the other to produce particles with diameters usually between the nano and micro scales. In the literature, the encapsulated substance can be found under several characterizations, such as active agent, internal phase, core material, etc. Similarly, the substance that is encapsulating may be found as membrane, wall material, coating material, carrier, external phase, shell material, matrix, etc. (Zuidam and Shimoni 2009). Within this chapter, the encapsulated substance and the substance that is encapsulating will be referred to as core material and shell material, respectively.

Encapsulation is applied to protect valuable compounds (such as polyphenols, micronutrients, enzyme, antioxidants, nutraceuticals, etc.) and, in parallel, to enhance final products with advanced characteristics, such as controlled release of the core material at targeted sites, increased bioavailability, and extended shelf life. Moreover, the application of encapsulation could alter the macroscale characteristics of products, such as texture, color, etc., and improve sensory characteristics of additives, which may not be attractive to consumers (unusual colors and textures, bad taste, etc.) (Bhandari et al. 1999, Ezhilarasi et al. 2013, Khare and Vasisht 2014).

Encapsulates can be categorized into two main types, the capsule and the sphere (Figure 7.1). Capsule, often called reservoir type, mononuclear, single-core, monocore, core-shell type, etc., presents the core material concentrated at the center of a sphere and the shell material around it. In the sphere type, also called multinuclear, polynuclear, etc., the core material is much more dispersed over the shell material. Core material in this type could be either homogenously distributed over the encapsulate or not. In addition, in this type of encapsulates the core material could also be present at the surface or have an additional coating layer (Teunou and Poncelet 2002, Zuidam and Shimoni 2009). Encapsulates might also be characterized by their particle size, such as nanoparticles, microcapsules, microreservoir, etc. Microencapsulates are particles that have a diameter between 3 and 800 µm, while nanoencapsulates are colloidal-sized particles with diameters ranging from 10 to 1000 nm (Ezhilarasi et al. 2013).

Nowadays, numerous encapsulation techniques have been developed and continuously modified in order to produce final products with desirable properties (Meyers 2014). Methods for encapsulation could be divided into two basic categories, chemical and physical, and the latter can also be subdivided into physicochemical and physicomechanical methods. Air-suspension coating, pan coating, and vacuum drying are physicomechanical methods that have been widely used as encapsulation techniques finding applications in many industries, such as food, cosmetic, nutraceutical, etc., due to their several advantages compared to other techniques (Patel et al. 2013). The aforementioned methods have been mainly used as microencapsulation methods (Bhandari 2015).

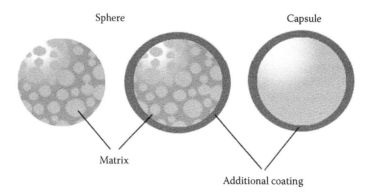

Sphere   Capsule

Matrix

Additional coating

FIGURE 7.1 Different morphologies of encapsulates.

Air-suspension coating, also referred to as *fluidized bed coating*, is typically used to coat solid particles (Risch 1995). During the process, the core material particles are circulated in a chamber with high-velocity air at a specific temperature in a batch process or a continuous setup. As they circulate, the shell material particles are atomized into the particle stream and deposited on their surface. The shell material properties are very important for efficient encapsulation during the process. It must have an appropriate viscosity to enable pumping and atomizing, to be thermally stable, and be able to form a film over a particle surface. The design of coating chamber and the control of the operating parameters are crucial since they affect the amount of coating applied (Jyothi et al. 2012, Patel et al. 2013, Risch 1995).

Pan coating has been widely used in the pharmaceutical industry for the preparation of controlled-release particles and is among the oldest industrial encapsulation techniques. The selected core material is coated in a coating pan while the shell material is applied slowly as a solution or as atomized spray. The excess amount of solvent is removed either by passing warm air in the coating pans or in a drying oven (Jyothi et al. 2012).

Vacuum drying has been traditionally used as a drying method in the food and pharmaceutical industry due to the low thermal operation conditions, in order to protect the thermosensitive compounds. However, over the years it has been also extensively used as an encapsulating technique. The core and shell materials are dissolved or dispersed in an appropriate solvent, usually water, and then dried in low temperatures (usually bellow 50°C) under vacuum. After drying, the brittle cake obtained can be broken into smaller pieces by grinding, if necessary (Zuidam and Shimoni 2009).

The optimum encapsulation technique is different for each application. The choice should be based on the nature of core and shell materials, the required size, and the desirable compositional and morphological characteristics of the final products. Table 7.1 presents an overview of air-suspension drying, pan coating, and vacuum drying as encapsulation techniques. Furthermore, the operating conditions in each

---

TABLE 7.1
Overview of Vacuum Drying, Air Suspension Drying, and Pan Coating as Encapsulation Methods

Technology	Applicable Core Material	Morphology	Approximate Particle Size (μm)[a]
Air suspension coating	Solids	Capsule	5–5,000
Pan coating	Solids	Capsule	600–5,000
Vacuum drying	Solids and liquids	Sphere (matrix)	20–50,000

Sources: Zuidam, N.J. et al., Overview of microencapsulates for use in food products or processes and methods to make them, in *Encapsulation Technologies for Active Food Ingredients and Food Processing*, eds. J. N. Zuidam and A. V. Nedovic, pp. 2, 5–6, 9–10, 22, Springer Science & Business Media., New York, 2009; Sanjoy Kumar Das, A.N. et al., *Int. J. Pharm. Sci. Technol.*, 6 (2), 1, 2011.

[a] The size of 5000 μm does not consist particle limitation as the methods are also applicable as macro-coating methods (Banker 1986, Sanjoy Kumar Das et al. 2011).

encapsulation technique affect the final encapsulates; thus, those factors need to be investigated and optimized (Ezhilarasi et al. 2013). A more analytical approach is presented in the following paragraphs.

7.2   AIR SUSPENSION COATING (FLUIDIZED BED)

Air suspension coating technology was originally developed for pharmaceutical applications, such as microencapsulation technique, to achieve controlled release of the encapsulated substances and desirable properties in the final products. In the last years, this technique has also been increasingly used in the food industry for a number of key applications (DeZarn 1995, Nedovic et al. 2011).

Although air suspension coating technology is generally expensive, pharmaceutical and cosmetic industries were able to counterbalance this cost by the high prices of their final products (Teunou and Poncelet 2002). On the other hand, processed food products are comparatively cheaper than cosmetics and pharmaceutical products, as the food industry's requirements of reduced cost and energy consumption are more demanding. Toward this direction, over the last years, several types of air suspension coaters have been developed with improved encapsulation yields, cost, and efficient energy consumption. In general, encapsulation at least doubles the cost of final products, but the higher production volumes, along with the newly developed technologies, have made many encapsulated products in the food industry standard items that are available at cost-effective prices (Dewettinck and Huyghebaert 1999, DeZarn 1995). As for the small increment in the final cost, it is acceptable by the consumers, since the new products present several improved properties such as extended shelf life, new functionalities, and more consistent product quality (Dewettinck and Huyghebaert 1999).

Thus, over the years, air suspension as an encapsulation technique has been widely used in several areas in the food industry. In the nutritional supplement market, it has been applied for supplying encapsulated versions of several bioactive compounds such as vitamins, minerals, etc. (DeZarn 1995). Other sectors in the food industry that apply air suspension encapsulation technique are the bakery and meat industry, which use it mostly to increase consumers' acceptability of the products with unacceptable organoleptic characteristics (Desai and Jin Park 2005, DeZarn 1995).

Air suspension techniques can generally be attributed to professor Dale E. Wurster's inventions (Arshady 1993, Dewettinck and Huyghebaert 1999). However, the air suspension process is known as "Wurster process" when the spray nozzle is located at the bottom of the fluidized bed of particles (Jyothi et al. 2012). The basic air suspension encapsulation process includes a dispersion of the powdered particles of core materials in a supporting air stream and the spray coating of the air-suspended particles in a batch-to-batch process or a continuous setup (Banker 1986, Dewettinck and Huyghebaert 1999, Nedovic et al. 2011, Werner et al. 2007, Zuidam and Shimoni 2009).

The coating core particles (encapsulates) are produced by spraying the shell material onto a fluidized powder bed (Hemati et al. 2003). Shell material atomized through nozzles into fluid bed cabin and covers the surface of core material particles. More specifically, the apparatus consists of different sections such as control panel, coating chamber, air distribution plate, and the nozzle for applying shell material coatings. Within the coating chamber of air suspension apparatus, particles are suspended on an upward-moving air stream. In the coating zone, the coating material is applied by spraying it on the moving core particles. The atomized shell material is sprayed at a specific temperature, and the shell material is gradually covered during each pass through the coating zone. Thus, in order to achieve the desirable shell material layer thickness, core particles should pass several times through the coating zone (Arshady 1993, Dewettinck and Huyghebaert 1999). Typically, the process duration should be 2 to 12 hours to achieve high encapsulation efficiency, in terms of the number of particles that have been coated (meaning that 98.5–99.8% of the particles are coated). Usually, in each encapsulated particle 5–50% of coating is applied, depending on the particle size of the core material and the final application of encapsulates (Dewettinck and Huyghebaert 1999, Teunou and Poncelet 2002). Upon reaching the top of the air stream, the core particles move into the outer, downward-moving column of air, which returns them to the fluidized bed. The supporting air stream helps in drying the product during encapsulation, thus having their coating nearly dried and hardened (Arshady 1993, Dewettinck and Huyghebaert 1999). Regarding the produced particle size, the air suspension encapsulation process is applicable for both microencapsulation and macroencapsulation coating processes (Sanjoy Kumar Das et al. 2011).

The design and operating parameters of the chamber affect the recirculating flow of the core particles through the coating zone, thus having an impact on the agglomeration and film formation of the particles, influencing the coating efficiency (Nedovic et al. 2011). The air suspension process has the capability of applying coatings that may be in a molten state or dissolved in an evaporable solvent, in the form of organic solutions, aqueous solutions, emulsions, dispersions, or hot melts, as long as the basic requirements are covered (Dewettinck and Huyghebaert 1999, Zuidam and Shimoni 2009). Specifically, the shell material must have an acceptable viscosity to be able to be pumped and atomized. Also, it must be thermally stable and be able to form a film over a particle surface. The temperature of the coated particles is also a crucial parameter. Coating of shell materials at low temperatures presents more pores and defects due to congealing that occurs prior to the completion of spreading. In addition, if the temperature is too close to shell materials' melting temperature, agglomeration phenomena occur. As the temperature of the coated core material particles approaches the shell materials' melting point, the fluidized bed becomes very viscous (Jozwiakowski et al. 1990, Dewettinck and Huyghebaert 1999).

The evaporation of the excess amount of water in encapsulated particles surface is controlled by several factors—mainly, the air flow; the conditions in the chamber such as humidity of the air inlet, the spray rate, the water content, and the temperature of the shell solution; the atomized air; the nozzle atomization pressure; and the properties of the core material (Coronel-Aguilera and San Martín-González 2015, Dewettinck and Huyghebaert 1999, Guignon et al. 2002, Teunou and Poncelet 2002, Zuidam and Shimoni 2009). The drying rate of the excess amount of water could be increased by efficient circulation of the coated core material particles. This is usually achieved in specific conditions of air flow rate in the center and in the inner column of periphery. These conditions also result in reduced agglomeration. The core material particles should ideally be spherical and dense, as well as have a narrow particle size distribution and good flow ability (Coronel-Aguilera and San Martín-González 2015, Zuidam and Shimoni 2009).

The nozzles typically used in fluidized bed coating are binary or pneumatic: liquid is supplied at a low pressure and is sheared into droplets by air. The droplet size and distribution are more controllable with this type of nozzle than with a hydraulic nozzle, especially at low–liquid flow rates (Dewettinck and Huyghebaert 1999, Filková and Mujumdar 1995). Moreover, the air used for atomization also contributes to evaporation of the coating solvent. This evaporation increases the droplet viscosity, thus inhibiting spreading and coalescence upon contact with the core material (Dewettinck and Huyghebaert 1998, 1999). Another factor affecting droplet viscosity is the distance that the droplets travel through the fluidization air before contacting the core particles. This problem is amplified by the use of organic solvents that have much lower heats of evaporation than water and polymers, whose viscosity is very sensitive to changes in solids concentration. Consequently, the nozzle should be positioned to minimize droplet travel distance (Dewettinck and Huyghebaert 1999).

One of the major side effects when using air suspension coating as an encapsulation method is the agglomeration phenomena. During coating, particles often collide, resulting in their agglomeration. In order to avoid agglomeration phenomena one thing that can be done is to increase the kinetic energy of the particles and/or to lower the moisture content in the cabin. However, lowering the moisture content simultaneously increases processing time and, consequently, the production cost. The selection of a proper core material is also a crucial parameter for preventing such phenomena (Dewettinck et al. 1998, 1999b, Dewettinck and Huyghebaert 1999).

7.2.1 AIR SUSPENSION COATING CONFIGURATIONS

There are three basic configurations of air suspension coating: top spray, bottom spray, and tangential spray or rotary fluidized bed coating, which basically differ in the position of the nozzle to be used for encapsulating the solid core material particles. These three fluidized bed processes present different advantages and disadvantages; however, in the food industry, the most commonly used configuration is the conventional top spray method. This fact can be attributed mainly to some basic characteristics such as its high versatility, the relatively high batch size, and the relative simplicity (Desai and Jin Park 2005, Dewettinck and Huyghebaert 1999). In addition, the continuous fluidized bed coaters have been developed to offer a continuous encapsulation process with the ability of adapting the system depending on the materials used and the final product requirements in order to be capable for use in a wide range of applications.

In *conventional top-spray method*, the particles are accelerated past the nozzle that sprays the shell material onto the randomly fluidized particles. The coated particles move through the coating zone into the expansion chamber, and then they fall back into the product container and continue cycling throughout the process (Jones 1988). One of the major disadvantages in such systems is that the distance the droplets travel before contacting the core particles cannot be adjusted, and premature droplet evaporation may occur resulting in coating imperfections (Jones 1988). However, the usage of top-spray systems has been reported for their ability to efficiently encapsulate small particles (2–100 µm) (Jones 1988, Thiel and Nguyen 1984).

The *bottom-spray method,* known as the Wurster system, can also be used for small particles in the size of 100 µm. The particles are recycled through the coating zone in seconds, as in the top-spray technique, but a major difference is that the fluidization process is much more controlled in the Wurster system (Jones 1988, Ortega-Rivas 2011). In addition, the distance of the droplets before contacting the core particles is extremely short, and the premature droplet evaporation is almost absent. The shell material droplets can spread out at the lowest viscosity, producing a very dense coat with a superior physical quality. However, this method also presents disadvantages. It has been reported that particles of core material that are coated in the bottom-spray mode do not display a uniform coating film thickness. This coating thickness variations could be explained by the differences in fluidization patterns, while this phenomenon does not occur in other configurations (Dewettinck and Huyghebaert 1999).

The newest configuration of air suspension encapsulation method is the *tangential spray, or rotary fluidized bed coating.* During fluidization, the process in the chamber, often described as spiraling helix, resulted from the combination of three forces. The particles move toward the wall of the chamber due to the centrifugal force caused by the rotating disc, air velocity through the gap provides upward acceleration, and finally the particles move inward and toward the disc once again due to the gravity. A nozzle is positioned beneath the surface of the rapidly tumbling bed, to spray the shell material tangentially to and simultaneously with the flow of particles (Dewettinck and Huyghebaert 1999, Jones 1988, Ortega-Rivas 2011, Teipel 2005).

7.2.2 Process Variables

Process variables along with some ambient parameters are very important in air suspension coating encapsulation technique. These variables refer not only to system-related parameters but also to selected properties of shell and core materials. For example, the evaporation rate of the shell material affects significantly the coating formation. However, evaporation rate is also affected by other parameters such as air velocity, temperature, and humidity.

The recommended spray rate is dependent on three factors: the evaporation capacity of air, the stickiness of the applied coating, and the speed of the particles that move through the coating zone (Jones 1988, Ortega-Rivas 2011). Another crucial parameter is the shell materials' droplet size, which should be adjusted to a size relative to the size of core material particles. Although droplet size depends on the characteristics of the shell solutions, such as viscosity and surface tension, atomizing air volume and pressure should also be adjusted since their values influence droplet size as well. Specifically, the higher these values are the smaller are the produced droplets. Also, for small particles (<250 µm), high atomization pressure values may be necessary in order for the droplets to retain their small size and avoid agglomeration phenomena. The atomization pressure influences not only the droplet size but also the droplet velocity and fluid bed temperature (Dewettinck and Huyghebaert 1998).

Product quality characteristics such as encapsulation efficiency, coating quality, and agglomeration phenomena, along with energy and cost consumption, are related to the thermodynamic operation point. Numerous process and ambient variables influence this point (Dewettinck and Huyghebaert 1999). Although process variables can easily be controlled toward the desirable results, environmental or ambient, variables are frequently not fully controlled causing a phenomenon called "weather effect." Climatic conditions can play a significant role in the air suspension coating process when using fluidizing air of low temperature. The basic problem is the variations in specific humidity of the air, which is very intense in some geographic locations during the year (Jones 1988). The simplest approach could be to dehumidify the air to achieve a desirable maximum dew point. However, this solution increases the operating cost of encapsulation as the selected dew point is lowered. This is not an issue for industries as the price of the final products is high enough to tolerate with an increment in operation cost, such as in the pharmaceutical industry. In addition, these industries have introduced even more expensive air-handling systems with dehumidifying and humidifying elements to ensure that the specific humidity of the process air is constant. It should be noted, however, that some process variables could be adjusted to set the thermodynamic operation point within limits when ambient conditions alter. This solution could be an alternative low-cost and efficient way to control the encapsulation process. For this scope, in order of importance, the most suitable process parameters that could be adjusted are spray rate, inlet air temperature, and the inlet air velocity (Dewettinck and Huyghebaert 1999).

As it is obvious, in order to minimize cost and energy consumption, a proper design and application of process controllers is essential in encapsulation processing operations that use air suspension coating (Alden et al. 1988, Haley and Mulvaney 1995). Toward this direction, many systems have been proposed. However, in the proposed systems, the thermodynamic operation point has been described to be characterized either only by temperature and not (relative) humidity or only the (relative) humidity and not the temperature (Alden et al. 1988, Dewettinck et al. 1999a, Watano et al. 1993). In the latest proposed models, the effects of major process and ambient variables on the thermodynamic operation point have been quantified by means of an appropriate thermodynamic model (Dewettinck et al. 1999a, Dewettinck and Huyghebaert 1999).

7.3 PAN COATING

Pan coating was one of the first encapsulation processes used to manufacture encapsulates on an industrial scale for commercial use in the pharmaceutical industry to produce controlled release capsules. This process is suitable for the encapsulation of solid core materials to produce particles that have size between micrometers and a few millimeters. In the pan coating process, the shell material is applied as a solution, or as an atomized spray, to the desired solid core material within the coating pan. Warm air is passed over the coated materials in the coating pan during the process to remove the excess amount of solvent from the surface of the encapsulates. Further, final solvent removal is accomplished in a drying oven, if necessary. The coating operation is repeated as many times as necessary to achieve the production of encapsulates with desired properties (Sahni and Chaudhuri 2011, Singh et al. 2010). The basic constrains in the usage of coating pan encapsulation technique refer to the core material particles that should have an adequate size (rather large particles) and high strength to avoid agglomeration phenomena and produce microencapsulates with desired properties.

The basic advantages when using pan coaters as an encapsulation method are the small process times and the large batch size (which, however, is comparable to air suspension coating method). On the other hand, the limited range of the size of pan material particles and several difficulties concerning the usage of this type of coaters, such as time-consuming cleaning process, replacement of entire pan to suit the different sizes of particles, etc., could prevent their selection as encapsulation method.

7.3.1 STANDARD PAN SYSTEM

The standard pan coating system consists of a circular metal pan mounted at a specific angle on a stand. The pan is rotated by a motor on its horizontal axis. Coating solutions (shell material) are applied to the core material by pouring or spraying the material onto the rotating bed. The use of atomizing systems to spray the liquid shell material onto the core material distributes faster the solution or suspension to the pan. Spraying can not only significantly reduce the operation time of the coating process but can also allow the continuous application of shell material in core particles (Banker 1986).

Even though pan pour coating methods have been used for many years in pharmaceutical industry, pan spray coating has predominated not only due to its higher encapsulation efficiency and reduced processing duration but also for the widest range of application and the better reproducibility of the products. In addition, encapsulates produced by pan pour methods are almost always followed by drying processes to remove the excess amount of solvents. Moreover, numerous problems concerning product quality such as agglomeration phenomena and product instability were observed when aqueous-based shell materials were applied due to unacceptably high latent moisture content.

The drying efficiency of standard pan coating systems has been significantly improved by the introduction of several modifications in the basic system such as Pellegrini pan, immersion sword, and the immersion tube systems. In Pellegrini systems, the air is uniformly distributed over the tablet bed surface through a baffled pan and a diffuser, while in some models the entire system is enclosed. Thus, further increment in their drying efficiency is achieved and they can be fully automated. The immersion sword systems introduce drying air in the pan through a perforated metal sword providing more efficient drying environment. Shell material solution is applied by an atomized spray system directed at the surface of the rotating core particles. In the immersion tube system, the heated air is delivered through a tube that is immersed in the particles bed. Spray nozzle, which is built at the top of the tube, supplies the system with the shell solution. In this type of systems, the heated air and the shell solution are applied simultaneously from the immersed tube, while the drying air flows upward and gets exhausted by a conventional duct. The major advantage of these systems besides their higher drying efficiency is the short processing time (Banker 1986).

7.3.2 Perforated Pan System

Perforated or partially perforated coating pans are among the most widely used encapsulation systems in the pharmaceutical industry. The type of perforation presents several differences in their performance when comparing for the same air volume and spray rate. Specifically, the partially perforated pans require a lower inlet temperature to maintain the same exhaust temperature, and the required energy is also lower. This lower energy consumption when using this type of pans could be attributed to the lower heat loss and/or the more efficient heat transfer. Process times are also shorter in partially perforated pans due to the higher spray rates that are usually used. However, no significant differences can be observed between the different types of perforations in terms of encapsulation efficiency (Melo Junior et al. 2010).

In both types of perforations, the pan rotates on its horizontal axis and drying air directed into the drum is passed through the particles bed and is exhausted through perforation in the drum (Banker 1986). However, several airflow configurations are applied depending on the type of perforated pan system.

Regardless of the selected method, in all the perforated pan systems, the basic operation procedure is the same. The shell solution material is applied to the surface of the rotating core material particles through spraying nozzles that are positioned inside the pan. Perforated pan coaters not only have high encapsulation capacity but are also efficient drying systems; thus, they can be completely automated for a wide range of encapsulation processes. Perforated pans are preferred over the conventional ones due to their higher encapsulation efficiency (Banker 1986).

7.3.3 Process Variables

Regardless of the selected pan coating method (conventional pan system, perforated pan systems), process variables need to be controlled to ensure consistent and desirable product quality, which is an essential characteristic of validated processes (Bauer et al. 1998, Melo Junior et al. 2010). The integrated automation in this type of equipment currently has several degrees, which allows each process parameter to be monitored and controlled, as well as continually recorded (Melo Junior et al. 2010). As it is obvious, many quality aspects of the final encapsulated products are greatly influenced by the combined effect of process parameter values.

More specifically, coating process parameters affect the spreading, penetration, and drying (i.e., evaporation of water) of the shell material on the core material surface and, subsequently, the surface roughness and the residual moisture in the coated particles (Sahni and Chaudhuri 2011).

The process parameters that affect the final products are pan-related variables, core material parameters, and spray-related parameters. Regarding pan variables, most important are those relative to pan design (pan diameter, pan depth, perforation, pan brim volume), pan rotation speed and pan load, inlet airflow and temperature, air properties, and exhaust temperature, while core material-related parameters are core material shape and size, baffle efficiency, and acceleration due to gravity. Concerning spray-related parameters most significant are the spray rate, atomizing air, solution parameters, nozzles to bed distance, and nozzle type and size (Banker 1986, Jyothi et al. 2012, Sahni and Chaudhuri 2011).

The selection of optimized pan operating conditions depends on the equipment availability, the type of core and shell materials, and the desirable characteristics of the produced encapsulates.

Pan variables mostly affect the mixing of the core material mass, which is a crucial parameter, since a uniform mixing is essential to produce homogeneous encapsulates with similar shape and shell material layer thickness. Consequently, the shape of the produced encapsulates could also affect the mixing.

Rotation speed of the pan affects not only mixing, as it was expected, but also the velocity of particles and, consequently, the time that they spend on the spraying zone. This duration affects the homogenous distribution of the shell material on the surface of each particle throughout the batch. The increment of pan speed produces encapsulates with decreased shell layer thickness and increased uniformity of the coating. However, if the speed is too high, it may affect the required drying time before the same particles are reintroduced to the spray zone resulting in rough coating appearance of the final products. High rotation speed can also cause excessive attrition and breakage of the particles. On the other hand, too slow speeds may cause over wetting in the particles surface resulting in the production of agglomerations or sticking of particles to the pan surface. Usually, the applying rotated speeds in large pan coaters when using nonaqueous shell materials are approximately 10–15 rpm, while, for aqueous based shell materials is slower, 3–10 rpm.

Variables of processed air should also be controlled for each encapsulation procedure. *Pan air temperature* has an upper limit value, which is determined from the thermal deterioration characteristics of core and shell materials. Temperature affects the drying efficiency of the coating pan and, consequently, the coating uniformity and the coating rate. Very high values of air temperature could affect the encapsulation efficiency, since it increases premature drying of the core material during spraying. *Air moisture content* can alter drying conditions, and thus, the quality of encapsulation. Special care should also be taken to minimize seasonal fluctuation of humidity content in air. In addition, supply and exhaust airflow ratio should be controlled to maintain all the components (dust, solvent, etc.) within the coating system.

The most important *spray-related parameters* to be controlled are flow rate of shell solution, nozzle type, and the degree of atomization (Dubey et al. 2011). These parameters are independent of each other, but however are usually directly affected by spraying air pressure. The shell solution is dispersed into droplets through spraying air pressure, which affects the distribution of droplets' size along with their spreading and penetration on core material's surface and the quality of encapsulation. An increment in pressure could lead to the production of encapsulates with denser and thinner shell material walls with lower surface roughness. However, in very high values, great spray loss is observed, the droplets of shell material are very fine and dried before reaching the core material surface, leading to inadequate encapsulation. On the other hand, if spraying air is insufficient, the spray loss is negligible but the thickness and density of shell material present great variations.

The *flow rate of shell material* should be adjusted according to the drying and mixing efficiency of the pan coating system, along with the core and shell material characteristics. In order to achieve the desired final product quality and to avoid problems related to over or under wetting, flow rate must be adjusted within a medium range of values. In aqueous-based shell materials, the most efficient encapsulation rate is accomplished when the flow rate value is adjusted to the rate of water evaporation from the encapsulates surface (Dubey et al. 2011).

Nozzle type is also a crucial parameter that affects the final properties of the produced encapsulates. The number of nozzles depends on the pan size. Specifically, in larger pans, a larger number of nozzles is required to cover all the core particles. The spray pattern should also be arranged in a medium width. If it is too wide, the encapsulation efficiency will be small, and large quantities of shell material will be wasted, while if it is too narrow, nonuniformed coated particles will be produced. The spray width can be adjusted by changing the distance between nozzles and core materials, the air pressure and/or air direction. Atomization of shell material (degree of atomization, size, and size distribution of droplets) is also a crucial parameter, but it is not easily controllable (Dubey et al. 2011).

7.4 VACUUM DRYING

Drying is an important process to preserve food components, and it is widely used in the food industry. Vacuum drying can be considered as the best dehydration method to preserve heat-sensitive ingredients such as aromatic compounds of herbs and spices, vitamins, enzymes, and several bioactive compounds, due to the low operation temperatures. However, vacuum drying can also be applied to encapsulate several substances in glassy carbohydrate–based matrices such as starches and maltodextrins. Nowadays, encapsulation by drying technology is of a growing interest and several drying techniques (spray drying, freeze drying, vacuum drying) have been used toward this direction. Using drying as an encapsulation method, excellent properties of protection, stabilization, solubility, and controlled release of encapsulated substances could be achieved. Moreover, drying improves the effectiveness of food additives, broadens the application range of food ingredients, enhances shelf life of final products, and in some cases could lower their cost. In addition, regardless of the selected encapsulation method, the drying procedure is essential after or during the encapsulation process. Thus, there are many reasons to apply drying technologies as encapsulation methods (Ray et al. 2016).

In general, the encapsulation using vacuum drying method starts with the dissolution, dispersion, or emulsification of core material and shell material in external (mostly aqueous) phase, which are then spread in a thin layer in a tray and dried under vacuum in low temperatures (Luan et al. 2006). The acceleration of the process could be achieved by the foaming of dispersion prior to the drying. At the end of the process, the obtained brittle cake is milled to the desired particle size (Kirk et al. 2008).

Vacuum oven consists of a jacketed vessel to withstand vacuum within the oven. The oven is connected through a condenser and a liquid receiver to a vacuum pump. Operating pressure is generally 0.03–0.06 atm, in which the boiling point of water is 25–30°C. The operating conditions of vacuum drying are significantly important for the stabilization of final encapsulated products. It is very crucial to investigate the relationship between the drying parameters and the stability of encapsulates in order to achieve efficient encapsulation of core ingredient, when using drying methods for encapsulation due to the additional stress caused during the processing of the encapsulates (Ezhilarasi et al. 2013).

Vacuum drying as an encapsulating method is in general fast and cheaper compared to the other, more popular techniques. The encapsulation procedure using vacuum drying presents high energy conservation, since only a small amount of energy is needed for drying. Vacuum drying processes also tend to work faster than other drying methods (Nedovic et al. 2011). However, there are several limitations concerning the morphology in the final encapsulated products. More specifically, thermal processing, during vaccum drying, affects the microstructure of the encapsulates by influencing their size, surface, pore size, leading to low release rates of the core material, however, higher than other drying methods such as freeze drying (Fang and Bhandari 2012).

7.4.1 VACUUM DRYERS

Vacuum drying is typically a batch operation due to the dryer's sealing requirements. However, it could perform as a continuous operation with some modifications to equipment, such as surge hoppers installation, in order to create a hybrid batch-continuous operation (Fuller 1999).

Vacuum dryer basically consists of the vacuum drying system, heating and circulating components, and vacuum/solvent recovery components. The dryer consists of an enclosed, thermal-jacketed vessel known as the drying chamber. The vessel is usually constructed from stainless steel with capacity typically from 1 to 15 m^3 (Fuller 1999).

Depending on the operation requirements, several heat sources can be used. For efficient vacuum drying of the thermal jacket, heating media and circulating components should be correctly specified to ensure that the media flow rates and pressure are compatible with the jacket. In addition, when using steam or pressurized hot water or oil in the jacket more analytical design is required.

Concerning vacuum and solvent-recovery components, usually a vacuum pump is used to reduce the atmospheric pressure in the dryer. The vacuum pump is primarily responsible for the vacuum level in the dryer, as long as the vessel is properly welded and the vacuum line is effectively sealed to the vessel. The most common type is a liquid ring vacuum pump. The sealing liquid in this pump can be water, oil, or a compatible solvent. The pump typically produces a vacuum in the range of 0.1 atm. For a very high vacuum (<0.01 atm), a rotary blower and air injectors should be used to boost the liquid ring pump's capability. The vapors that exit the dryer get transferred through the vacuum line as the wet material dries, and then the vapors are captured by a condensing system located between the vacuum pump and the dryer. The system typically includes a pre-condenser and a condensate receiver tank. At first, the pre-condenser is chilled to condense the vapor, and then the condensed water or solvent is captured in the condensate receiver tank. Finally, a pump removes the condensed water or solvent from the tank.

In some cases, if the solvent is toxic or hazardous and would pose an environmental hazard if discarded without special treatment, the solvent can be reused for the sealant in the liquid ring vacuum pump. However, this requires equipping the vacuum with a condensing system that has a closed-loop sealant arrangement (Fuller 1999).

7.4.1.1 Tumble and Agitated Vacuum Dryers

The two principal types of vacuum dryers are tumble and agitated dryers. However, since most of the batch type dryers can be operated under vacuum, many conventional types of dryers could be considered as vacuum dryers. For example, a tray dryer could become a vacuum dryer when the trays are placed in a vacuum chamber and a heat transfer medium is circulated through the hollow shelves.

Regardless of the selected type of vacuum dryer, the variables that can be controlled in batch vacuum dryers are absolute pressure (vacuum), the rate of tumbling or agitation, and the temperature of the heat transfer medium (Fuller 1999, Liptak 1998, Mujumdar 2006).

The air in all vacuum dryers is practically removed by the vacuum; thus, the vapor space is filled with water vapor. The pressure of this water vapor depends on the vapor pressure of water at the operating temperature. This difference between the absolute pressure inside the vacuum dryer and the vapor pressure of water at the operating temperature is the driving force for evaporation (Fuller 1999).

Tumble vacuum dryers, also known as the rotating batch vacuum dryers, consist mainly of a rotating vessel that imparts a tumbling motion. These types of dryers, most commonly, are in double cone shape. During the encapsulation procedure, the vessel rotates about the trunnion's axis. Heat for the drying process is supplied by a thermal jacket, which encloses the entire vessel. The thermal jacket is supported by two trunnions, while a small, stationary vacuum line in one trunnion extends into the vessel. A small vacuum filter is located at the line's inlet and a disaggregation bar is usually mounted on one trunnion to remove the evaporated moisture from the drying chamber (Baker 1997). During rotation, the material cascades inside the chamber, gently tumbling and folding to bring the material into contact with the heated walls. This action makes the dryer especially suitable for handling friable and fragile materials that cannot withstand shear from agitation. However, the material may stick and create a layer in the walls of the chambers, thus reducing heat transfer. These problems could be solved by applying very slow or very high rotating speed. In addition, the disaggregation bar can be operated periodically to break up undesired agglomerates (Baker 1997, Fuller 1999, Liptak 1998).

Agitated vacuum dryer consists of a stationary cylindrical or trough-shaped vessel with a low-speed rotor mounted at the bottom of the vessel. The rotor speed could adjust through either a belt/sheave pulley system or an AC frequency inverter, which ensures constant contact of the dried material with the heated surfaces, which results in a homogeneous temperature in the product. In some models, the rotating shaft and agitators can also be heated with hot media.

The dryer's agitator configurations are based on various blender designs, including ribbon, paddle, and plow mixers. A large vacuum line is connected to the jacketed vacuum filter at the top of the dryer vessel. During operation, the material blends as the shaft-mounted agitators rotate, and the material comes into contact with the dryer's heated walls. The important features of this type of dryers include shorter drying times, no agglomeration phenomena due to the movement of the product, temperature sensor to control heat exchange, continual product contact with the vessel's jacketed surfaces, final moisture lower than 0.1%, ability to work in explosive or corrosive environments (through the use of special manufacturing alloys and appropriate controls), and the ability to reduce the temperature of the product after drying by the circulation of cooling fluid in the jacket (Fuller 1999, Guedu 1993, Liptak 1998, Mujumdar 2006).

Both tumble and agitated dryers are efficient, cost effective, and easy to use. However, discharge and cleaning could be performed easier in the tumble dryer because it has no internal shaft or agitators. The tumble dryer also has a significantly lower capital cost than the agitated dryer. On the other hand, the agitated dryer provides greater efficiency, and by providing a higher overall heat transfer coefficient and effective heat transfer surface area than in the tumble dryer allows better heat transfer to the material. The agitated unit's larger vacuum line and larger filter area (in the jacketed vacuum filter housing) also provide a vapor flow advantage for an application where the vaporization of unbound moisture is especially fast during constant rate drying. Also, in the stationary vessel in agitated dryers, hard piping and fixed connections, which are not possible in the tumble dryer, can be used in order to simplify installation and maintenance (Mujumdar 2006).

The most suitable vacuum dryer for each application depends mainly on the materials' moisture content along with the particle characteristics that affect the moisture flow and the required temperature difference between the material and the heating media. In addition, a very crucial parameter to choose the best vacuum dryer for each application is the determination of the required batch size (Fuller 1999, Mujumdar 2006, Newman 2015). In addition, moisture content of the material, the rate of vaporization, and the vaporization rate in each drying cycle along with the desirable capacity will determine the components and the solvent recovery system requirements to ensure the dryer achieves the temperature, pressure, and the solvent recovery needed. Material characteristics, characteristics of the removed liquid, and drying requirements are also important criteria to be considered (Fuller 1999, Mujumdar 2006, Newman 2015).

7.4.2 Process Variables

Process variables in vacuum drying procedure, vacuum pressure, operating temperature, and drying duration affect moisture, effective diffusivity, and, consequently, the morphology of the encapsulates. Size and morphology (mass, layer thickness, core material–shell material ratio, initial moisture content, shell material type, etc.) of the thin layer in trays are also the parameters that influence encapsulates' properties and encapsulation efficiency.

Encapsulated particles having the same final composition, with different microstructures, lead to different release profiles and encapsulation ratio (Gabas et al. 2007, Luan et al. 2006, Šumić et al. 2015). In general, the increment in porosity and diameter of microparticles can lead to a decrease in the encapsulation efficiency and to an increment in the initial release and the release rate of the core material (Chan 2011, Luan et al. 2006, Newman 2015). In addition, the efficiency of protection and controlled release of the encapsulated substance mainly depend on the composition and the structure of the shell material (Ray et al. 2016).

Concerning food stuff, the highest values of porosity are observed in lower vacuum pressure and operational temperature (Krokida et al. 1997). However, further examination should be performed when applying vacuum drying as an encapsulating method due to the complex properties of the initial emulsion/solution (Gabas et al. 2007).

REFERENCES

Alden, M., P. Torkington, and A. C. R. Strutt. 1988. Control and instrumentation of a fluidized-bed drier using the temperature-difference technique I. Development of a working model. *Powder Technology* 54 (1):15–25. doi: http://dx.doi.org/10.1016/0032-5910(88)80044-5.

Arshady, R. 1993. Microcapsules for food. *Journal of Microencapsulation* 10 (4):413–435. doi: 10.3109/02652049309015320.

Baker, C. G. J. 1997. *Industrial Drying of Foods*. London, U.K.: Chapman & Hall.

Banker, G. S. 1986. The theory and practice of industrial pharmacy. *Journal of Pharmaceutical Sciences* 59 (10):412–429, 348–350. doi: 10.1002/jps.2600591046.

Bauer, K. H., K. Lehmann, H. P. Osterwald, and G. Rothgang. 1998. *Coated Pharmaceutical Dosage Forms*, pp. 63–119. Stuttgart, Germany: Medpharm Scientific.

Bhandari, B. 2015. Handbook of industrial drying, fourth edition edited by AS. Mujumdar. *Drying Technology* 33 (1):128–129. doi: 10.1080/07373937.2014.983704.

Bhandari, B. R., B. R. D'Arcy, and I. Padukka. 1999. Encapsulation of lemon oil by paste method using beta-cyclodextrin: Encapsulation efficiency and profile of oil volatiles. *Journal of Agricultural and Food Chemistry* 47 (12):5194–5197.

Chan, E.-S. 2011. Preparation of Ca-alginate beads containing high oil content: Influence of process variables on encapsulation efficiency and bead properties. *Carbohydrate Polymers* 84 (4):1267–1275. doi: http://dx.doi.org/10.1016/j.carbpol.2011.01.015.

Coronel-Aguilera, C. P., and M. Fernanda San Martín-González. 2015. Encapsulation of spray dried β-carotene emulsion by fluidized bed coating technology. *LWT—Food Science and Technology* 62 (1 Part 1):187–193. doi: 10.1016/j.lwt.2014.12.036.

Desai, K. G. H., and H. Jin Park. 2005. Recent developments in microencapsulation of food ingredients. *Drying Technology* 23 (7):1361–1394. doi: 10.1081/DRT-200063478.

Dewettinck, K., L. Deroo, W. Messens, and A. Huyghebaert. 1998. Agglomeration tendency during top-spray fluidized bed coating with gums. *LWT—Food Science and Technology* 31 (6):576–584. doi: http://dx.doi.org/10.1006/fstl.1998.0421.

Dewettinck, K., A. De Visscher, L. Deroo, and A. Huyghebaert. 1999a. Modeling the steady-state thermodynamic operation point of top-spray fluidized bed processing. *Journal of Food Engineering* 39 (2):131–143. doi: http://dx.doi.org/10.1016/S0260-8774(98)00144-7.

Dewettinck, K., and A. Huyghebaert. 1998. Top-spray fluidized bed coating: Effect of process variables on coating efficiency. *LWT—Food Science and Technology* 31 (6):568–575. doi: http://dx.doi.org/10.1006/fstl.1998.0417.

Dewettinck, K., and A. Huyghebaert. 1999. Fluidized bed coating in food technology. *Trends in Food Science & Technology* 10 (4–5):163–168. doi: http://dx.doi.org/10.1016/S0924-2244(99)00041-2.

Dewettinck, K., W. Messens, L. Deroo, and A. Huyghebaert. 1999b. Agglomeration tendency during top-spray fluidized bed coating with gelatin and starch hydrolysate. *LWT—Food Science and Technology* 32 (2):102–106. doi: http://dx.doi.org/10.1006/fstl.1998.0507.

DeZarn, T. J. 1995. Food ingredient encapsulation. In S.J. Risch, and G.A. Reineccius (eds.), *Encapsulation and Controlled Release of Food Ingredients*, pp. 74–86. Washington, DC: American Chemical Society.

Dubey, A., R. Hsia, K. Saranteas, D. Brone, T. Misra, and F. J. Muzzio. 2011. Effect of speed, loading and spray pattern on coating variability in a pan coater. *Chemical Engineering Science* 66 (21):5107–5115. doi: http://dx.doi.org/10.1016/j.ces.2011.07.010.

Ezhilarasi, P. N., P. Karthik, N. Chhanwal, and C. Anandharamakrishnan. 2013. Nanoencapsulation techniques for food bioactive components: A review. *Food and Bioprocess Technology* 6 (3):628–647. doi: 10.1007/s11947-012-0944-0.

Fang, Z., and B. Bhandari. 2012. 4-Spray drying, freeze drying and related processes for food ingredient and nutraceutical encapsulation. In N. Garti, and D.J. McClements (eds.), *Encapsulation Technologies and Delivery Systems for Food Ingredients and Nutraceuticals*, pp. 73–109. Cambridge, U.K.:Woodhead Publishing.

Filková, I., and A. S. Mujumdar. 1995. Industrial spray drying systems. In A.S. Mujumdar (ed.)., *Handbook of Industrial Drying*, Vol. 1. New York: Marcel Dekker.

Fuller, W. O. 1999. How to choose a vacuum dryer. *Powder and Bulk Engineering* 10:55–71.

Gabas, A. L., V. R. N. Telis, P. J. A. Sobral, and J. Telis-Romero. 2007. Effect of maltodextrin and arabic gum in water vapor sorption thermodynamic properties of vacuum dried pineapple pulp powder. *Journal of Food Engineering* 82 (2):246–252. doi: http://dx.doi.org/10.1016/j.jfoodeng.2007.02.029.

Guedu, S. A. 1993. Agitated vacuum dryer from guedu. *Filtration & Separation* 30 (8):698. doi: http://dx.doi.org/10.1016/0015-1882(93)80311-J.

Guignon, B., A. Duquenoy, and E. D. Dumoulin. 2002. Fluid bed encapsulation of particles: Principles and practice. *Drying Technology* 20 (2):419–447. doi: 10.1081/drt-120002550.

Haley, T. A., and S. J. Mulvaney. 1995. Advanced process control techniques for the food industry. *Trends in Food Science & Technology* 6 (4):103–110. doi: http://dx.doi.org/10.1016/S0924-2244(00)88992-X.

Hemati, M., R. Cherif, K. Saleh, and V. Pont. 2003. Fluidized bed coating and granulation: Influence of process-related variables and physicochemical properties on the growth kinetics. *Powder Technology* 130 (1–3):18–34. doi: http://dx.doi.org/10.1016/S0032-5910(02)00221-8.

Jones, D. M. 1988. Controlling particle size and release properties. In S.J. Risch, and G.A. Reineccius (eds.), *Flavor Encapsulation*, pp. 158–176. Washington, DC: American Chemical Society.

Jozwiakowski, M. J., D. M. Jones, and R. M. Franz. 1990. Characterization of a hot-melt fluid bed coating process for fine granules. *Pharmaceutical Research: An Official Journal of the American Association of Pharmaceutical Scientists* 7 (11):1119–1126. doi: 10.1023/a:1015972007342.

Jyothi, Sri. S., A. Seethadevi, K. Suria Prabha, P. Muthuprasanna, and P. Pavitra. 2012. Microencapsulation: A review. *International Journal of Pharma and Bio Sciences* 3 (1):509–531.

Khare, A. R., and N. Vasisht. 2014. Chapter 14: Nanoencapsulation in the food industry: Technology of the future. In A.G. Gaonkar, N. Vasisht, A.R. Khare, and R. Sobel (eds.), *Microencapsulation in the Food Industry*, pp. 151–155. San Diego, CA: Academic Press.

Kirk, R. E., D. F. Othmer, and A. Seidel. 2008. *Food and Feed Technology*. Hoboken, NJ: Wiley-Interscience.

Krokida, M. K., N. P. Zogzas, and Z. B. Maroulis. 1997. Modelling shrinkage and porosity during vacuum dehydration. *International Journal of Food Science & Technology* 32 (6):445–458. doi: 10.1111/j.1365-2621.1997.tb02119.x.

Liptak, B. 1998. Optimizing dryer performance through better control. Part 1. *Chemical Engineering* 105 (2):96–104.

Luan, X., M. Skupin, J. Siepmann, and R. Bodmeier. 2006. Key parameters affecting the initial release (burst) and encapsulation efficiency of peptide-containing poly(lactide-co-glycolide) microparticles. *International Journal of Pharmaceutics* 324 (2):168–175. doi: http://dx.doi.org/10.1016/j.ijpharm.2006.06.004.

Melo Junior, V. P. de, R. Löbenberg, and N. A. Bou-Chacra. 2010. Statistical evaluation of tablet coating processes: Influence of pan design and solvent type. *Brazilian Journal of Pharmaceutical Sciences* 46:657–664.

Meyers, M. A. 2014. Chapter 34—Flavor release and application in chewing gum and confections. In A.G. Gaonkar, N. Vasisht, A.R. Khare, and R. Sobel (eds.), *Microencapsulation in the Food Industry*, pp. 443–453. San Diego, CA: Academic Press.

Mujumdar, A. S. 2006 (ed.). Principles, classification, and selection of dryers. *Handbook of Industrial Drying*, 3rd edn. Boca Raton, FL: CRC Press.

Nedovic, V., A. Kalusevic, V. Manojlovic, S. Levic, and B. Bugarski. 2011. An overview of encapsulation technologies for food applications. *Procedia Food Science* 1:1806–1815. doi: http://dx.doi.org/10.1016/j.profoo.2011.09.265.

Newman, A. 2015. *Pharmaceutical Amorphous Solid Dispersions*. Hoboken, NJ: John Wiley & Sons.

Ortega-Rivas, E. 2011. Fluidization. In *Unit Operations of Particulate Solids*, pp. 249–279. Boca Raton, FL: CRC Press.

Patel, M. K., K. N. Patel, K. M. Patel, and K. J. Patel. 2013. Microencapsulation: A vital technique in novel drug delivery system. *International Journal of Medicine and Pharmaceutical Research* 1 (1):117–126.

Ray, S., U. Raychaudhuri, and R. Chakraborty. 2016. An overview of encapsulation of active compounds used in food products by drying technology. *Food Bioscience* 13:76–83. doi: http://dx.doi.org/10.1016/j.fbio.2015.12.009.

Risch, S. J. 1995. Encapsulation: Overview of uses and techniques. In S.J. Risch and G.A. Reineccius (eds.), *Encapsulation and Controlled Release of Food Ingredients*, pp. 2–7. Washington, DC: American Chemical Society.

Sahni, E., and B. Chaudhuri. 2011. Experiments and numerical modeling to estimate the coating variability in a pan coater. *International Journal of Pharmaceutics* 418 (2):286–296. doi: http://dx.doi.org/10.1016/j.ijpharm.2011.05.041.

Sanjoy Kumar Das, A. N., S. R. N. David, R. Rajabalaya, H. K. Mukhopadhyay, T. Halder, M. Palanisamy, J. Khanam, and A. Nanda. 2011. Microencapsulation techniques and its practice. *International Journal of Pharmaceutical Science and Technology* 6 (2):1–23.

Singh, M. N., K. S. Y. Hemant, M. Ram, and H. G. Shivakumar. 2010. Microencapsulation: A promising technique for controlled drug delivery. *Research in Pharmaceutical Sciences* 5 (2):65–77.

Šumić, Z. M., A. N. Tepić, S. D. Jokić, and R. V. A. Malbaša. 2015. Optimization of frozen wild blueberry vacuum drying process. *Hemijska Industrija* 69 (1):8.

Teipel, U. 2005. *Energetic Materials: Particle Processing and Characterization*. Weinheim, Germany: Wiley-VCH Verlag GmbH & Co. KGaA.

Teunou, E., and D. Poncelet. 2002. Batch and continuous fluid bed coating—Review and state of the art. *Journal of Food Engineering* 53 (4):325–340. doi: http://dx.doi.org/10.1016/S0260-8774(01)00173-X.

Thiel, W. J., and L. T. Nguyen. 1984. Fluidized bed film coating of an ordered powder mixture to produce microencapsulated ordered units. *Journal of Pharmacy and Pharmacology* 36 (3):145–152. doi: 10.1111/j.2042-7158.1984.tb06928.x.

Watano, S., T. Harada, K. Terashita, and K. Miyanami. 1993. Development and application of moisture control system with IR moisture sensor to aqueous polymeric coating process. *Chemical and Pharmaceutical Bulletin* 41 (3):580–585.

Werner, S. R. L., J. R. Jones, A. H. J. Paterson, R. H. Archer, and D. L. Pearce. 2007. Air-suspension particle coating in the food industry: Part I—State of the art. *Powder Technology* 171 (1):25–33. doi: http://dx.doi.org/10.1016/j.powtec.2006.08.014.

Zuidam, N. J., and E. Shimoni. 2009. Overview of microencapsulates for use in food products or processes and methods to make them. In J. N. Zuidam, and A. V. Nedovic (eds.), *Encapsulation Technologies for Active Food Ingredients and Food Processing*, pp. 2, 5–6, 9–10, 22. New York: Springer Science & Business Media.

8 Industry-Relevant Encapsulation Technologies for Food and Functional Food Production

Drvenica Ivana, Đorđević Verica,
Trifković Kata, Balanč Bojana, Lević Steva,
Bugarski Branko, and Nedović Viktor

CONTENTS

8.1 INTRODUCTION

Since the discovery of the first encapsulation technique in the 1950s (Srivastava et al. 2013), it was hardly believed that encapsulation would become an essential technology in food production and processing today. Its great importance in the food sector was initially underestimated by neglecting the existence of strong interconnections of different industries on each other and the impact brought about by this technology. Along with the developed high performance of agriculture after the last world war, encapsulation has provided improvements in agricultural production methods through achievements such as slow fertilizer release and inoculation of soil by immobilized rhizobacteria (Ivanova et al. 2006; Han et al. 2009; Zhang et al. 2014; Qiao et al. 2016), as well as improvement of live-stock farming by supplementing feed with vitamins or antimicrobial agents dif-ferent from antibiotics (herbal extracts, peptides, copper, zinc) (Thacker 2013). As in a chain reaction, increments in agricultural and livestock farming pro-ductivities brought about the so-called abundance of food. This surplus in food, permitting us to live nowadays without worrying "how to survive next winter," consequently has driven people to demand from the food they eat not to be solely the source of individual nutrients and energy, but to provide safety/quality, ease/ convenience, health, and—why not?—fun/pleasure (Poncelet et al. 2011). With this increasing demand for functional foods, such as fortified foods, nutraceuti-cals, and other health-based foods, especially in developed countries such as the United States, the United Kingdom, and Germany, improving technology along with R&D investments and easy usage techniques by manufacturers are likely to drive the market for food encapsulation; thus, in 2015 the food encapsulation market, valued at US$29.36 billion, is projected to reach US$41.74 billion by 2021, at a compound annual growth rate of 6.0% from 2016 (Marketsandmarkets 2016). Besides the well-known leading companies dealing with food encapsula-tion, such as Cargill, Incorporated (United States), Friesland Campina Kievit (the Netherlands), Royal DSM (the Netherlands), Kerry Group (Ireland), Symrise AG (Germany), every day one witnesses an increase in the number of new companies offering food industry concepts, ideas, and complete solutions for encapsulated components for food products.

Despite the numerous definitions of encapsulation under different terms in each particular field of application, in the terminology favored by the food indus-try, it is a process by which one material or mixture of materials is coated with or entrapped within another material or system. The coated (entrapped) material most often represents liquid—but could also be a solid particle—and is referred to as core material or fill or internal phase, while material that forms the coating is referred to as the wall material, matrix, or coating (Wandrey et al. 2009; Fang and Bhandari 2010; Burgain et al. 2011). Produced encapsulates can be of different morphology types, that is, reservoir (capsule), matrix, and coated matrix (Zuidam and Shimoni 2010) in the nanometer (nanoencapsulation), micrometer (microen-capsulation), or millimeter scale (Lakkis 2007; Burgain et al. 2011). While the reservoir type has a so-called layer around the core material, the matrix type has the active agent dispersed over the carrier material and can also be found on the surface. A combination of these two types of encapsulates gives a third encapsu-late called coated matrix (Ray et al. 2016). Although nanoencapsulation technol-ogy in the agro-food sector is the fastest growing field in nano research (Dasgupta et al. 2015), undefined legal aspects related to the handling and consumption of nano-sized materials in food applications and scarce information of risks associ-ated to consumption of nano-sized food product limits mass production of nano-foods (Quintanilla-Carvajal et al. 2010). This chapter will provide an overview of encapsulation technologies that are already adopted for industrial-scale food production or have strong prospects for implementation, regardless of the size of resulting encapsulates.

What kinds of changes and benefits have brought about the introduction of encapsulation as a tool in food industry? Modern work and lifestyle have gradually moved us from traditional cooking/food products to industrial and pre-cooked food (fast food and catering). In order to be cost effective, production of these kinds of foods implies usage of powders and mixes (simple to handle, with long shelf life) to a greater extent than hydrated food products/ingredients (costly to transport, store, and handle) (Poncelet et al. 2011). The inevitable side effects of the dehydration process on texture, flavor, and solubility of rehydrated food could be overcome by the employment of different encapsulation methods. Contrary to traditional foods that contain many probiotics, industrial food must be "enriched" with these beneficial microbes. Unfortunately, the efficient ones are very susceptible and cannot survive under adverse conditions in food and during gastrointestinal transit. Today, encapsulation methods allow us to efficiently deliver probiotics in food, even to combine them with prebiotic to promote their growth (Khalf et al. 2010; Mitropoulou et al. 2013; Sathyabama et al. 2014). Different types of probiotic encapsulates have been successfully implemented mainly in dairy products such as yogurts, cheeses, ice cream, mayonnaise, fermented sausages, and some fermented plant-based products (Manojlović et al. 2010). Industrial processes, such as pasteurization, drying, and long-term storage can negatively affect content of inherent aroma and vitamins in food. For this particular application, as well as for controlled release of aroma (e.g., during cooking) from spices and herbs incorporated in food, encapsulation became of great importance for food process engineers. Furthermore, encapsulation technology emerged as a competitive tool in the designing of new properties and functions of commonly used ingredients in the food industry. It is well known that mixing of food-valuable hydrophobic liquids, such as unsaturated fatty acids, some vitamins, and plant extracts in hydrophilic food powder, represents a big challenge, but can easily be overcome by encapsulation technologies. In addition to protection, encapsulation allows transformation of ingredients to free-flowing, easily suspendable (e.g., hydrophobic vitamins in fruit juices), and dispersible (e.g., cocoa in cold milk) powders (Poncelet et al. 2011). Introduction of encapsulation technologies in the food sector provided a solution to a great extent for taste masking and oxidation limiting; for example, although consumption of unsaturated fatty acids may be beneficial to health, it is sometimes followed by an unpleasant or completely unacceptable taste when they are in free form and oxidized. Furthermore, incorporating flavors or coloring materials in the coating through encapsulation helps make the functional food pleasant to consume and easy to differentiate between both a drug and a food.

Encapsulation strongly contributed to innovative development in the food industry. It provided a biocatalyst immobilization system to process food in a more safe and efficient manner, such as cheese ripening, wine, beer, and other fermented beverage productions (Champagne et al. 2010; Kourkoutas et al. 2010; Verbelen et al. 2010). Many companies have launched original functional food products or solutions to active ingredients application in food and beverages by means of encapsulation technologies. Good examples are manufacturers such as Polaris®, in the field of nutritional marine oils (POLARIS 2016), having products like Megakrill® powder (microencapsulated powdered krill oil), and Omegacaps® basic C (microencapsulated/compressible powder) rich in eicosapentaenoic and docosahexaenoic acid or Mane® (MANE Flavor & Fragrance Manufacturer 2016) in the field of flavor industry for all types of applications—tea, confectionery, bakery, instant beverages, instant dessert (N-CAPTURE™). Used encapsulation technologies have provided not just health benefits or pleasure through the related products' consumption but a fun experience too, like Clearly Canadian company® did it with a soft drink called Orbitz® containing colored encapsulates of different aromas and/or vitamins.

The main criterion for application of an encapsulation technology in food industry is to be easily scalable to industrial level with an adequate production rate. Secondly, the additional cost of an encapsulation technology introduction in food/functional food production has to be economically viable. This is particularly important if some sophisticated requirements for food/functional food products cannot be achieved solely through the employment of one technology, that is, where two or more encapsulation processing steps are required to get encapsulates with superior characteristics or to produce multiencapsulated formulations (e.g., fluid-bed coating of pre-encapsulates produced by spray drying [to provide extended release]).

Also, food grade materials suitable for encapsulation purposes are limited. Newly introduced food grade encapsulation materials could strongly contribute to sophisticated application of encapsulation technologies as a tool to satisfy specific needs of the food industry today. A good example is—Q-Naturale SF launched by Ingredion Inc. (2016) in 2012, a sugar-free noncariogenic encapsulation matrix that enables manufacturers to address an increasing consumer demand for sugar-free gums and beverages. This newly introduced encapsulating material, defined as Quillaja extract (nondigestible dextrin, polyol) possesses better characteristics than gum arabic; it is optimized for spray drying encapsulations, and offers even 40% flavor (whereas most flavor spray drying is carried out at 20% flavor load), increasing manufacturing throughput twice. In addition, this ingredient raises the oil/encapsulant ratio from 1:4 to 1:1.5, meaning that, for the same amount of flavor, only 37.5% encapsulant will be used and half the powder product needs to be handled (Ingredion Incorporated 2016).

As discussed, at the beginning of this century (Gouin 2004) the applicability of an encapsulation process in food industry needs to be in equilibrium between the final price of the product and its benefits. This means that in the case of an expensive ingredient, the cost-in-use must be lower than the nonencapsulated ingredient. However, if encapsulation delivers the ingredient with a particular property, then the cost-in-use can be slightly higher than the nonencapsulated ingredient (Estevinho et al. 2013). For example, considering that microencapsulated components of one food product represent 1%–5% of the final product, the maximum cost for microencapsulation process in the food industry was estimated to be approximately £ 0.1/kg (Gouin 2004). All this said, it is not surprising that only some technologies have really found their application in the food sector, although literature data provide many more that are still at the laboratory stage of development. Thus, food industry relevant encapsulation

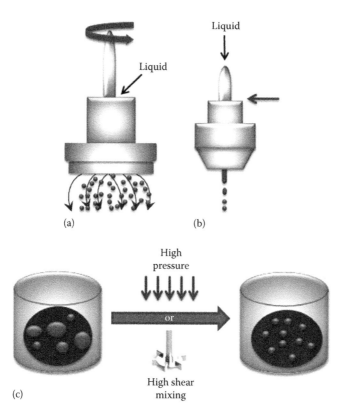

Liquid

Liquid

(a)   (b)

High
pressure

or

High shear
mixing

(c)

FIGURE 8.1 A simplified schematic representation of food industry relevant encapsulation technologies: (a) spraying, (b) dripping, and (c) emulsion-based encapsulation technologies.

technologies can be the simplest classified by the equipment needed to perform the process (Figure 8.1). These represent

1. Spraying-based technologies (spray drying, spray cooling, and spray coating)
2. Dripping-based technologies (electrostatic extrusion, coaxial airflow, vibrating jet/nozzle, jet cutting, and spinning disk atomization)
3. Emulsion-based technologies (high shear mixing, high pressure homogenization, microfluidization, ultrasonic homogenization, and membrane emulsification)

The objective of this chapter is to provide a general concept and benefits of existing food industry encapsulation technologies, their comparison and examples of the use of encapsulates for food and functional food development and improvement. Although the actual production of food encapsulates accounts for thousands of tons per year (with an annual growth of around 10%), the development of an encapsulated product still represents a challenge requiring an expert multidisciplinary approach. Comprehension of key factors (active ingredient properties, wall materials,

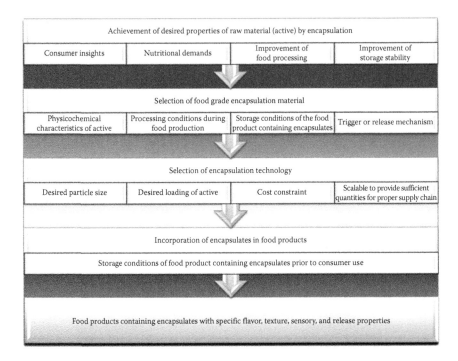

FIGURE 8.2 Schematic representation of main issues to be considered with the aim to adopt certain encapsulation technology for food and functional food production.

interactions between the core, matrix, and the environment), stability of the micro-encapsulated ingredient in storage (refrigerated or ambient), its release mechanisms (fracturation, dissolution, pressure, heat and shear), and legal and regulatory requirements for addition into foods should be considered. A series of stages are needed to produce a final encapsulated product (Figure 8.2). Thus, from an industrial point of view, current trends and future prospects for food industry encapsulation technologies will be discussed.

8.2 SPRAYING-BASED TECHNOLOGIES

Spraying-based technologies are one of the oldest approaches for encapsulates production in the food industry (Poncelet et al. 2011). They include two main methods: spray drying if the production of encapsulate occurs by encapsulating material drying, and spray cooling if encapsulate is formed by matrix cooling down. Herein, each of them will be discussed separately.

8.2.1 SPRAY DRYING

Spray drying is a well-known technology used in food industry for large-scale drying of heat-sensitive liquid, slurry, or pasty food into powders (fruit juices, instant coffee, whey, milk products), that is, food that is not always perceived as an encapsulate (Zuidam and Shimoni 2010; Jangam and Mujumdar 2013). Through achievement of fast elimination of liquid and decrease in water content and activity, spray drying allows avoiding of product quality damaging by biological/chemical factors, ensures microbiological stability of the products, and reduces storage and transportation costs (Gharsallaoui et al. 2007). Besides, it is the most commonly used encapsulation technology too, since it represents a relatively simple and continuous cost-effective process for encapsulation of aromas, essential oils, colorants, and different extracts (Liu and Liu 2009; Drosou et al. 2016). Encapsulation by spray drying provides protection, stabilization, and controlled release of bioactives in powder form (Zuidam and Shimoni 2010; Jangam and Mujumdar 2013), with particle size in the range from a few microns to tens of microns with narrow size distribution (Drosou et al. 2016). Basically, spray drying is realized in three major steps: (1) production and homogenization of feed solution (or dispersion) that is composed of active compound and carrier material, (2) dispersion (atomization) of feed solution into the droplets, and (3) drying of droplets to dry mass (usually in the form of powder) (Desai and Park 2005; Zuidam and Shimoni 2010). The drying (dehydration) of droplets occurs in the drying chamber where hot air is introduced, causing rapid water evaporation. Industrial equipment has evaporating rates in the range of 0.1–12 T/h (Haffner et al. 2016). Further, the air stream carry formed particles into a cyclone where solid material is removed from spent air (Zuidam and Shimoni 2010; Anandharamakrishnan and Padma Ishwarya 2015). Separation of fine particles (those past the cyclone) is completed in the filters that are arranged at the end of the drying system (Anandharamakrishnan and Padma Ishwarya 2015). Atomization of feed liquid can be achieved by two main systems: (1) by compressed air using special nozzle system, (2) by mechanical atomization using rotating disc that disperses feed liquid under high rotational speed (Zuidam and Shimoni 2010). Other atomization techniques such as ultrasonic, multiple fluid channel, and electrohydrodynamic atomization are also available for production of fine engineered particles or for atomization of medium with specific properties (e.g., highly viscous fluids) (Anandharamakrishnan and Padma Ishwarya 2015).

There are several operating conditions affecting the quality of final spray-dried encapsulate: feed temperature, inlet and outlet air temperature, flow rate, emulsion properties, types of encapsulating materials, carrier to active ingredient ratio in the liquid dispersion, total solids content, viscosity of the atomizing fluid, stability, and droplet size (Drosou et al. 2016). Since feed temperature alters feed viscosity (fluidity) and its capacity to be homogenously dispersed, it is considered as one of the most significant parameters influencing encapsulation efficiency and morphology of final spray-dried encapsulates (Drosou et al. 2016). Although an ideal feed temperature cannot be defined, general recommendation for feed temperature employed for encapsulation of ingredients in food industry is in the range from 40°C to 60°C. This can be explained by the lower feed viscosity and the higher drying rate, which results in a quicker formation of the dry layer of carrier material around the droplet (Brückner et al. 2007; Drosou et al. 2016). As expected, there are inconsistent literature data regarding optimal feed temperature for the same bioactive ingredient, for example, D-limonene, depending on used carrier material, setup, and operating conditions (Paramita et al. 2009; Yamamoto et al. 2012). Recent studies have shown that the increase of inlet air temperature up to limits of thermo stability of actives (accompanied with high concentration of active substance) contributes to rapid formation of a semipermeable membrane on the droplet surface and consequently better encapsulation efficiency, especially in the case of volatile oils (Fernandes et al. 2013; Botrel et al. 2014a; de Souza et al. 2015). Recent review by Drosou and coworkers gives a good summary of optimized inlet and outlet temperature for the encapsulation of different food ingredients by spray drying (Drosou et al. 2016).

One of the main advantages of spray drying is the wide range of particle sizes (10–400 μm) that can be obtained, as well as a huge number of potentially used materials as carriers by this technology. Depending on active compound properties and final product requirements, carrier materials can be selected from a wide variety of natural and synthetic polymers (Zuidam and Shimoni 2010). Nevertheless, almost all spray drying processes in the food industry require the use of water soluble carrier with film forming, good emulsification properties, and low viscosity (Botrel et al. 2014b). In this way, the most used carriers in spray drying encapsulation in the food sector are carbohydrates, including modified and hydrolyzed starches, cellulose derivatives, gums and cyclodextrins, as well as proteins including whey proteins, caseinates, and gelatin (Drosou et al. 2016). The choice of adequate carrier material has to be in balance with the physicochemical properties of an active substance. For instance, natural gum such as alginate efficiently stabilizes flavor emulsion (Lević et al. 2015), but the high viscosity of alginate limits its use in the spray drying processes. Thermal pre-treatment (e.g., autoclaving) of alginate can reduce its viscosity in order to ensure efficient droplets formation and encapsulation of lipophilic compound (Tan et al. 2009). On the other hand, modified starches and gum arabic are found to be very suitable for spray drying of liquid aromas. For example, gum arabic exhibits relatively low viscosity even at high concentrations (above 30%) and provides stable emulsions. However, gum arabic is a naturally occurring plant exudate and its production is affected by climatic factors, which affect availability, quality, and cost of such material. Modified starches are more available and could be modified in order to fulfill specific demands (Porzio 2007). In order to overcome general lack of active surface properties of polysaccharides (except gum arabic), very often a mixture of proteins and carbohydrates is employed in spray drying encapsulation of hydrophobic core actives (Botrel et al. 2014a).

A very important processing parameter is core to carrier material concentration too. Although encapsulation efficiency in each particular spray drying process depends on physicochemical properties of the active ingredient and carrier material, there is a trend that higher core loads lead to poorer retention (lower encapsulation) due to greater proportions of core material close to the drying surface, which shorten the diffusion path length at the air–particle interface (Drosou et al. 2016). This is confirmed in the case of linoleic acid, lycopene, and oils such as lemon myrtle oil, peppermint and fish oil as reviewed by Drosou et al. (2016). The majority of published works reported a typical core-to-wall material ratio of 1:4 as being optimal for various wall materials, such as gum arabic and modified starches (Jafari et al. 2008).

Since spray drying in the food industry has made a great contribution in microencapsulation of oils, a significant processing parameter is the physical characteristics of emulsions (droplet size, droplet surface charge, viscosity, and other interfacial properties) from which particles are produced (Drosou et al. 2016). With an aim to have successful encapsulation by spray drying, which means high retention of volatiles and low content of surface oil (Jafari et al. 2007a, 2008), well-documented recommendations include use of emulsion with reduced size of droplets and increased viscosity (Huynh et al. 2008; Jafari et al. 2008; Tonon et al. 2011).

Despite its advantage in providing rapid dryness without significant increase of the products surface temperature, the most known limitation in spray drying represents low energy efficiency due to usage of high-temperature drying and water evaporation. A good example for illustration is the production of skim milk powder by spray drying, where energy consumption accounts for 5300 kJ/kg for water evaporation. This amount of energy could be reduced if two- (4500 kJ/kg of water) or three-stage (4000 kJ/kg of water) spray driers are used. The data are based on spray drying of previously concentrated raw milk (50% of solids) followed by spray drying to 3%–4% (wet basis) of moisture content (Kudra 2008). However, in comparison to freeze drying as technology, which includes the water evaporation process too, the food industry considers spray drying as a more energy efficient process due to lower energy consumption, shorter process time, and lower porosity of final dried materials (Zuidam and Shimoni 2010). Another potential problem in spray drying can be wall deposition and loss of the final product, as very often seen in the processing of fruits and vegetables (e.g., tomato, garlic, and spinach) into powdered products. These spray-dried powders could be added into the product such as soup premixes or snacks, as well as be mixed in baby foods, beverages, bakery products, etc. (Phinix International 2016). Goula and Adamopoulos (2008) proposed that addition of maltodextrin into tomato pulp, combined with the use of dehumidified air as drying medium, may reduce losses during spray drying.

Based on all the aforementioned, there is a constant demand for improving spray driers' performances and the quality of spray-dried products on an industrial scale. SANOVO TECHNOLOGY GROUP introduced the new horizontal drier configuration (Gentle-Air Spray Drying System-Box Dryer) as an alternative to the conventional spray drying process. The main features of this design are reducing building height and heat recovery. Energy saving is achieved by a special energy recovery unit that provides pre-heating of the inlet air by the hot exhaust air from the drying chamber. Also, the drying unit is insulated with special polyurethane foam panels that minimize heat loss. The system provides a new design of dispersion-atomization process and special filter section with a bag filter integrated in the drying chamber that reduces dust emission and product loss (SANOVO TECHNOLOGY GROUP 2016).

Keeping all this in mind, prior to the application of spray drying as an encapsulation technology in the food industry, the following should be considered: energy efficiency, adequate drying chamber of setup (to reduce wall deposition), and atomization technique (to provide particles of desired sizes and shapes).

One of the most important applications of spray drying technology in food processing is flavor encapsulation. Almost 10 years ago, that is, in 2008, the ratio of encapsulated flavors in total flavor market was 20%–25%, while 90% of all encapsulated flavors were produced by spray drying (Zuidam and Heinrich 2010). Along with the constant improvement of spray drying devices and progressive increase in research interest in this encapsulation technology, the portion of encapsulated flavor on the global market nowadays is expected to be much higher. Flavors encapsulated by spray drying possess better oxidative stability than unencapsulated ones, and could be stored up to 1 year and is easy to handle (Gharsallaoui et al. 2007). In order to control the flavor release, a combination of spray drying and other encapsulation was proposed with different flavors distributed between the core (solid, paste, liquid) and the solid external layer, which provides two levels of release (Đorđević et al. 2015). For instance, an oil-in-water emulsion with coffee aroma was sprayed on coffee powder to form a dry coating layer enriched in aroma, with release in hot water (Buffo et al. 2002). The critical step in flavor encapsulation by spray drying is the formation of stable feed solution that is composed of solvent (i.e., water), adequate type and concentration of carrier material (one or more compounds), and active ingredient (i.e., flavor). For example, by combining maltodextrin with mono- and disaccharides prior to spray drying it is possible to achieve flavor oxidative stability for up to 1 year (Bošković et al. 1992). Also, spray-dried encapsulated Swiss cheese bioaroma produced by *P. freudenreichii* revealed better retention of short chain organic acids flavoring agents (propionic and acetic acid) when 50% modified starch and 50% maltodextrin was used at 175°C in comparison with the formulations with modified starch concentrations of 100% or 14.5% and air inlet temperature of 180°C and 163°C (da Costa et al. 2015). At the same time, the main disadvantage of spray drying for flavor encapsulation is the application of relatively high drying temperatures that can cause flavor degradation in some cases.

Also, spray drying can be used for masking of unpleasant taste of active compound. Favaro-Trindade et al. (2010) showed that gelatin and soy protein isolate mixtures successfully reduced the bitter taste of casein hydrolysate. This is also important for commercial coffee and tea extracts produced by spray drying. Instant coffee and tea are very popular products that can be produced by spray drying. However, some negative aspects of such products, such as bitter taste may harm their attractiveness to consumers. Belščak-Cvitanović et al. (2015b) showed that green tea extract microencapsulates with modified starch, inulin, and carrageenan exhibited the lowest bitterness and astringency and the highest flavor intensity. Similarly, in order to mask the strong and unpleasant odor of propolis, which generally compromises its beneficial antioxidant property and acceptability in foodstuff, Spinelli et al. (2015) have microencapsulated propolis with gum arabic and chemically modified starch by using spray drying. Obtained encapsulate of spray-dried propolis was incorporated in fish burgers and resulted in increased phenolics and antioxidant activity and also a good acceptability of fish burger by consumers (Spinelli et al. 2015).

Despite numerous research reports on spray-dried probiotics, the production of spray-dried probiotic bacteria is scarce on a large commercial scale (Haffner et al. 2016) and still represents a huge challenge due to the cells' heat sensitivity. Although spray drying can provide high production rates and relatively low operating costs, that is, 4–7 times cheaper than freeze drying (Chávez and Ledeboer 2007), most of the probiotic strains do not survive in large quantities after the heat stress and dehydration. Nevertheless, a growing number of recent studies are providing complete research assay for efficient industrial exploitation, like the one by Dimitrellou et al. (2016); they conducted evaluation of *L. casei* ATCC 393 in spray-dried form under simulated gastrointestinal conditions and during fermented milk production and storage. Ganeden Biotech® and Tipton Mills® in 2012 revealed the first instant coffee with probiotic bacteria under commercial name Ganeden BC30 (Ganedenbc30.com 2016). The main feature of this product is that bacterial cells are cultivated in a manner to produce spores. Spores are further spray- or freeze dried and directly applied into various products. Using spores instead of the vegetative cells provides better survivability under heat treatments during food preparation (Endres 2012). As we pointed out before, one of the main obstacles to wider application of spray-dried encapsulates in the food products is low water resistance of carrier materials. This is even more important in the case of probiotic bacteria, where the encapsulate must resist not only food processing but also gastrointestinal conditions. Picot and Lacroix (2003) reported production of water insoluble microcapsules using an emulsification/spray drying process. Multiple carriers such as anhydrous milk fat, micronized skim milk powder, and whey protein isolate were combined in the several preparation process steps prior to spray drying. Encapsulates obtained thus exhibited relative stability toward water and could be used for probiotic encapsulation. In a most recent report by Guerin et al. (2017), they developed matrices with innovative functionalities by spray drying encapsulation of *L. rhamnosus* GG in milk proteins insoluble in water and enzymatic cleavage of milk proteins by chymosin before atomization. The innovative functionalities for food industry are reflected during microparticle reconstitution with water: (1) when rehydrated in cold (8°C) water, produced microparticles may be interesting for ferment production, as the release of encapsulated bacteria can be easily achieved by a careful choice of reconstitution temperature, and (2) when dispersed in warm (40°C) water, these spray-dried microparticles may provide bacteria vectorization in high moisture content food products (such as milk, fermented drink, juice, yogurt, etc.) (Guerin et al. 2017) (Figure 8.3).

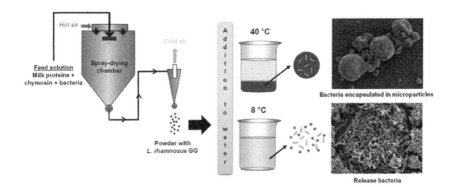

FIGURE 8.3 Production of innovative microparticles containing *L. rhamnosus* GG by spray drying. (Reprinted from *J. Food Eng.*, 193, Guerin, J., Petit, J., Burgain, J. et al., *Lactobacillus rhamnosus* GG encapsulation by spray drying: Milk proteins clotting control to produce innovative matrices, 10–19. Copyright 2017, with permission from Elsevier.)

The food industry is showing an interest in the encapsulation of oils as rich sources of unsaturated fatty acids (Stajić et al. 2014). Transformation of edible oils into powders is relatively simple through spray drying, while particle sizes of spray-dried encapsulates are more suitable for food applications in comparison to other encapsulation technologies. However, spray drying of oil requires preparation of stable emulsion using various carrier materials. Besides the type of carrier used for encapsulation, spray drying of oils usually exhibits low yield at high oil loadings (Tan et al. 2005). This could be overcome by combining various materials such as alginate and starch (Tan et al. 2009), maltodexrin and acacia gum (Fuchs et al. 2006), or emulsion stabilized by β-lactoglobulin prior to spray drying (Mezzenga and Ulrich 2010). Company Connoils already offers a huge number of vegetable and fish oils in powdered form suitable for application in pet food, cosmetics, and special and functional food products (CONNOILS—Connect with Connoils for Your Nutritional Health & Beauty Ingredients 2016). Also, Watson company offers spray-dried sunflower oil with relatively high oil content of 0.3 g of oil/gram. The same company also produces spray-dried vitamins encapsulated in the matrix of gum arabic or modified starch. For this purpose, cold water dispersible starch was found to be a suitable medium for spray drying. The same carrier is also used for encapsulation of beta-carotene, which is at the same time of natural color (yellow or orange), but also a rich source of vitamin A (Watson Inc. 2016).

8.2.2 Spray Cooling

Utilizing the spray cooling process is a growing trend in the food industry. Although the terminology defining spray cooling can be inconsistent (also referred to as prilling, spray chilling, and spray congealing), the general principle of microencapsulation methodology covered by these terms is more important and to be explained here. A more precise description for spray cooling is that it presents a mix of three methods—melt extrusion, melt coating, and melt agglomeration (Becker et al. 2015), rather than just one of melt agglomeration methods as listed by some authors (Eliasen et al. 1999; Becker et al. 2015). Also, the spray cooling technique may be considered as a mixture of hot melt technology (coating or agglomeration) and spray drying technique (Okuro et al. 2013a). Regardless of the precise definition by some authors, the most appropriate term for this encapsulation technology and resulting encapsulates in food industry is prilling (Oxley 2012; Dubey and Windhab 2013), that is, prills as products with particle sizes of 500–2000 μm (Pivette et al. 2009; Vervaeck et al. 2013). Independent of the names and definitions, products of this encapsulation technology are particles consisting of solidified (via crystallization) lipid carrier material and active ingredient dispersed all over the volume of the particle. Perfectly spherical produced particles thus represent fat-based systems of matrix type (microsphere) or a more core/shell-like structure (microcapsules) depending on the nozzle type and fluid delivery used (Becker et al. 2015). For the food industry, employed lipid carriers in spray cooling have to be GRAS—generally recognized as safe—typically nonallergenic, and to provide release of active ingredient in the gastrointestinal tract after digestion (lipid metabolism) (Favaro-Trindade et al. 2015); thus, generally recommended are fatty acids, alcohols, triacylglycerols, and waxes (e.g., palm oil, beeswax, cocoa butter, and kernel oil) (Okuro et al. 2013a).

The first required step in the spray cooling process is the production of stable, homogenous melt dispersion of the ingredient intended to be encapsulated, and some additional excipients, such as soy lechtine as an emulsifier for water-in-fat emulsion production (Okuro et al. 2013a). This could be done by a few common techniques, such as ultrasound homogenizer or high shear mixer. Since each of them could have a strong effect on the resulting particle size, rheology, and crystallinity, they have to be controlled (Okuro et al. 2013a; Becker et al. 2015). The following step in spray cooling is atomization, which could be done by various atomization units (different on mechanism, liquid channeling, quantity, and characteristics of obtained products), such as spinning disk, vibrating nozzle, pneumatic nozzle, ultrasonic devices, or dual-fluid nozzle (Favaro-Trindade et al. 2015). Regardless of the used atomization procedure, particles are then solidified by cooling below the melting point of the matrix material in a large chamber, that is, prilling tower. Cooling in chamber can be provided at room temperature, with cold air, liquid nitrogen, or a carbon dioxide ice bath. If the chamber is at room temperature, the melting point of the encapsulation material is between 45°C and 122°C as opposed to 32°C–42°C if the chamber is cooled (Okuro et al. 2013a). Phase of solidifying is crucial for the final characteristics of products, since an incomplete recrystallization or inadequate time of flight may lead to deformation or sintered lipid blocks in the product vessel (Becker et al. 2015).

Compared to widespread spray drying technology in the food industry, spray cooling technology is operationally similar to spray drying. However, there is a difference between these processes in terms of direction of energy flow. With regard to spray drying, the energy is applied to the droplet forcing the solvent evaporation. In contrast, in spray cooling encapsulation, the energy is removed from the droplet driving carrier solidification (Okuro et al. 2013a), so there is no mass transfer (Favaro-Trindade et al. 2015). Regarding particle sizes and morphology, spray cooling provides spherical larger ones (20–200 µm) with a smooth surface, compared to those geometrically shaped (5–150 µm) with irregular surface (due to solvent evaporation) produced by spray drying (Okuro et al. 2013a). Generally, in the group of spraying-based technologies, the operational highlights of spray cooling encapsulation include easy scale up, controlled particle size, free flowing powders, and reduced energy costs. Also, as described in Favaro-Trindade et al. (2015), spray cooling is considered as a fast and clean technology (because solvents are not used) for continuous production of large quantities of encapsulates in a single-step operation (because post-processing such agglomeration or coating is not usually necessary).

There are some general recommendations regarding process conditions for encapsulation by spray cooling according to Cordeiro et al. (2013) and Okuro et al. (2013a): the temperature of the melt should be higher by 10°C than the melting temperature of the carrier; optimal feed viscosity should be lower than 500 mPas (if a particular nozzle design is not available) and the load of active ingredient should be up to 30%, that is, 50% when some adaptations in atomizing step, such as usage of pneumatic atomizing nozzle could give this possibility (Albertini et al. 2008; Favaro-Trindade et al. 2015).

Although spray cooling technology can provide low-cost, continuous mass production of encapsulates under mild temperature conditions and fine tuning of release kinetics of encapsulated material due to lipophilic encapsulating material (Haffner et al. 2016), the last one is concomitantly the biggest industrial hurdle. In other words, lipophilic carrier used by this encapsulation technology could contribute to final organoleptic and nutritional value of such food product, two very important parameters of "modern" food.

Thanks to the mild processing parameters of spray cooling, its main application in the food sector represents probiotics encapsulation. Besides protection of probiotics from unfavorable conditions during processing, storage, interaction with the food matrix, packaging, and the environment (temperature, moisture, and presence of oxygen), encapsulation by spray cooling in lipid matrix is particularly beneficial due to its ability to provide delivery of probiotics in intestine (their action site) through lipid matrix digestion (Okuro et al. 2013a; Favaro-Trindade et al. 2015). Even early studies (Rutherford et al. 1994) resulted in a patented product by DuPont Pioneer (Haffner et al. 2016) in the animal feed sector. This patent implies microspheres of fatty acid matrix (stearic acid) containing probiotics (*Enterococcus faecium, Lactobacilli*, and yeast) produced by spray cooling using a rotary disc; this product can be stored for a period of 3–6 months, and cell viability is maintained even if there is exposure to some moisture or antibiotics. More recent investigations also present promising results regarding potential of spray cooling in probiotics encapsulation. Thus, in models simulating fermented and nonfermented drinks, it has been shown that bifidobacteria (*B. longum*) encapsulated in cocoa butter matrix possess prolonged viability during storage (Lahtinen et al. 2007). Also, in studies of Pedroso et al. (2012) on encapsulation of *B. animalis* subsp. *lactis* and *L. acidophilus* by spray cooling in fat produced with hydrogenated palm and kernel oil, they reported better resistance of encapsulated probiotics to simulated gastric and intestinal fluids in comparison to free microorganisms and viability during storage (Pedroso et al. 2012). The same lipid carrier was used in the case of symbiotic microparticles encapsulating *L. acidophilus, L. rhamnosus,* and prebiotics (inulin and polydextrose) (mean size 62.4 ± 2.8 to 69.6 ± 5.1 μm) produced by spray cooling revealing similar results regarding viability and susceptibility of probiotics to enteric conditions (Okuro et al. 2013b). The formulation containing polydextrose demonstrated cell viability with 6.9 or 6.8 log CFU/g at relative humidity of 11% after 120 days of storage at −18°C or +7°C, respectively (Okuro et al. 2013b), which is the minimum concentration of viable cells required by the U.S. FDA (Haffner et al. 2016).

Spray cooling is gaining attention in the production of nutraceuticals too. Spray cooled ferric pyrophosphate and ferrous sulfate as iron food fortificants within a matrix of hydrogenated palm oil exhibited high stability against oxidation and consequent deterioration (Biebinger et al. 2009). This technology has proved to be effective in increasing protection to different vitamins as well, like vitamin D (Paucar et al. 2016), vitamin C (Matos et al. 2015a; Sartori et al. 2015; Alvim et al. 2016; Sartori and Menegalli 2016), and vitamin E (Gamboa et al. 2011). To illustrate an example of vitamin D, microparticles obtained from vegetable fat and 1% of beeswax by spray cooling increased vitamin D stability at 25°C, since 86.3% of the vitamin was detected after 65 days compared with nonencapsulated vitamin under the same conditions (60.8%) (Paucar et al. 2016). Furthermore, promising results have been demonstrated in the case of antioxidants, such as gallic acid (Consoli et al. 2016) and lycopene (Pelissari et al. 2016) by encapsulation in lipid microparticles by spray cooling.

A good example of spray cooling technology adopted in the production of food-stuffs is vinegar entrapped in melted, partially hydrogenated cottonseed oil and 90% pure alpha form glycerol monostearate. This encapsulate produced by spray cooling has a form of free flowing powder with a strong vinegar odor and can be added to a dry barbeque sauce mix to provide a vinegar note when the mix is dispersed in hot water (Morgan and Blagdon 1993). Furthermore, often used food encapsulates produced by spray cooling technology are hydrolyzed vegetable proteins in partially hydrogenated cottonseed oil as carrier. The obtained sprayed material in the form of free flowing powder is used for making gravy with a typical "beefy" taste (Morgan and Blagdon 1993).

8.2.3 SPRAY COATING

In some cases, additional protection of encapsulates in the food industry is required by the formation of protective layer(s) on the material surface, and it is usually made by a spraying technology called spray coating. Very often, this technology is confused with spray drying/cooling. Spray coating consists of fluidizing a powder (in a fluid bed or a pan) and spraying a coating solution on the nebulized particles (Poncelet et al. 2011). Thus, while spray drying and cooling technologies produce particles, spray coating is used for recovering particles allowing the final morphology of reservoir-type microcapsules (Okuro et al. 2013a). The coatings can be solidified by drying in the case of polymer, or cooling when a fat-based coating is desired. Since the spray coating allows adjustment of the thickness and hydrophility/lipophility of the coats, fine tuning of the release profile of active ingredient can be achieved, as well as greater flexibility for incorporation of these particles in food products. In the same way, through spray coating technology the encapsulated ingredient is better protected compared to encapsulates produced by spray cooling. When a larger quantity of coated material is required, fluid bed coating has an advantage compared to traditional pan coating.

Generally, there are three major technical designs of fluid bed coaters: (1) fluid bed coaters with top spraying of coating material, (2) fluid bed coaters with bottom spraying of coating material, and (3) fluid bed coaters with tangential spraying of coating material. For food processes, a fluid bed coater with top spraying of coating material is more suitable, since it is a relatively simple configuration with high capacity and versatility. Also, development of continual fluid bed coaters could be the turning point for wider spreading of fluidized bed processes in food industry (Dewettinck and Huyghebaert 1999).

When particles of obtained encapsulates are too small and one is interested in their aggregation into bigger particles, fluid bed coating could be applied even without additional material for agglomeration. Such an example could be spray-dried powders containing oil as active ingredient and maltodextrin and acacia gum as carriers

(Turchiuli et al. 2005). After spray drying, the particle size is less than 50 μm, but after agglomeration in the air fluidized bed with spraying of water, their size can increase up to 150 μm. Coated oil exhibited good oxidation stability, while overall agglomerated particles showed satisfactory technological characteristics from the viewpoint of food engineering (Fuchs et al. 2006).

8.3 DRIPPING-BASED TECHNOLOGIES

It is a common belief that dripping (extrusion) technologies are not suitable for industrial level production. Although these are simple and low-cost technologies that can be performed under mild conditions (pressure and temperature are generally lower than 100 psi and 118°C, so they can be used for heat-sensitive products) (Rohman 2007) and despite the fact that microbeads produced by extrusion technologies have a less porous structure than those produced by, for example, spray drying (Rodríguez et al. 2016), the cost of utilization of classic extrusion technologies on an industrial level is still twice as high than in the case of spray drying. In addition, the size of obtained particles may be too bulky for application on certain food products, as it can be detectable in the mouth (Kaushik et al. 2015).

However, some major advances related to scale-up of extrusion technologies and their utilization for encapsulation of bioactives on an industrial level has been introduced. Namely, the production of droplets on a large-scale level can be achieved by alternative extrusion technologies, for example, multiple-nozzle systems, spinning disc atomization, or jet-cutting (de Vos et al. 1997, 2010; Kailasapathy 2002). It is important to point out that extrusion technologies enable actual encapsulation of bioactives versus their immobilization (de Vos et al. 2010). In addition, these technologies are very gentle as they do not utilize any harmful solvents, and it is possible to conduct the encapsulation procedure under both aerobic and anaerobic conditions; all listed determine the extrusion technologies as preferable for living cells encapsulation. This is particularly advantageous when anaerobic microbes are encapsulated in order to be applied in the food industry. The encapsulation can be achieved by simply placing the extrusion setup in the sterile cabinet where oxygen is replaced with nitrogen (de Vos et al. 2010). Besides cells, extrusion technologies can be used for the encapsulation of various bioactives, such as enzymes, aromas and flavors, vitamins, proteins, antioxidants, etc.

The procedure behind dripping technologies is relatively simple: aqueous solution of polymer and active substance(s) is added dropwise into the gelling solution, where the beads are left to harden for an adequate period of time. The tool used for dripping distinguishes the difference between the technologies; in that respect, five different extrusion technologies can be determined: electrostatic extrusion, coaxial air flow extrusion, vibrating jet/nozzle extrusion, spinning disk atomization, and jet-cutting (Figure 8.4) (Zuidam and Shimoni 2010). Usually, by means of these techniques, particles in the size range of 0.2–5 mm can be produced, depending not only on the chosen dripping technology but also on the viscosity and rheological behavior of polymer solution (Nedović et al. 2006; Wandrey et al. 2009). In addition, the particle size can be further controlled by manipulation of process parameters essential for the particular technique.

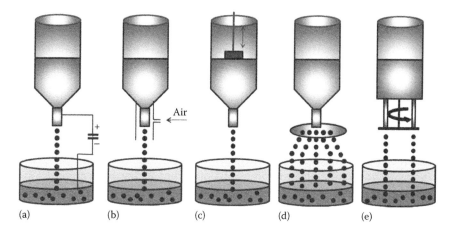

(a) (b) (c) (d) (e)

FIGURE 8.4 Different dripping encapsulation technologies. Extrusion of polymer/active substance solution into the gelling solution by means of: (a) electrostatic potential, (b) coaxial air flow, (c) vibrating jet/nozzle, (d) spinning disk, and (e) jet cutter.

8.3.1 ELECTROSTATIC EXTRUSION

Electrostatic extrusion technique implies droplet generation by means of electrostatic potential. Namely, applied electrostatic forces cause the disruption of a liquid surface at the top of a needle, resulting in the formation of a charged stream of small droplets (Nedović et al. 2006). The breakup of droplets under the applied electrostatic potential can be done in two modes: (1) first is the dripping mode, where low current is applied, so the polymer-actives solution is gently pressed through the nozzle/needle, and (2) the second mode, which implies higher electric current, so the polymer solution is pushed at a higher velocity through the nozzle, forming a smooth stable jet of droplets. From the production rate aspect, mode (2) is superior, since higher productivity can be achieved, resulting in microbeads sized from 1 to 15 μm (Tran et al. 2011). In that respect, the main advantage of electrostatic extrusion technique is that it enables the production of particles that are small in diameter with very uniform size distribution (Bugarski et al. 1994).

It is important to point out that applied electric potential in the encapsulation process does not influence cell integrity and viability; thus, the electrostatic extrusion technology can be successfully used for the encapsulation of cells (Bugarski et al. 1993, 2004; Bezbradica et al. 2004; Manojlovic et al. 2006; Lopez-Rubio et al. 2012). The most frequently, if not almost exclusively, used carrier material in these encapsulation procedures is alginate hydrogel ionically crosslinked with calcium ions. It has been shown that this matrix enables successful encapsulation of various cell types (Bugarski et al. 1993, 1997, 1999; Goosen et al. 1997; Pjanovic et al. 2000; Rosinski et al. 2002); it has also been utilized for brewing yeast encapsulation aimed at continuous beer production (Nedović et al. 2001, 2002).

As it is well known, the most abundantly used cells in the contemporary functional food domain of the food industry are probiotic cells (Burgain et al. 2011).

There has been some anticipation related to the probiotic functional food market, where it was projected to reach a value of US$ 28.8 billion by 2015 (Hernández-Rodríguez et al. 2014). However, extensive utilization of probiotics in functional food products is limited by their instability and loss of viability during food processing, product storage, or consumption, as well as through the human gastrointestinal tract upon intake (Đorđević et al. 2015). In order to stabilize them, encapsulation via electrostatic extrusion can be applied (Obradović et al. 2015; Krunić et al. 2016). Subsequent to encapsulation, an increase in cell viability, fermentative activity, as well as antioxidant characteristic ascribed to probiotics has been reported (Krunić et al. 2016).

In recent researches, electrostatic extrusion technique has been proposed as a means of stabilization and preservation of plant antioxidants (Fang and Bhandari 2010). Plant polyphenols can be easily encapsulated within the Ca-alginate matrix (Belščak-Cvitanović et al. 2010, 2011; Stojanović et al. 2012; Trifković et al. 2012; Ćujić et al. 2016). However, a problem of actives loss (in case of low-molecular weight polyphenolic compounds) by diffusion through the porous matrix network has arisen. Several solutions to this problem can be applied. Some authors proposed manufacturing of empty polymer beads, while encapsulation is carried out subsequently, by placing the beads in the polyphenols solution (so-called post-loading entrapment) (Chan et al. 2009; Stojanović et al. 2012; Trifković et al. 2014); following this procedure, the actives entrapment can be enhanced up to sixfold (Chan et al. 2009). Another solution to the problem of actives percolation can be found in the eradication of driving force for actives diffusion by simply retaining the concentration of the actives in the gelling solution at the same level as in the polymer-actives solution; with this approach it is important to evaluate its impact on the encapsulation efficiency (Stojanović et al. 2012). Finally, complexation of alginate with oppositely charged polymers, such as chitosan, carrageenan, pectin, PVA, can be used to overcome the surplus diffusion of actives (Belščak-Cvitanović et al. 2015a).

From the operational point of view, it is important to highlight the influence of process parameters on the final features of obtained encapsulates, especially on the beads size. Since the polymer solution has to be extruded through a needle/nozzle with a small inner diameter, its viscosity is first to be monitored and adjusted. One of the major obstacles to wider use of extrusion technologies in the food sector is their inability to process highly viscous polymer solutions when it comes to electrostatic extrusion technology—even when higher voltage is applied, highly viscous polymer solutions are difficult to handle. For alginate solution, numerous studies showed the impact of increased viscosity on microbeads size, precisely on enlargement of beads diameter (Poncelet et al. 1999; Jayasinghe and Edirisinge 2004; Manojlovic et al. 2006), until some critical concentration after which it is no longer possible to push the solution through the needle. Also, process parameters such as flow rate of polymer solution (Poncelet et al. 1999; Nedović et al. 2006), intensity of voltage applied (Poncelet et al. 1994; Lewinska et al. 2004), as well as needle size and spacing in the system (Bugarski et al. 1994; Bezbradica et al. 2004; Nedović et al. 2006) affect the final bead characteristics.

8.3.2 EXTRUSION WITH COAXIAL AIRFLOW, VIBRATING JET/NOZZLE, JET CUTTING, AND SPINNING DISK ATOMIZATION

Coaxial air flow technique involves droplet formation under the influence of compressed air; in this way, detachment of drops from the nozzle is faster when compared to simple dripping, where detachment of droplets is governed only by gravitational force (Prusse et al. 2008). Using this technique, it is possible to produce microbeads of around 200 μm in diameter with a uniform size distribution; what is more important from an operational point of view is that the process is highly reproducible and can be performed under mild as well as sterile conditions (Heinzen et al. 2004). However, the coaxial air flow technology has a major drawback in the low production rates that can be achieved (Đorđević et al. 2015).

Nevertheless, some recent studies highlighted the effectiveness of coaxial air flow technology, particularly in the case of hydrophobic compounds encapsulation (Ghayempour and Mortazavi 2013; Moghaddam et al. 2015). Using such a method, Tylingo et al. (2016) encapsulated cold pressed canola essential oil, a rich source of fatty acids, within the chitosan matrix, to be used as a functional food additive (Figure 8.5). Special emphasis has been laid to the possibility of coaxial air flow technique in forming capsules with a separate oily core and coat material. The additional advantage is that it can be used regardless of the hydrophilic-hydrophobic nature of actives to be encapsulated. The authors argue the possibility of process scale-up to industrial level by proposing the use of battery consisting of a series of presented encapsulation heads, while the media could be pneumatically forced.

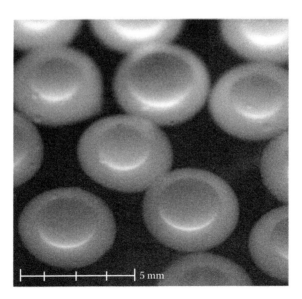

FIGURE 8.5 Encapsulated cold-pressed canola essential oil within the chitosan matrix by coaxial airflow extrusion. (Reprinted from *React. Funct. Polym.*, 100, Tylingo, R., Mania, S., and Szwacki, J., A novel method for drop in drop edible oils encapsulation with chitosan using a coaxial technique, 64–72. Copyright 2016, with permission from Elsevier.)

In order to overcome the low production capacity problem, extrusion technologies based on controllable droplets, a jet breakup can be used. This is achievable by (1) vibrating the nozzle at a defined frequency, (2) cutting device, or (3) rotating a disk in a specified direction, and in that respect three different extrusion methods can be identified: (1) vibrating jet/nozzle extrusion, (2) jet cutting technology, and (3) spinning disk atomization (Whelehan and Marison 2011). Prusse et al. (2008) compared and evaluated these technologies for the production of microbeads on a large scale. It was shown that polymer viscosity has a detrimental effect on the performances of each technology; accordingly, low viscous alginate solutions (up to 2% w/w) made possible the process with all technologies and spherical beads were obtained. However, processing of alginate solutions of higher viscosity (higher than 3% w/w) was not possible with the vibrating nozzle technology. In addition, microbeads obtained via the coaxial air flow technology using highly concentrated alginate solutions had a nonuniform shape, where the sphericity was lost and the beads were deformed to egg-like and drop-like shapes. As regarding the productivity of compared technologies, the following conclusions can be drawn: in comparison to coaxial air flow technique and electrostatic extrusion, which exhibited low production rates and therefore are limited to lab scale applications, vibration technology showed production rates higher by up to 50 times, thus meeting the requirements of large-scale applications (especially when multi nozzle systems are utilized). The same conclusion can be applied for jet-cutting technology (Prusse et al. 2008).

An alternate extrusion technology that can be used for encapsulation in the food industry is coextrusion where, as a tool for dripping, a concentric nozzle is utilized. In this way, spherical beads with a hydrophobic (lipophilic) core and a hydrophilic or hydrophobic shell can be produced (Zuidam and Shimoni 2010). Thus, by using the so-called submerged coextrusion (dropping of oil and wall material through a vibrating concentric nozzle in a stream of cooling oil) (Beindorff and Zuidam 2010), fish oil (rich source of polyunsaturated fatty acid) in a load of 70%–95% can be encapsulated in different polysaccharide particles of a size between 0.8 and 8 mm; this technology is currently held by Morishita Jintan, Japan.

Despite the fact that a lot of research activities are related to the encapsulation of bioactives by extrusion technologies on an industrial scale, not much of these research findings are actually transferred to food industry processes. However, some of the food products are successfully enriched with encapsulated bioactive ingredients via extrusion processes. For example, the problem of probiotic cell survival during the process of cheese ripening can be solved by their encapsulation via extrusion. It has been shown that various types of cheese can be fortified with different strains of probiotic cultures on a large scale: cheddar cheese (Amine et al. 2014), white-brined cheese (Ozer et al. 2009), Iranian white cheese (Zomorodi et al. 2011; Mirzaei et al. 2012), Kasar cheese (Ozer et al. 2008), and mozzarella cheese (Ortakci et al. 2012). It was proven that encapsulation provides higher cell survival in comparison to free cell incorporation in cheeses without significant impact on the appearance, color, texture, or overall sensorial perceptibility among the experimental cheeses.

Although the application of living probiotic cells to yogurt can be sufficient to enable a therapeutic level (Burgain et al. 2011; Mousa et al. 2014), the utilization of encapsulated forms exhibit advantages reflected in: physical protection of probiotics against inhibitory metabolic products or severe conditions in the gastrointestinal tract, as well as enhancement of probiotics survival during fermentation (Ziar et al. 2012). In that respect, process of incorporation of microencapsulated probiotic cells of *Lactobacillus* species in yogurts were investigated by numerous researchers (Iyer and Kailasapathy 2005; Urbanska et al. 2007; Krasaekoopt and Tandhanskul 2008; Ortakci and Sert 2012). The addition of encapsulates into yogurt did not affect its safety and tolerability (Jones et al. 2012), what's more, sensorial properties of the product seemed to have improved in terms of creamier consistency, due to the polysaccharides' (carrier material for encapsulation) presence in the encapsulates, as judged by some panelists (Tolve et al. 2016). In addition, encapsulation by extrusion technologies offers protection to the sensitive probiotic cells in gastrointestinal conditions. As already mentioned, in order to exhibit health beneficial effects, probiotics have to be present at a certain concentration (10^7 cfu g^{-1} or mL^{-1}) after the passage through the gastrointestinal tract. Ziar et al. (2012) showed improved survival of encapsulated probiotics in comparison to free ones. This is supported by investigations of Urbanska et al. (2007) and Ortakci and Sert (2012), who demonstrated improved viability of probiotics, as well as enhanced storage stability over 1 month of storage at 4°C.

Similar results for extrusion technologies based on the encapsulated form of probiotics were demonstrated in food products such as ice-creams and processed meat products as well. It was reported that microencapsulation of *Lactobacillus* species by extrusion technologies, followed by addition of encapsulates in ice-creams, increases the survivability rate of probiotics in the storage period of 180 days at −23°C by 30% in comparison to nonencapsulated forms, while the sensorial characteristics of ice-creams remained unchanged (Karthikeyan et al. 2014). By addition of encapsulated probiotic cells in dry and fermented sausages, Muthukumarasamy and Holley (2007) proved increased survival of lactic acid bacteria cells, but on the other hand decreased inhibitory activity against *Escherichia coli*.

Encapsulates produced via extrusion technologies are not exclusively used as functional food ingredients; they can be utilized in active food packaging as well. Maresca et al. (2016) investigated encapsulation of antimicrobial peptide nisin, the only bacteriocin recognized as safe by FAO/WHO and the Expert Committee on Food Additives, and permitted for use as preservative in food products. The reported findings signify the importance of nisin encapsulation—protection by nisin against protease degradation, thus achieving prolonged bioavailability and enhanced stability. This implies that extrusion technologies can be successfully used for the production of functional food preservatives.

8.4 EMULSION-BASED TECHNOLOGIES

This part of the chapter will focus on encapsulation technologies for production of emulsion-based delivery systems. The most important methods will be discussed regarding the influence of process parameters on particle characteristics as well as their food industry applications. As already mentioned, the success of an encapsulation process is often linked to the know-how of the formulation or of the chemistry to achieve stabilization (Poncelet et al. 2011). Although food science researchers provide a lot of results with important implications for the design of emulsion-based delivery systems, they are not yet of as much industrial importance as, for example, spray-dried food additives. The main reasons are many engineering problems related to the scale-up of the processes for production of emulsion based encapsulates, high costs of materials used for them (surfactants and cosurfactants) and time- and cost-consuming methods for their preparation. Another issue is how to avoid consumption of synthetic emulsifiers, whose addition to food products is limited to low amounts. Thus, food industry is searching for natural "label friendly" surfactants to replace conventional compounds (Yang and McClements 2013). Proteins from animal and vegetable sources have been considered as natural emulsifiers due to their amphiphilic character, polymeric structure, and electrical charge characteristics (Burgos-Díaz et al. 2016). Emulsion-based delivery systems are primarily being used by the pharmaceutical industry to carry active agents to specific locations within the gastrointestinal tract and release them at a controlled rate. The food industry has been developing similar edible systems (conventional emulsions, multiple emulsions, multilayer emulsions, solid lipid particles, and filled polymer particles) for food components, such as antimicrobials, antioxidants, nutraceuticals, and flavors. In addition, encapsulation of bioactives (e.g., polyphenols) in delivery systems of nanometric size (nanoemulsions, liposomes, solid lipid nanoparticles) contributes more significantly to the improvement of its bioavailability (Sessa et al. 2011).

A number of different homogenization processes and devices have been developed for the preparation of emulsions in the food industry, of which the most important are high shear mixing, agitation by rotor-stator homogenizer, ultrasound-assisted homogenization, membrane emulsification, microfluidization, and high pressure homogenization (Schultz et al. 2004; van der Graaf et al. 2005). The homogenization process is the main step to emulsion preparation, having high impact on droplet size distribution, the parameter that affects many physicochemical properties of emulsions (e.g., stability), microbial safety (bacteria growth is reduced in case of smaller droplet diameter due to the lack of nutrients inside of the droplets), and product quality (e.g., taste if flavoring compound is in the dispersed phase) (Charcosset 2009). The choice of the appropriate homogenization method depends on the quantity of the sample to be homogenized, materials used, desired size distribution, and expected physicochemical properties and, of course, on the process costs. Therefore, the choice has to be made carefully bearing in mind the required characteristics of the final emulsion along with performances of the available devices.

High shear mixing has been frequently used for the preparation of various emulsion-based systems (e.g., conventional emulsions, multiple emulsions), which can contain an active compound. The oil and aqueous phase are agitated by high shear stirrers (20–2000 rpm). This high-speed rotation generates longitudinal, radial, and rotational velocity gradients in the liquids that further decrease the droplet size.

The emulsion droplet size depends on the design of the stirrer. The increase in agitation speed and duration of mixing decrease the droplet size. Average size of droplets in high-shear mixed emulsions is commonly between 2 and 10 μm (McClements 1999). One has to be circumspect with this type of homogenization due to the possibility of temperature increase in the emulsion. In case of temperature-sensitive compounds (e.g., ω-3 fatty acids, carotenoids), it is necessary to control the temperature during the entire process. Furthermore, if lipid crystallization does occur (e.g., during preparation of solid lipid emulsions from saturated fats), then the homogenizer will be blocked and potentially damaged.

Rotor-stator systems, such as toot-disc high-speed homogenizers and colloid mills, can be used in continuous but also in discontinuous mode. In the discontinuous process it is usual to use gear-rim dispersion machines or agitators of different geometry (Schultz et al. 2004). The breakup energy is in the form of inertia and shearing forces of turbulent flow. Droplet diameters >2 μm are produced in this way. Somewhat smaller droplets (~1 μm) can be formed in colloid mills operating continuously, which basically apply the same kinds of forces. Colloid mills are used for medium- and high-viscosity emulsions and are able to give high product throughputs (McClements 1999). There are different colloid mills on the market, but the operating principle remains the same. The rapid rotation of the rotor provides a shear stress in the gap located between the rotor and stator. The shear stress causes the reduction in droplet size. The size of the gap, the rotation speed, the processing time, flow rate, as well as the number of passes, all affect the final droplet distribution (McClements 1999). The operating conditions have to be carefully chosen in order to obtain a stable emulsion. For example, the production of double emulsions (e.g., W1/O/W2) requires application of two homogenization cycles. The homogenization conditions used in the second stage are usually less intense than those in the first stage (to avoid disruption or expulsion of the W1 droplets within the oil phase). A requirement for narrow size distribution usually increases the costs in the manufacturing process.

Ultrasound homogenization is a high energy and long-lasting process (Jafari et al. 2007b). Ultrasonic sonotrodes (probes) provide high energy input, thus creating cavitation and microturbulences. This method is mainly applied at laboratory scale to produce emulsions with a droplet diameter of around 0.1 μm. The intensity, duration, and frequency of the generated ultrasound waves are the main factors affecting droplet size and emulsion quality. The common range of frequency is between 20 and 50 kHz, and the homogenization efficiency decreases with increasing of frequency. The size distribution of droplets produced by ultrasound homogenization is usually wide and depends on pre-emulsion preparation procedure (Jafari et al. 2007b). A leakage of an active compound may occur during the process due to high shear forces causing a disruption of the large liposomes (Isailović et al. 2013).

Microfluidization system is capable of producing very small droplets, even smaller than 0.1 μm. The system contains separate streams of oil and aqueous phase, which are accelerated to high velocity, mixed, and broken into small droplets. If the emulsion recirculates through the device a number of times, the droplet size can be decreased. Microfluidization works in continuous mode and it is suitable for low to medium viscous materials. It is an efficient, cost-effective, and solvent-free method for the fabrication of liposomes having high encapsulation efficiency of an active compound. The method can provide a few hundred liters of aqueous liposomes per hour on a continuous basis (Gouin 2004). In addition, microfluidization can be used for the production of conventional emulsions containing a high drug loading (Yang and McClements 2013). Jafari et al. (2007b) reported that this process was more efficient for the production of small droplets (~0.5 μm) with a narrow distribution in comparison with ultrasonication.

High pressure systems are common in the food industry. The emulsion mixture is forced through a narrow orifice creating a great shearing action, which can provide an extremely fine emulsion. The usual pressure is between 5×106 and 3.5×107 Pa. The forces predominantly derive from inertia shear, turbulent flow, as well as from cavitation. High pressure systems are mostly suitable for low- and intermediate-viscosity emulsions (McClements 1999). The average droplet size is around 0.1 μm, but the size and size distribution are difficult to control. The size of the droplets in an O/W emulsion produced by this type of homogenizer can be decreased by increasing the processing pressure or number of passes (McClements et al. 2007). High pressure homogenization can be used for the production of beverage emulsions. Beverage emulsions are unique compared to all other food dispersion systems because they are very dilute (20 mg/L dispersed oil phase in finished product) (Given 2009). The production of beverage emulsions requires two-stage high pressure homogenization (2000–5000 psi) to reduce particle size to between 0.2 and 2.0 μm diameter. However, the quality of the emulsion obtained by high pressure homogenization can vary from one batch to another. Poor reproducibility is one of the scale-up difficulties (van der Graaf et al. 2005). When extremely high pressures are used, high local temperatures are being developed; thus the emulsion should be kept cooled.

Membrane emulsification process is a relatively new method. It works either in batch or continuous mode. Low pressure is used to force the disperse phase through the pores of microporous glass or ceramic membranes into the continuous phase (Charcosset 2009). Disperse phase flux and the size of membrane pores are critical to the final droplet size. This process provides narrow size distribution and droplets down to 0.2 μm. Membrane emulsification is alluring due to easier control of droplet size and droplet size distribution as well as low energy consumption. The process is characterized by mild shear stresses convenient for both types of conventional emulsions (W/O and O/W), and for multiple emulsion production (van der Graaf et al. 2005; Muschiolik 2007). Matos et al. (2015b) compared membrane emulsification using tubular membranes in the cross-flow system (Figure 8.6a) with conventional emulsification (by mechanical agitation) for fabrication of a double emulsion (W1/O/W2) containing either trans-resveratrol or vitamin B12; membrane emulsification gave emulsions with uniform droplets (~60 μm) of controlled mean size and span values below 1, while mechanical agitation led to polydisperse emulsions (Figure 8.6b). However, during membrane emulsification process there is a risk of fouling of the pores by the emulsion components, which may slow down the process (Nazir et al. 2014; Schroën et al. 2015). The use of a "dynamic membrane" for emulsification has been suggested to overcome the problem of pore clogging (van Der Zwan et al. 2008; Nazir et al. 2010). The system consists of a metal sieve that supports a bed of small glass beads. It can be used for conventional and double food-grade emulsions (Eisinaite et al. 2016), operated at high throughput. The device can be easily cleaned after emulsification.

Emulsions have typically a mean droplet diameter within food systems somewhere between 100 nm and 100 μm. If dispersed phased droplet sizes are less than 100 nm, these (nano) emulsions are called microemulsions (McClements 2012). These have beneficial properties for a design of functional food such as optical clarity, high stability to gravitational separation, flocculation and coalescence, and improved absorption and bioavailability of bioactive components (Shamsara et al. 2015; Mehrnia et al. 2016). The encapsulation efficiency depends on the materials used (continued and dispersed phases and surfactants) as well as on the technique applied for fabrication of the emulsion. For example, Matos et al. (2015b) showed considerably higher encapsulation efficiency of trans-resveratrol and vitamin B12 in double (W1/O/W2) emulsions by membrane emulsification in comparison with those obtained by mechanical agitation. The shelf life and physicochemical properties of emulsion-based delivery systems actually depends on the most appropriate stabilizer(s). Manipulation of the droplet size distribution, interfacial properties, and

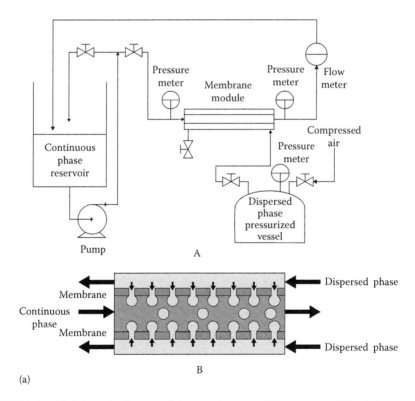

FIGURE 8.6 (a) Schematic diagram of the membrane emulsification unit (A) and the membrane module (B). *(Continued)*

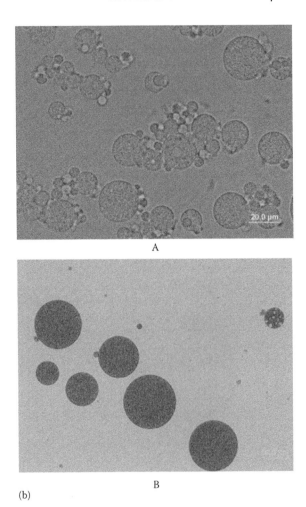

(b)

FIGURE 8.6 (*Continued*)   (b) Confocal image of W1/O/W2 emulsion containing resveratrol produced by mechanical agitation (A) and membrane emulsification (B). (Reprinted from *J. Food Eng.*, 166, Matos, M., Gutiatos, G., Iglesias, O., Coca, J., Pazos, C., Enhancing encapsulation efficiency of food-grade double emulsions containing resveratrol or vitamin B12 by membrane emulsification, 212–220. Copyright 2015, with permission from Elsevier.)

physical state can be performed to create delivery systems with specific functional performances appropriate for different types of active components and food matrices (McClements et al. 2007). For example, ω-3 fatty acids and carotenoids being susceptible to chemical degradation should be protected from contact with prooxidants (e.g., transition metals). This can be achieved by engineering the properties of the interfacial layer around the droplets (e.g., making it cationic so that it repels cationic transition metal ions). The main problem of food emulsions is gravitational instability, which reflects as creaming or sedimentation. Furthermore, bioactives are easily released from conventional emulsions due to the small size of dispersed droplets and thin interfacial layer (nanometric size). In multiple emulsions, the bioactive components can potentially be located in different molecular and physical environments. Multiple emulsions have better release properties; thus, bioactives can be trapped inside an inner phase and be released at a controlled rate or in response to specific environmental stimuli. For example, Giroux et al. (2016) have shown that the kinetics of peptides (azocasein hydrolysates) released from double (W1/O/W2) emulsions can be controlled by adjusting the oil phase composition. On the other hand, multiple emulsions are rarely used in food products because they are easily broken during storage or food processing. A novel approach to increase stability against coalescence of inner oil droplets of O1/W/O2 emulsions is to incorporate a hydrocolloid (e.g., pectin) as stabilizer in the water phase that separates the inner oil droplets from the outer continuous oil phase; a hydrocolloid can fixate oil droplets in its gel network and prevent the droplets from coalescing (Schmidt et al. 2015).

Despite the numerous research studies in academia on emulsion-based encapsulation technologies and resulting encapsulates, only a small number of these are really adopted by the food industry. Very interesting examples of emulsion techniques application in the food industry are probiotics encapsulates. Unilever™ holds an invention of W/O/W edible emulsion for salad dressing containing probiotics in the internal phase (Beck et al. 2002). The main benefit arises from the fact there is no volume restriction in respect to the concentration of probiotics that can be added to the internal phase of this emulsion (Haffner et al. 2016). Also, world-renowned company Abbott GmbH holds the patent on production of stable emulsion containing any acidophilic and/or bifidobacteria (Mazer and Kessler 2014), which could be added further into a powdered nutritional product designed for infants, children, or adults. Another good commercial example is emulsion of micronized ferric pyrophosphate, patented by Taiyo Kagaku Co. Ltd., Japan, under the brand name SunActive®. This liquid form of iron (12%) is stabilized by emulsifiers, such as polyglycerol esters and hydrolyzed lecithin and can be used in the fortification of milk based drinks; up to 94% of iron absorption can be achieved in humans by this formulation (Fidler et al. 2004; Moretti et al. 2006). More recent studies have confirmed that for large-scale milk production fortified with the desired high percentage of microencapsulated iron, the emulsification process gave better results compared to liposome, fatty acid esters, and freeze-drying methods (Gupta et al. 2015) and provide sensory properties similar to unfortified milk.

Liposomes are spherical-shell structures consisting of a phospholipid bilayer (or many such bilayers) enclosing an aqueous liquid core within which the hydrophilic solutes can be entrapped, while lipophilic compounds can be dissolved in lipids. In the food industry, liposomes and nanoliposomes are emerging carrier vehicles of functional compounds such as vitamins, enzymes, food antimicrobials, essential oils, bioactive peptides, and polyphenols. In the 1980s and 1990s, a lot of research work was done on liposomes for enzyme encapsulation in food and nutrition, primarily for acceleration of cheese ripening. The encapsulation efficiency varies in quite a wide range, for example, 1%–36% for enzymes and 80%–99% for liposoluble vitamins. In general, it depends on the type of active compound (lipophilic or hydrophilic and molecular weight), the composition of the bilayer (saturated versus unsaturated phospholipids, and presence of sterols)

(Lee et al. 2005a,b), the structure of the bilayer (unilamellar vs. multilamellar), and the method of liposome preparation can also influence the same compound. For example, high degrees of encapsulation have been obtained by freeze drying and dehydration/rehydration cycles (Emami et al. 2016).

Liposomes and solid lipid particle emulsion as emulsion-based delivery systems got more attention in the domain of nutraceuticals than in common foodstuff and fortified, that is, functional foods. There are a number of reports proving enhanced stability of liposomal formulations of commonly used nutraceuticals, such as vitamin C (Wechtersbach et al. 2012; Yang et al. 2012), vitamin D (Banville et al. 2000), and vitamin E (Liu and Park 2009; Ko and Lee 2010; Marsanasco et al. 2011) in comparison to solutions of free vitamins during storage or processing conditions (i.e., heat treatment). For example, Wechtersbach et al. (2012) found increased half-life of liposomal vitamin C of as much as up to 300-fold in model systems relative to that of free vitamin. By providing retarded release of actives, liposomes can improve the bioavailability of antioxidants (Takahashi et al. 2009) and reduce their cytotoxicity (Isailović et al. 2013). Enhanced antimicrobial (e.g., *Citrus limon*), antifungal (e.g., *Eucalyptus camaldulensis*), and other biological activities (e.g., *Artemisia arborescens* L.) of essential oils incorporated into liposomes have also been reported (Sinico et al. 2005; Gortzi et al. 2007; Moghimipour et al. 2012). However, there are some successful reports regarding liposomes implementation in foodstuff. For example, it was reported that the recovery of vitamin D in cheddar cheese was significantly higher when the vitamin was added using multilayer liposomes compared to other methods (commercial water-soluble emulsion of vitamin D or homogenization of vitamin D in a portion of cream) (Banville et al. 2000). Similarly, β-carotene-loaded multilayer liposomes were incorporated in yogurt (Toniazzo et al. 2014). Before the application, the liposomes were fortified and were able to protect the β-carotene from degradation for 95 days; the physicochemical test showed no significant changes in the texture of yogurt with the added liposomes (Toniazzo et al. 2014).

In general, liposomes are prone to physical and chemical deterioration in an aqueous environment. Encapsulated antioxidants (essential oils and polyphenols) protect lipid bilayers from oxidation to some extent (Liolios et al. 2009; Detoni et al. 2012; Balanč et al. 2015). Liposomal formulations undergo digestion in the GI tract through bile juice containing bile salts, which work as surfactants. Bile salts interact with liposomes and burst them by extracting lipids from the liposomal membrane, which results in the release of active compounds. Surface coating of liposomes with polymers could avoid the direct interaction of bile salts with lipid membrane (Gültekin-Özguven et al. 2016) and also secure long-term physical and oxidative stability (Gibis et al. 2012). A possible approach to secure stability of liposomes and to make them industrially applicable might be also their conversion into dry forms (by freeze drying, fluidized bed-coating, or spray drying). Thus, one smart commercial example is Curcusome® (Nanolife Tonics 2016) functional beverage containing freeze-dried liposomal formulation of curcumin/quercetin/piperine in release caps. In such a method, the destabilizing effect of process parameters on liposomes, such as low pH or high temperatures during beverage production, is avoided by releasing liposomal powder from the cap into the beverage right at consumption. The same company is producing liposomal formulation of potent antioxidant resveratrol as well, under the commercial name Resverasome®.

Solid lipid particle emulsions consist of solid lipid droplets coated with an emulsifier and dispersed in an aqueous continuous phase. The morphology and packing of the crystals within the lipid phase should be controlled in order to obtain particular functional attributes. This kind of emulsion offers the possibility to control the delivery of bioactives, for example, at a specific temperature by designing a solid lipid phase to melt at that temperature. Solid lipid particle emulsions are prepared at elevated temperatures (hot homogenization process), which may cause chemical degradation of heat-sensitive compounds. Another disadvantage is that the lipid phase is usually highly saturated (so as to have a sufficiently high melting point to form emulsions), which may have an unfavorable impact on health. In addition, the expulsion of the encapsulated lipophilic component during storage may occur unless the appropriate lipid carrier and processing conditions are selected for the preparation. Apart from the homogenization methods used for preparation of solid lipid emulsions, spray drying and spray chilling can be used for manufacturing solid lipid particles (Tulini et al. 2016; Wang et al. 2016). Solid lipid particles have poor stability in gastrointestinal conditions. Under strong acidic condition (i.e., in the stomach) they easily aggregate to a bulky structure due to the protonation of carboxyl groups and neutralization of surface charge. Surface modifications, for instance coating with (crosslinked) polymeric layers, may confer new physicochemical properties to them and provide enhanced functionalities (Wang et al. 2016). In addition, coating may help to avoid severe agglomeration of lipid nanoparticles into large aggregates during the drying process. The literature shows that in in vitro models the solid lipid particles are degraded by lipase, but at a slower rate than liquid lipid droplets (Bonaire et al. 2008; Nik et al. 2012); this means that digestibility of the crystalline lipid phase in comparison to liquid lipid phase of conventional emulsions is slower and this reflects as reduced bioavailability of the encapsulated lipophilic component. Food scientists have studied solid lipid particles to improve physical and chemical stability of β-carotene during storage (Helgason et al. 2009; Qian et al. 2013) and to enhance cellular uptake and oral bioavailability of curcumin (Kakkar et al. 2011; Sun et al. 2013; Wang et al. 2016).

Structured delivery systems, fabricated from natural lipids and polymers, are the topic of the latest research studies as they are the most efficient in improving the oral bioavailability of poorly water-soluble nutraceuticals. They consist of polysaccharide particles filled with emulsified lipids (lipid particles, vesicles, liposomes, or droplets) in which the active compound of interest is entrapped. These complex systems, as being incorporating the two (polymer-based and the lipid-based) systems, join their advantages and avoid disadvantages. For example, being protected by a polymer membrane, the lipid-based systems become more stable; otherwise, they are prone to oxidation and aggregation, which generally constrains their application, especially in the food industry. On the other hand, most of the polymers regarded as safe for food actually do not provide controlled release to the encapsulates made by common encapsulation technologies. In fact, more designable drug release kinetics could be achieved by hydrogel-microencapsulated liposomes compared with that by hydrogel or liposomes alone. For example, by modulating the electrostatic interactions between emulsifier-coated lipid droplets and the biopolymer matrix within hydrogel particles, it is possible to control the release characteristics (Zeeb et al. 2015). Recently, Balanč et al. (2016) have designed Ca-alginate microparticles filled with liposomes (made of a commercial phospholipids mixture) with incorporated resveratrol. The authors applied the proliposome method for the fabrication of liposomes and electrostatic extrusion for the production of Ca-alginate microbeads (~400 μm), and the resulting encapsulation efficiency was up to 91%. Release studies showed the overall resistance to mass transfer one order of magnitude higher than the resistance ascribed solely to the liposomal membrane. Chitosan coating around Ca-alginate microbeads (~7 μm thick in dry state) prolonged the release of resveratrol even more. However, from the application point of view, it should be borne in mind that liposomes interfered with the thermal behavior of alginate in the temperature region above 220°C (according to DSC studies) and that the presence of liposomes decreased the strength of the beads in comparison to placebo beads (according to mechanical tests on compression). Other studies have shown that complex encapsulation systems are convenient for implementation in food products as they can survive different processes associated with food production. Thus, Gültekin-Özguven et al. (2016) have designed a delivery system consisting of spray-dried maltodextrin microparticles (~5 μm) with entrapped black mulberry (*Morus nigra*) extract-loaded liposomes; the liposomes were prepared by high pressure homogenization and then coated with chitosan using the layer-by-layer depositing method in order to protect them from the action of the bile salts during digestion, thus providing enhanced bioaccessibility of anthocyanins. Such structured extract powder was implemented in dark chocolate at various alkalization degrees (pH 4.5, 6, 7.5) and conching temperatures of 40°C, 60°C, and 80°C. Chocolate was fortified with encapsulated anthocyanins maximum 76.8%, depending on conching temperature and pH.

8.5   CONCLUSION

With the constant improvements in encapsulation technologies and their availability nowadays, as well as consumers' increasing interest in functional foods, the development of products based on encapsulates has become one of the target market strategies for the food industry. Although there are plenty of possibilities as to how to encapsulate an active ingredient, the suitability of one technology for food industry actually depends on many factors; the most important are: the type of active compound of interest for encapsulation, required size and size distribution of produced encapsulates, acceptability of encapsulating material as food grade, compatibility with the food matrix in which the encapsulates are going to be incorporated, their rheological and textural impacts, as well as the final return on investment.

This chapter gives an overview on the most relevant technologies used for food encapsulates manufacture on an industrial scale with regard to the general principles behind these technologies, the most important characteristics of these, variations, advantages, and limitations. Spraying-based technologies (spray drying, spray cooling, and spray coating), which are already well-established at the industrial scale, are expected to play a dominant role in the food sector in the future as well, especially due to persistent upgrading of spraying equipment. Although extrusion-based technologies are simple, low cost, and under mild conditions can produce encapsulates with less porous structure than those produced by, for example, spray drying, the cost of utilization of classic extrusion technologies on an industrial level is still twice as high than in the case of spray drying. However, probiotics encapsulated by extrusion technologies on a large scale and implemented in different varieties of cheese provide higher cell survival in comparison to free cell incorporation, without significant impact on their appearance, color, texture, and overall sensorial perceptibility. Moreover, encapsulation by extrusion offers protection to the sensitive probiotic cells in gastrointestinal conditions as demonstrated in the case of yogurts with incorporated probiotics encapsulates. Emulsion-based technologies provide encapsulates that could deliver active agents to specific locations within the gastrointestinal tract and release them at a controlled rate, but the main restrictions to a wider use in the food sector are the high cost of the materials used (surfactants and cosurfactants), time- and cost-consuming methods for their preparation, as well as limitation of synthetic emulsifiers addition to food products. Nevertheless, food industry has been developing suitable edible systems (conventional emulsions, multiple emulsions, multilayer emulsions, solid lipid particles, and filled polymer particles) for food components, such as antimicrobials, antioxidants, and flavors. Up-to-date research has shown that liposomes can be used to significantly accelerate the production of some foodstuffs and to develop new food products with improved shelf life. Since very often, increasingly sophisticated demands by consumers could be impossible to achieve if only one encapsulation technology is employed, it is more likely that two or more encapsulation processing steps are required to get encapsulates with superior characteristics. This indicates that all encapsulation technologies described herein (as relevant ones for food industry) are constantly being subjected to changes and improvements by research in academia.

ACKNOWLEDGMENT

This work was supported by the Ministry of Education, Science and Technological Development, Republic of Serbia (Projects No. III46010).

REFERENCES

Albertini, B., Passerini, N., Pattarino, F., Rodriguez, L. 2008. New spray congealing atomizer for the microencapsulation of highly concentrated solid and liquid substances. *European Journal of Pharmaceutics and Biopharmaceutics* 69:348–357.

Alvim, I.D., Stein, M.A., Koury, I.P., Dantas, F.B.H., Cruz, C.L.C.V. 2016. Comparison between the spray drying and spray chilling microparticles contain ascorbic acid in a baked product application. *LWT: Food Science and Technology* 65:689–694.

Amine, K.M., Champagne, C.P., Raymond, Y. 2014. Survival of microencapsulated *Bifidobacterium longum* in Cheddar cheese during production and storage. *Food Control* 37:193–199.

Anandharamakrishnan, C., Padma Ishwarya, S., eds. 2015. Introduction to spray drying. In: *Spray Drying Techniques for Food Ingredient Encapsulation*, pp. 1–36. Malden, MA: Wiley-Blackwell.

Balanč, B., Ota, A., Đorđević, V. et al. 2015. Resveratrol-loaded liposomes: Interaction of resveratrol with phospholipids. *European Journal of Lipid Science and Technology* 117:1615–1626.

Balanč, B., Trifković, K., Đorđević, V. et al. 2016. Novel resveratrol delivery systems based on alginate-sucrose and alginate-chitosan microbeads containing liposomes. *Food Hydrocolloids* 61:832–842.

Banville, C., Vuillemard, J., Lacroix, C. 2000. Comparison of different methods for fortifying Cheddar cheese with vitamin D. *International Dairy Journal* 10:375–382.

Beck, N.T., Franch, G., Geneau, D.L. 2002. Edible emulsion comprising live micro-organisms and dressings or side sauces comprising said edible emulsion. Patent Number US 20020076467 A1.

Becker, K., Salar-Behzadi, S., Zimmer, A. 2015. Solvent-free melting techniques for the preparation of lipid-based solid oral formulations. *Pharmaceutical Research* 32:1519–1545.

Beindorff, C.M., Zuidam, N.J. 2010. Microencapsulation of fish oil. In: *Encapsulation Technologies for Active Food Ingredients and Food Processing*, eds. N.J. Zuidam and V.A. Nedović, pp. 161–186. New York: Springer.

Belščak-Cvitanović, A., Komes, D., Karlović, S. et al. 2015a. Improving the controlled delivery formulations of caffeine in alginate hydrogel beads combined with pectin, carrageenan, chitosan and psyllium. *Food Chemistry* 167:378–386.

Belščak-Cvitanović, A., Lević, S., Kalušević, A. et al. 2015b. Efficiency assessment of natural biopolymers as encapsulants of green tea (*Camellia sinensis* L.) bioactive compounds by spray drying. *Food and Bioprocess Technology* 8:2444–2460.

Belščak-Cvitanović, A., Stojanović, R., Dujmić, F. et al. 2010. Encapsulation of polyphenols from *Rubus idaeus* L. leaves extract by electrostatic extrusion. In: *Fifth Central European Congress on Food*, Bratislava, Slovakia, pp. 8–14.

Belščak-Cvitanović, A., Stojanović, R., Manojlović, V. et al. 2011. Encapsulation of polyphenolic antioxidants from medicinal plant extracts in alginate–chitosan system enhanced with ascorbic acid by electrostatic extrusion. *Food Research International* 44:1094–1101.

Bezbradica, D., Matić, G., Obradović, B.M., Nedović, V.A., Leskošek-Čukalović, I., Bugarski, B.M. 2004. Immobilization of brewing yeast in PVA/alginate micro beads using electrostatic droplet generation. *Hemijska Industrija* 58:118–120.

Biebinger, R., Zimmermann, M.B., Al-Hooti, S.N. et al. 2009. Efficacy of wheat-based biscuits fortified with microcapsules containing ferrous sulfate and potassium iodate or a new hydrogen-reduced elemental iron: A randomised, double-blind, controlled trial in Kuwaiti women. *British Journal of Nutrition* 102:1362–1369.

Bonaire, L., Sandra, S., Helgason, T., Decker, E.A., Weiss, J., McClements, D.J. 2008. Influence of lipid physical state on the in vitro digestibility of emulsified lipids. *Journal of Agricultural and Food Chemistry* 56:3791–3797.

Bošković, A.M., Vidal, M.S., Saleeb, Z.F. 1992. Spray-dried fixed flavorants in a carbohydrate substrate and process. Patent Number US 5124162 A.

Botrel, D.A., Borges, S.V., Fernandes, R.V.D.B., Lourenço Do Carmo, E. 2014a. Optimization of fish oil spray drying using a protein: Inulin system. *Drying Technology* 32:279–290.

Botrel, D.A., de Barros Fernandes, R.V., Borges, S.V., Yoshida, M.I. 2014b. Influence of wall matrix systems on the properties of spray-dried microparticles containing fish oil. *Food Research International* 62:344–352.

Brückner, M., Bade, M., Kunz, B. 2007. Investigations into the stabilization of a volatile aroma compound using a combined emulsification and spray drying process. *European Food Research and Technology* 226:137–146.

Buffo, R.A., Probst, K., Zehentbauer, G., Luo, Z., Reineccius, G.A. 2002. Effects of agglomeration on the properties of spray-dried encapsulated flavours. *Flavour and Fragrance Journal* 17:292–299.

Bugarski, B., Li, Q., Goosen, M.F.A., Poncelet, D., Neufeld, R.J., Vunjak, G. 1994. Electrostatic droplet generation: Mechanism of polymer droplet formation. *AIChE Journal* 40:1026–1031.

Bugarski, B., Obradovic, B., Nedovic, V.A., Poncelet, D. 2004. Immobilisation of cells and enzymes using electrostatic droplet generator. In: *Fundamentals of Cell Immobilisation Biotechnology*, Focus on Biotechnology, Vol. 8a, eds. V. Nedović and R.G. Willaert, pp. 277–294. Dordrecht, the Netherlands: Kluwer Academic Publishers.

Bugarski, B., Smith, J., Wu, J., Goosen, M.F.A. 1993. Methods for animal cell immobilization using electrostatic droplet generation. *Biotechnology Techniques* 7:677–682.

Bugarski, B., Vunjak, G., Goosen, M.F.A. 1999. Principles of bioreactor design for encapsulated cells. In: *Cell Encapsulation Technology and Therapeutics*, eds. W.M. Kuhtreiber, R.P. Lanza, and W.L. Chick, pp. 395–416. Boston, MA: Birkhauser.

Bugarski, M.B., Sajc, L., Plavsic, M., Goosen, M.F.A., Jovanovic, G. 1997. Semipermeable alginate-Pill microcapsule as a bioartificial pancreas. In: *Animal Cell Technology, Basic and Applied Aspects*, eds. K. Funatsu, Y. Shirai, and T. Matsushita, pp. 479–486. Dordrecht, the Netherlands: Kluwer Academic Publishers.

Burgain, J., Gaiani, C., Linder, M., Scher, J. 2011 Encapsulation of probiotic living cells: From laboratory scale to industrial applications. *Journal of Food Engineering* 104:467–483.

Burgos-Díaz, C., Wandersleben, T., Marqués, A.M., Rubilar, M. 2016. Multilayer emulsions stabilized by vegetable proteins and polysaccharides. *Current Opinion in Colloid & Interface Science* 25:51–57.

Champagne, C.P., Lee, B.H., Saucier, L. 2010. Immobilization of cells and enzymes for fermented dairy or meat products. In: *Encapsulation Technologies for Active Food Ingredients and Food Processing*, eds. N.J. Zuidam and V.A. Nedović, pp. 345–366. New York: Springer.

Chan, E.S., Lee, B.B., Ravindra, P., Poncelet, D. 2009. Prediction models for shape and size of Ca-alginate macrobeads produced through extrusion–dripping method. *Journal of Colloid and Interface Science* 338:63–72.

Charcosset, C. 2009. Preparation of emulsions and particles by membrane emulsification for the food processing industry. *Journal of Food Engineering* 92:241–249.

Chávez, B.E., Ledeboer, A.M. 2007. Drying of probiotics: Optimization of formulation and process to enhance storage survival. *Drying Technology* 25:1193–1201.

CONNOILS—Connect with Connoils for Your Nutritional Health & Beauty Ingredients. 2016. Connoils.Com; http://www.connoils.com. Accessed October 15, 2016.

Consoli, L., Grimaldi, R., Sartori, T., Menegalli, F.C., Hubinger, M.D. 2016. Gallic acid microparticles produced by spray chilling technique: Production and characterization. *LWT: Food Science and Technology* 65:79–87.

Cordeiro, P., Temtem, M., Winters, C. 2013. Spray congealing: Applications in the pharmaceutical industry. *Chimica Oggi-Chemistry Today* 31:69–73.

Ćujić, N., Trifković, K., Bugarski, B., Ibrić, S., Pljevljakušić, D., Šavikin, K. 2016. Chokeberry (*Aronia melanocarpa* L.) extract loaded in alginate and alginate/inulin system. *Industrial Crops and Products* 86:120–131.

da Costa, J.M.G., Silva, E.K., Hijo, A.A.C.T. et al. 2015. Microencapsulation of Swiss cheese bioaroma by spray drying: Process optimization and characterization of particles. *Powder Technology* 274:296–304.

Dasgupta, N., Ranjan, S., Mundekkad, D., Ramalingam, C., Shanker, R., Kumar, A. 2015 Nanotechnology in agro-food: From field to plate. *Food Research International* 69:381–400.

de Souza, V.B., Thomazini, M., Balieiro, J.C.C., Fávaro-Trindade, C.S. 2015. Effect of spray drying on the physicochemical properties and color stability of the powdered pigment obtained from vinification byproducts of the Bordo grape (*Vitis labrusca*). *Food and Bioproducts Processing* 93:39–50.

de Vos, P., de Haan, B.J., Van Schilfgaarde, R. 1997. Upscaling the production of encapsulated islets. *Biomaterials* 18:1085–1090.

de Vos, P., Faas, M.M., Spasojevic, M., Sikkema, J. 2010. Encapsulation for preservation of functionality and targeted delivery of bioactive food components. *International Dairy Journal* 20:292–302.

Desai, H.G.K., Park, J.H. 2005. Recent developments in microencapsulation of food ingredients. *Drying Technology* 23:1361–1394.

Detoni, C.B., de Oliveira, D.M., Santo, I.E. et al. 2012. Evaluation of thermal-oxidative stability and antiglioma activity of *Zanthoxylum tingoassuiba* essential oil entrapped into multi-and unilamellar liposomes. *Journal of Liposome Research* 22:1–7.

Dewettinck, K., Huyghebaert, A. 1999. Fluidized bed coating in food technology. *Trends in Food Science & Technology* 10:163–168.

Dimitrellou, D., Kandylis, P., Petrović, T. et al. 2016. Survival of spray dried microencapsulated *Lactobacillus casei* ATCC 393 in simulated gastrointestinal conditions and fermented milk. *LWT: Food Science and Technology* 71:169–174.

Đorđević, V., Balanč, B., Belščak-Cvitanović, A. et al. 2015. Trends in encapsulation technologies for delivery of food bioactive compounds. *Food Engineering Reviews* 7:452–490.

Drosou, C.G., Krokida, M.K., Biliaderis, C.G. 2016. Encapsulation of bioactive compounds through electrospinning/electrospraying and spray drying: A comparative assessment of food related applications. *Drying Technology* 35:139–162. http://dx.doi.org/10.1080/07 373937.2016.1162797.

Dubey, N.B., Windhab, E.J. 2013. Iron encapsulated microstructured emulsion-particle formation by prilling process and its release kinetics. *Journal of Food Engineering* 115:198–206.

Eisinaite, V., Juraite, D., Schroën, K., Leskauskaite, D. 2016. Preparation of stable food-grade double emulsions with a hybrid premix membrane emulsification system. *Food Chemistry* 206:59–66.

Eliasen, H., Kristensen, H.G., Schaefer, T. 1999. Electrostatic charging during a melt agglomeration process. *International Journal of Pharmaceutics* 184:85–96.

Emami, S., Azadmard-Damirchi, S., Peighambardoust, S.H., Valizadeh, H., Hesari, J. 2016. Liposomes as carrier vehicles for functional compounds in food sector. *Journal of Experimental Nanoscience* 11:737–759. DOI: 10.1080/17458080.2016.1148273.

Endres, R.J. 2012. Agency response letter GRAS notice no. GRN 000399. Silver Spring, MD: U.S. Food and Drug Administration.

Estevinho, B.H., Rocha, F., Santos, L., Alves, A. 2013. Microencapsulation with chitosan by spray drying for industry applications: A review. *Trends in Food Science & Technology* 31:138–155.

Fang, Z., Bhandari, B. 2010. Encapsulation of polyphenols: A review. *Trends in Food Science & Technology* 21:510–523.

Favaro-Trindade, C.S., Okuro, P.K., Matos Jr, F.E. 2015. Encapsulation via spray chilling/ cooling/congealing. In: *Handbook of Encapsulation and Controlled Release*, ed. M.K. Mishra, pp. 71–88. Boca Raton, FL: CRC Press.

Favaro-Trindade, C.S., Santana, A.S., Monterrey-Quintero, E.S., Trindade, M.A., Netto, F.M. 2010. The use of spray drying technology to reduce bitter taste of casein hydrolysate. *Food Hydrocolloids* 24:336–340.

Fernandes, R.V.D.B., Borges, S.V., Botrel, D.A., Silva, E.K., Costa, J.M.G., Queiroz, F. 2013. Microencapsulation of rosemary essential oil: Characterization of particles. *Drying Technology* 31:1245–1254.

Fidler, M.C., Walczyk, T., Davidsson, L. et al. 2004. A micronised, dispersible ferric pyrophosphate with high relative bioavailability in man. *British Journal of Nutrition* 91:107–112.

Fuchs, M., Turchiuli, C., Bohin, M. et al. 2006. Encapsulation of oil in powder using spray drying and fluidized bed agglomeration. *Journal of Food Engineering* 75:27–35.

Gamboa, O.D., Gonçalves, L.G., Grosso, C.F. 2011. Microencapsulation of tocopherols in lipid matrix by spray chilling method. *Procedia Food Science* 1:1732–1739.

Ganedenbc30.com. 2016. The probiotic that delivers. https://www.ganedenprobiotics.com/innovative-ideas/probiotics-for-hot-beverages. Accessed October 15, 2016.

Gharsallaoui, A., Roudaut, G., Chambin, O., Voilley, A., Saurel, R. 2007. Applications of spray-drying in microencapsulation of food ingredients: An overview. *Food Research International* 40:1107–1121.

Ghayempour, S., Mortazavi, S.M. 2013. Fabrication of micro-nanocapsules by a new electrospraying method using coaxial jets and examination of effective parameters on their production. *Journal of Electrostatics* 71:717–727.

Gibis, M., Vogt, E., Weiss, J. 2012. Encapsulation of polyphenolic grape seed extract in polymer-coated liposomes. *Food & Function* 3:246–254.

Giroux, H.J.., Robitaille, G., Britten, M. 2016. Controlled release of casein-derived peptides in the gastrointestinal environment by encapsulation in water-in-oil-in-water double emulsions. *LWT: Food Science and Technology* 69:225–232.

Given, P.S. 2009. Encapsulation of flavors in emulsions for beverages. *Current Opinion in Colloid & Interface Science* 14:43–47.

Goosen, M.F.A., Mahmud, E.S.C., M-Ghafi, A.S., M-Hajri, H.A., Al-Sinani, Y.S., Bugarski, M.B. 1997. Immobilization of cells using electrostatic droplet generation. In: *Immobilization of Enzymes and Cell*, ed. G.F. Bickerstaff, pp. 167–174. Totowa, NJ: Humana Press.

Gortzi, O., Lalas, S., Tsaknis, J., Chinou, I. 2007. Enhanced bioactivity of *Citrus limon* (Lemon Greek cultivar) extracts, essential oil and isolated compounds before and after encapsulation in liposomes. *Planta Medica* 73:184.

Gouin, S. 2004. Microencapsulation: Industrial appraisal of existing technologies and trends. *Trends in Food Science & Technology* 15:330–347.

Goula, M.A., Adamopoulos, G.K. 2008. Effect of maltodextrin addition during spray drying of tomato pulp in dehumidified air: I. Drying kinetics and product recovery. *Drying Technology* 26:714–725.

Guerin, J., Petit, J., Burgain, J. et al. 2017. *Lactobacillus rhamnosus* GG encapsulation by spray-drying: Milk proteins clotting control to produce innovative matrices. *Journal of Food Engineering* 193:10–19.

Gültekin-Özguven, M., Karadağ, A., Duman, Ş, Özkal, B., Özçelik, B. 2016. Fortification of dark chocolate with spray dried black mulberry (*Morus nigra*) waste extract encapsulated in chitosan-coated liposomes and bioaccessability studies. *Food Chemistry* 201:205–212.

Gupta, C., Chawla, P., Arora, S. 2015. Development and evaluation of iron microcapsules for milk fortification. *CyTA: Journal of Food* 13:116–123.

Haffner, F.B., Diab, R., Pasc, A. 2016. Encapsulation of probiotics: Insights into academic and industrial approaches. *AIMS Materials Science* 3:114–136.

Han, X., Chen, S., Hu, X. 2009. Controlled-release fertilizer encapsulated by starch/polyvinyl alcohol coating. *Desalination* 240:21–26.

Heinzen, C., Berger, A., Marison, I. 2004. Use of vibration technology for jet break-up for encapsulation of cells and liquids in monodisperse microcapsules. In: *Fundamentals of Cell Immobilisation Biotechnology*, eds. V. Nedovic and R. Willaert, pp. 257–275. Dordrecht, the Netherlands: Kluwer Academic Publishers.

Helgason, T., Awad, T.S., Kristbergsson, K., Decker, E.A., McClements, D.J., Weiss, J. 2009. Impact of surfactant properties on oxidative stability of b-carotene encapsulated within solid lipid nanoparticles. *Journal of Agricultural and Food Chemistry* 57:8033–8040.

Hernández-Rodríguez, L., Lobato-Calleros, C., Pimentel-González, D., Vernon-Carter, E. 2014. *Lactobacillus plantarum* protection by entrapment in whey protein isolate: κ-carrageenan complex coacervates. *Food Hydrocolloids* 36:181–188.

Huynh, T.V., Caffin, N., Dykes, G.A., Bhandari, B. 2008. Optimization of the microencapsulation of lemon myrtle oil using response surface methodology. *Drying Technology* 26:357–368.

Ingredion Incorporated. 2016. http://www.ingredion.us/Ingredients/foodbeverage/encapsulationemulsification.html. Accessed October 15, 2016.

Isailović, B., Kostić, I., Zvonar, A. et al. 2013. Resveratrol loaded liposomes produced by different techniques. *Innovative Food Science & Emerging Technologies* 19:181–189.

Ivanova, E., Teunou, E., Poncelet, D. 2006. Alginate based macrocapsules as inoculants carriers for production of nitrogen fixing biofertilizers. *Chemical Industry and Chemical Engineering Quarterly* 12:31–39.

Iyer, C., Kailasapathy, K. 2005. Effect of co-encapsulation of probiotics with prebiotics on increasing the viability of encapsulated bacteria under in vitro acidic and bile salt conditions and in yogurt. *Journal of Food Science* 70:M18–M23.

Jafari, S.M., Assadpoor, E., He, Y., Bhandari, B. 2008. Encapsulation efficiency of food flavours and oils during spray drying. *Drying Technology* 26:816–835.

Jafari, S.M., He, Y., Bhandari, B. 2007a. Encapsulation of nanoparticles of d-limonene by spray drying: Role of emulsifiers and emulsifying techniques. *Drying Technology* 25:1069–1079.

Jafari, S.M., He, Y., Bhandari, B. 2007b. Production of sub-micron emulsions by ultrasound and microfluidization techniques. *Journal of Food Engineering* 82:478–488.

Jangam, S.V., Mujumdar, A.S. 2013. Recent developments in drying technologies for foods. In: *Advances in Food Process Engineering Research and Applications*, eds. S. Yanniotis, P. Taoukis, N.G. Stoforos, and V.T. Karathanos, pp. 153–172. New York: Springer.

Jayasinghe, S.N., Edirisinge, M.J. 2004. Electrically forced jets and microthreads of high viscosity dielectric liquids. *Journal of Aerosol Science* 35:233–243.

Jones, M.L., Martoni, C.J., Tamber, S., Parent, M., Prakash, S. 2012. Evaluation of safety and tolerance of microencapsulated *Lactobacillus reuteri* NCIMB 30242 in a yogurt formulation: A randomized, placebo-controlled, double-blind study. *Food and Chemical Toxicology* 50:2216–2223.

Kailasapathy, K. 2002. Microencapsulation of probiotic bacteria: Technology and potential applications. *Current Issues in Intestinal Microbiology* 3:39–48.

Kakkar, V., Singh, S., Singla, D., Kaur, I.P. 2011. Exploring solid lipid nanoparticles to enhance the oral bioavailability of curcumin. *Molecular Nutrition & Food Research* 55:495–503.

Karthikeyan, N., Elango, A., Kumaresan, G., Gopalakrishnamurty, T.R., Raghunath, B.V. 2014. Enhancement of probiotic viability in ice cream by microencapsulation. *International Journal of Environmental Science and Technology* 3:339–347.

Kaushik, P., Dowling, K., Barrow, C.J., Adhikari, B. 2015. Complex coacervation between flaxseed protein isolate and flaxseed gum. *Food Research International* 72:91–97.

Khalf, M., Dabour, N., Kheadr, E., Fliss, I. 2010. Viability of probiotic bacteria in maple sap products under storage and gastrointestinal conditions. *Bioresource Technology* 101:7966–7972.

Ko, S., Lee, S.C. 2010. Effect of nanoliposomes on the stabilization of incorporated retinol. *African Journal of Biotechnology* 9:6158–6161.

Kourkoutas, Y., Manojlović, V., Nedović, V.A. 2010. Immobilization of microbial cells for alcoholic and malolactic fermentation of wine and cider. In: *Encapsulation Technologies for Active Food Ingredients and Food Processing*, eds. N.J. Zuidam and V.A. Nedović, pp. 327–344. New York: Springer.

Krasaekoopt, W., Tandhanskul, A. 2008. Sensory and acceptance assessment of yogurt containing probiotic beads in Thailand. *Kasetsart Journal* 42:99–106.

Krunić, T.Ž., Obradović, N.S., Bulatović, M.L., Vukašinović-Sekulić, M.S., Trifković, K.T., Rakin, M.B. 2016. Impact of carrier material on fermentative activity of encapsulated yoghurt culture in whey based substrate. *Hemijska Industrija* 71:41–48. DOI: 10.2298/HEMIND150717016K.

Kudra, T. 2008. Energy aspects in food dehydration. In: *Advances in Food Dehydration*, ed. C. Ratti, pp. 423–446. Boca Raton, FL: CRC Press.

Lahtinen, S.J., Ouwehand, A.C., Salminen, S.J., Forssell, P., Myllärinen, P. 2007. Effect of starch and lipid based encapsulation on the culturability of two *Bifidobacterium longum* strains. *Letters in Applied Microbiology* 44:500–505.

Lakkis, J.M. 2007. Confectionery products as delivery systems for flavours, health and oral-care actives. In: *Encapsulation and Controlled Release Technologies in Food Systems*, ed. J.M. Lakkis, pp. 171–200. Hoboken, NJ: John Wiley & Sons.

Lee, S.C., Kim, J.J., Lee, K.E. 2005a. Effect of β-sitosterol in liposome bilayer on the stabilization of incorporated retinol. *Food Science and Biotechnology* 14:604–607.

Lee, S.C., Lee, K.E., Kim, J.J., Lim, S.H. 2005b. The effect of cholesterol in the liposome bilayer on the stabilization of incorporated retinol. *Journal of Liposome Research* 15:157–166.

Lević, S., Lijaković, I.P., Đorđević, V. et al. 2015. Characterization of sodium alginate/D-limonene emulsions and respective calcium alginate/D-limonene beads produced by electrostatic extrusion. *Food Hydrocolloids* 45:111–123.

Lewinska, D., Rosinski, S., Werynski, A. 2004. Influence of process conditions during impulsed electrostatic droplet formation on size distribution of hydrogel beads. *Artificial Cells, Blood Substitutes, and Biotechnology* 32:41–53.

Liolios, C., Gortzi, O., Lalas, S., Tsaknis, J., Chinou, I. 2009. Liposomal incorporation of carvacrol and thymol isolated from the essential oil of *Origanum dictamnus* L. and *in vitro* antimicrobial activity. *Food Chemistry* 112:77–83.

Liu, C.-P., Liu, S.-D. 2009. Low-temperature spray drying for the microencapsulation of the fungus *Beauveria bassiana*. *Drying Technology* 27:747–753.

Liu, N., Park, H.J. 2009. Chitosan-coated nanoliposome as vitamin E carrier. *Journal of Microencapsulation* 26:235–242.

Lopez-Rubio, A., Sanchez, E., Wilkanowicz, S., Sanz, Y., Lagaron, J.M. 2012. Electrospinning as a useful technique for the encapsulation of living bifidobacteria in food hydrocolloids. *Food Hydrocolloids* 28:159–167.

MANE Flavor & Fragrance Manufacturer. 2016. http://www.mane.com/encapsulation. Accessed October 15, 2016.

Manojlovic, V., Djonlagic, J., Obradovic, B., Nedovic, V., Bugarski, B. 2006. Immobilization of cells by electrostatic droplet generation: A model system for potential application in medicine. *International Journal of Nanomedicine* 1:163–171.

Manojlović, V., Nedović, V.A., Kailasapathy, K., Zuidam, N.J. 2010. Encapsulation of probiotics for use in food products. In: *Encapsulation Technologies for Active Food Ingredients and Food Processing*, eds. N.J. Zuidam and V.A. Nedović, pp. 269–302. New York: Springer.

Maresca, D., De Prisco, A., La Storia, A., Cirillo, T., Esposito, F., Mauriello, G. 2016. Microencapsulation of nisin in alginate beads by vibrating technology: Preliminary investigation. *LWT: Food Science and Technology* 66:436–443.

Marketsandmarkets. 2016. Food encapsulation market by shell material, technology, region—2021. Marketsandmarkets.Com; http://www.marketsandmarkets.com/Market-Reports/food-encapsulation-advanced-technologies-and-global-market-68.html. Accessed October 15, 2016.

Marsanasco, M., Marquez, A.L., Wagner, J.R., del Alonso, V.S., Chiaramoni, N.S. 2011. Liposomes as vehicles for vitamins E and C: An alternative to fortify orange juice and offer vitamin C protection after heat treatment. *Food Research International* 44:3039–3046.

Matos Jr, F.E., Albertini, B., Favaro-Trindade, C.S. 2015a. Development and characterization of solid lipid microparticles loaded with ascorbic acid and produced by spray congealing. *Food Research International* 67:52–59.

Matos, M., Gutiatos, G., Iglesias, O., Coca, J., Pazos, C. 2015b. Enhancing encapsulation efficiency of food-grade double emulsions containing resveratrol or vitamin B12 by membrane emulsification. *Journal of Food Engineering* 166:212–220.

Mazer, T., Kessler, T. 2014. Methods for extruding powered nutritional products using a high shear element. Patent Number WO 2014093832 A1.

McClements, D.J. 1999. *Food Emulsions: Principles, Practice, and Techniques*. Boca Raton, FL: CRC Press/Taylor & Francis.

McClements, D.J. 2012. Nanoemulsions *versus* microemulsions: Terminology, differences, and similarities. *Soft Matter* 8:1719–1729.

McClements, D.J., Decker, E.A., Weiss, J. 2007. Emulsion-based delivery systems for lipophilic bioactive components. *Journal of Food Science* 72:109–124.

Mehrnia, M.A., Jafari, S.M., Makhmal-Zadeh, B.S., Maghsoudlou, Y. 2016. Crocin loaded nano-emulsions: Factors affecting emulsion properties in spontaneous emulsification. *International Journal of Biological Macromolecules* 84:261–267.

Mezzenga, R., Ulrich, S. 2010. Spray-dried oil powder with ultrahigh oil content. *Langmuir* 26:16658–16661.

Mirzaei, H., Pourjafar, H., Homayouni, A. 2012. Effect of calcium alginate and resistant starch microencapsulation on the survival rate of *Lactobacillus acidophilus* La5 and sensory properties in Iranian white brined cheese. *Food Chemistry* 132:1966–1970.

Mitropoulou, G., Nedovic, V., Goyal, A., Kourkoutas, Y. 2013. Immobilization technologies in probiotic food production. *Journal of Nutrition and Metabolism* 2013:716861. http://dx.doi.org/10.1155/2013/716861.

Moghaddam, M.K., Mortazavi, S.M., Khayamian, T. 2015. Preparation of calcium alginate microcapsules containing n-nonadecane by a melt coaxial electrospray method. *Journal of Electrostatics* 73:56–64.

Moghimipour, E., Aghel, N., Mahmoudabadi, A.Z., Ramezani, Z., Handali, S. 2012. Preparation and characterization of liposomes containing essential oil of *Eucalyptus camaldulensis* leaf. *Journal of Natural Pharmaceutical Products* 7:117–122.

Moretti, D., Zimmermann, M.B., Wegmüller, R., Walczyk, T., Zeder, C., Hurrell, R.F. 2006. Iron status and food matrix strongly affect the relative bioavailability of ferric pyrophosphate in humans. *American Journal of Clinical Nutrition* 83:632–638.

Morgan, R., Blagdon, P. 1993. Methods of encapsulating liquids in fatty matrices, and products thereof. Patent Number US 5204029 A.

Mousa, A., Liu, X.M., Chen, Y.Q., Zhang, H., Chen, W. 2014. Evaluation of physiochemical, textural, microbiological and sensory characteristics in set yogurt reinforced by microencapsulated *Bifidobacterium bifidum* F-35. *International Journal of Dairy Technology* 49:1673–1679.

Muschiolik, G. 2007. Multiple emulsions for food use. *Current Opinion in Colloid & Interface Science* 12:213–220.

Muthukumarasamy, P., Holley, R.A. 2007. Survival of *Escherichia coli* O157:H7 in dry fermented sausages containing micro-encapsulated probiotic lactic acid bacteria. *Food Microbiology* 24:82–88.

Nanolife Tonics. 2016. Liposome based nutritional supplements. http://nanolifetonics.com/. Accessed October 15, 2016.

Nazir, A., Boom, R., Schroën, K. 2014. Influence of the emulsion formulation in premix emulsification using packed beds. *Chemical Engineering Science* 116:547–557.

Nazir, A., Schroën, K., Boom, R. 2010. Premix emulsification: A review. *Journal of Membrane Science* 362:1–11.

Nedović, A.V., Obradovic, B., Leskosek, I., Pesic, R., Bugarski, B. 2001. Electrostatic generation of alginate microbeads loaded with brewing yeast. *Process Biochemistry* 37:17–22.

Nedović, V., Manojlovic, V., Pruesse, U., Bugarski, B., Djonlagic, J., Vorlop, K. 2006. Optimization of the electrostatic droplet generation process for controlled microbead production-single nozzle system. *Chemical Industry and Chemical Engineering Quarterly* 12:53–57.

Nedović, V.A., Obradovic, B., Poncelet, D., Goosen, M.F.A., Leskosek-Cukalovic, I., Bugarski, B. 2002. Cell immobilisation by electrostatic droplet generation. *Landbauforschung Volkenrode SH* 241:11–18.

Nik, A.M., Langmaid, S., Wright, A.J. 2012. Digestibility and β-carotene release from lipid nanodispersions depend on dispersed phase crystallinity and interfacial properties. *Food & Function* 3:234–245.

Obradović, N.S., Krunić, T.Ž., Trifković, K.T. et al. 2015. Influence of chitosan coating on mechanical stability of biopolymer carriers with probiotic starter culture in fermented whey beverages. *International Journal of Polymer Science* 2015:1–8. http://dx.doi.org/10.1155/2015/732858.

Okuro, P.K., Baliero, J.C.C., Liberal, R.D.C.O., Favaro-Trindade, C.S. 2013b. Co-encapsulation of *Lactobacillus acidophilus* with inulin or polydextrose in solid lipid microparticles provides protection and improves stability. *Food Research International* 53:96–103.

Okuro, P.K., Matos Jr, F.E., Favaro-Trindade, C.S. 2013a. Technological challenges for spray chilling encapsulation of functional food ingredients. *Food Technology and Biotechnology* 51:171–182.

Ortakci, F., Broadbent, J.R., McManus, W.R., McMahon, D.J. 2012. Survival of microencapsulated probiotic *Lactobacillus paracasei* LBC-1e during manufacture of Mozzarella cheese and simulated gastric digestion. *Journal of Dairy Science* 95:6274–6281.

Ortakci, F., Sert, S. 2012. Stability of free and encapsulated *Lactobacillus acidophilus* ATCC 4356 in yogurt and in an artificial human gastric digestion system. *Journal of Dairy Science* 95:6918–6925.

Oxley, J.D. 2012. Spray cooling and spray chilling for food ingredient and nutraceutical encapsulation. In: *Encapsulation Technologies and Delivery Systems for Food Ingredients and Nutraceuticals*, eds. N. Garti and D.J. McClements, pp. 110–130. Philadelphia, PA: Woodhead Publishing.

Ozer, B., Kirmaci, H.A., Senel, E., Atamer, M., Hayaloglu, A. 2009. Improving the viability of *Bifidobacterium bifidum* BB-12 and *Lactobacillus acidophilus* LA-5 in white-brined cheese by microencapsulation. *International Dairy Journal* 19:22–29.

Ozer, B., Uzun, Y.S., Kirmaci, H.A. 2008. Effect of microencapsulation on viability of *Lactobacillus acidophilus* LA-5 and *Bifidobacterium bifidum* BB-12 during Kasar cheese ripening. *International Journal of Dairy Technology* 61:237–244.

Paramita, V., Iida, K., Yoshii, H., Furuta, T. 2009. Effect of feed liquid temperature on the structural morphologies of d-limonene microencapsulated powder and its preservation. *Journal of Food Science* 75:E39–E45.

Paucar, O.C., Tulini, F.L., Thomazini, M., Balieiro, J.C.C., Pallone, E.M.J.A., Favaro-Trindade, C.S. 2016. Production by spray chilling and characterization of solid lipid microparticles loaded with vitamin D3. *Food and Bioproducts Processing* 100:344–350. http://dx.doi.org/10.1016/j.fbp.2016.08.006.

Pedroso, D.L., Thomazini, M., Heinemann, R.J.B., Favaro-Trindade, C.S. 2012. Protection of *Bifidobacterium lactis* and *Lactobacillus acidophilus* by microencapsulation using spray chilling. *International Dairy Journal* 26:127–132.

Pelissari, J.R., Souza, V.B., Pigoso, A.A., Tulini, F.L., Thomazini, M., Rodrigues, C.E.C., Urbano, A., Favaro-Trindade, C.S. 2016. Production of solid lipid microparticles loaded with lycopene by spray chilling: Structural characteristics of particles and lycopene stability. *Food and Bioproducts Processing* 98:86–94.

Phinix International. 2016. Welcome to Phinix International. Phinixinternational.Com; http://www.phinixinternational.com. Accessed October 15, 2016.

Picot, A., Lacroix, C. 2003. Production of multiphase water-insoluble microcapsules for cell microencapsulation using an emulsification/spray-drying technology. *Journal of Food Science* 68:2693–2700.

Pivette, P., Faivre, V., Daste, G., Ollivon, M., Lesieur, S. 2009. Rapid cooling of lipid in a prilling tower. *Journal of Thermal Analysis and Calorimetry* 98:47–55.

Pjanovic, R., Goosen, M.F.A., Nedovic, V., Bugarski, M.B. 2000. Immobilization /encapsulation of cells using electrostatic droplet generation. *Minerva Biotecnologica* 12:241–248.

POLARIS. 2016. Lipides nutritionnels. http://www.polaris.fr/. Accessed October 15, 2016.

Poncelet, D., Bugarski, B., Amsdem, B.G., Zhu, J., Neufeld, R.J., Goosen, M.F.A. 1994. A parallel plate electrostatic droplet generator: Parameters affecting microbead size. *Applied Microbiology and Biotechnology* 42:251–255.

Poncelet, D., Neufeld, R.J., Goosen, M.F.A., Burgarski, B., Babak, V. 1999. Formation of microgel beads by electric dispersion of polymer solutions. *AIChE Journal* 45:2018–2023.

Poncelet, D., Picot, A., El Mafadi, S. 2011. Encapsulation: An essential technology for functional food applications. *Innovations in Food Technology* 22:32–34.

Porzio, M. 2007. An in-depth look at the steps in spray drying and the different options available to flavorists. *Perfumer and Flavorist* 32:34–39.

Prusse, U., Bilancetti, L., Bucko, M. et al. 2008. Comparison of different technologies for alginate beads production. *Chemical Papers* 62:364–374.

Qian, C., Decker, E.A., Xiao, H., McClements, D.J. 2013. Impact of lipid nanoparticle physical state on particle aggregation and b-carotene degradation: Potential limitations of solid lipid nanoparticles. *Food Research International* 52:342–349.

Qiao, D., Liu, H., Yu, L., Bao, X., Simon, G.P., Petinakis, E., Chen, L. 2016. Preparation and characterization of slow-release fertilizer encapsulated by starch-based superabsorbent polymer. *Carbohydrate Polymers* 147:146–154.

Quintanilla-Carvajal, M.X., Camacho-Díaz, B.H., Meraz-Torres, L.S. et al. 2010. Nanoencapsulation: A new trend in food engineering processing. *Food Engineering Reviews* 2:39–50.

Ray, S., Raychaudhuri, U., Chakraborty, R. 2016. An overview of encapsulation of active compounds used in food products by drying technology. *Food Bioscience* 13:76–83.

Rodríguez, J., Martín, M.J., Ruiz, M.A., Clares, B. 2016. Current encapsulation strategies for bioactive oils: From alimentary to pharmaceutical perspectives. *Food Research International* 83:41–59.

Rohman, S. 2007. *Handbook of Food Preservation*. Boca Raton, FL: CRC Press.

Rosinski, S., Lewinska, D., Migaj, M., Wozniewicz, B., Werynski, A. 2002. Electrostatic microencapsulation of parathyroid cells as a tool for the investigation of cell's activity after transplantation. *Landbauforschung Volkenrode SH* 241:47–50.

Rutherford, W.M., Allen, J.E., Schlameus, H.W. et al. 1994. Process for preparing rotary disc fatty acid microspheres of microorganisms. Patent Number US 5292657.

SANOVO TECHNOLOGY GROUP. 2016. https://www.sanovogroup.com/products/powder-processing/spray-drying-egg-products/. Accessed October 15, 2016.

Sartori, T., Consoli, L., Hubinger, M.D., Menegalli, F.C. 2015. Ascorbic acid microencapsulation by spray chilling: Production and characterization. *LWT: Food Science and Technology* 63:353–360.

Sartori, T., Menegalli, F.C. 2016. Development and characterization of unripe banana starch films incorporated with solid lipid microparticles containing ascorbic acid. *Food Hydrocolloids* 55:210–219.

Sathyabama, S., Ranjith Kumar, M., Bruntha Devi, P., Vijayabharathi, R., Brindha Priyadharisini, V. 2014. Co-encapsulation of probiotics with prebiotics on alginate matrix and its effect on viability in simulated gastric environment. *LWT: Food Science and Technology* 57:419–425.

Schmidt, U.S., Bernewitz, R., Guthausen, G., Schuchmann, H.P. 2015. Investigation and application of measurement techniques for thedetermination of the encapsulation efficiency of O/W/O multipleemulsions stabilized by hydrocolloid gelation. *Colloids and Surfaces A: Physicochemical and Engineering Aspects* 475:55–61.

Schroën, K., Bliznyuk, O., Muijlwijk, K., Sahin, S., Berton-Carabin, C.C. 2015. Microfluidic emulsification devices: From micrometer insights to large-scale food emulsion production. *Current Opinion in Food Science* 3:33–40.

Schultz, S., Wagner, G., Urban, K., Ulrich, J. 2004. High-pressure homogenization as a process for emulsion formation. *Chemical Engineering & Technology* 27:361–368.

Sessa, M., Tsao, R., Liu, R., Ferrari, G., Donsì, F. 2011. Evaluation of the stability and antioxidant activity of nanoencapsulated resveratrol during in vitro digestion. *Journal of Agricultural and Food Chemistry* 59:12352–12360.

Shamsara, O., Muhidinov, Z.K., Jafari, S.M. et al. 2015. Effect of ultrasonication, pH and heating on stability of apricot gum-lactoglobuline two layer nanoemulsions. *International Journal of Biological Macromolecules* 81:1019–1025.

Sinico, C., De Logu, A., Lai, F. et al. 2005. Liposomal incorporation of *Artemisia arborescens* L. essential oil and in vitro antiviral activity. *European Journal of Pharmaceutics and Biopharmaceutics* 59:161–168.

Spinelli, S., Conte, A., Lecce, L., Incoronato, A.L., Del Nobile, M.A. 2015. Microencapsulated propolis to enhance the antioxidant properties of fresh fish burgers. *Journal of Food Process Engineering* 38:527–535.

Srivastava, Y., Semwal, A.D., Sharma, K. 2013. Application of various chemical and mechanical microencapsulation techniques in food sector: A review. *International Journal of Food and Fermentation Technology* 3:1–13.

Stajić, S., Živković, D., Tomović, V. et al. 2014. The utilisation of grapeseed oil in improving the quality of dry fermented sausages. *International Journal of Food Science and Technology* 49:2356–2363.

Stojanović, R., Belscak-Cvitanovic, A., Manojlovic, V., Komes, D., Nedovic, V., Bugarski, B. 2012. Encapsulation of thyme (*Thymus serpyllum* L.) aqueous extract in calcium alginate beads. *Journal of the Science of Food and Agriculture* 92:685–696.

Sun, J., Bi, C., Chan, H.M., Sun, S., Zhang, Q., Zheng, Y. 2013. Curcumin-loaded solid lipid nanoparticles have prolonged *in vitro* antitumor activity, cellular uptake and improved in vivo bioavailability. *Colloids and Surfaces B: Biointerfaces* 111:367–375.

Takahashi, M., Uechi, S., Takara, K., Asikin, Y., Wada, K. 2009. Evaluation of an oral carrier system in rats: Bioavailability and antioxidant properties of liposome-encapsulated curcumin. *Journal of Agricultural and Food Chemistry* 57:9141–9146.

Tan, H.L., Chan, L.W., Heng, P.W.S. 2005. Effect of oil loading on microspheres produced by spray drying. *Journal of Microencapsulation* 22:253–259.

Tan, H.L., Chan, W.L., Heng, S.W.P. 2009. Alginate/starch composites as wall material to achieve microencapsulation with high oil loading. *Journal of Microencapsulation* 26:263–271.

Thacker, P.A. 2013. Alternatives to antibiotics as growth promoters for use in swine production: A review. *Journal of Animal Science and Biotechnology* 4:35. DOI: 10.1186/2049-1891-4-35.

Tolve, R., Galgano, F., Caruso, M. et al. 2016. Encapsulation of health-promoting ingredients: Applications in foodstuffs. *International Journal of Food Science and Nutrition* 67:888–918. http://dx.doi.org/10.1080/09637486.2016.1205552.

Toniazzo, T., Berbel, I.F., Cho, S., Favaro-Trindade, C.S., Moraes, I.C., Pinho, S.C. 2014. β-carotene-loaded liposome dispersions stabilized with xanthan and guar gums: Physicochemical stability and feasibility of application in yogurt. *LWT: Food Science and Technology* 59:1265–1273.

Tonon, R.V., Grosso, C.R.F., Hubinger, M.D. 2011. Influence of emulsion composition and inlet air temperature on the microencapsulation of flaxseed oil by spray drying. *Food Research International* 44:282–289.

Tran, V.T., Benoit, J.P., Venier-Julienne, M.C. 2011. Why and how to prepare biodegradable, monodispersed, polymeric microparticles in the field of pharmacy? *International Journal of Pharmaceutics* 407:1–11.

Trifković, K.T., Milašinović, N.Z., Djordjević, V.B. et al. 2014. Chitosan microbeads as carriers for encapsulation of thyme (*Thymus serpyllum* L.) polyphenols. *Carbohydrate Polymers* 111:901–907.

Trifković, K.T., Milašinović, N.Z., Isailović, B.D. et al. 2012. Encapsulation of *Thymus serpyllum* L. aqueous extract in chitosan and alginate-chitosan microbeads. In: *Sixth Central European Congress on Food*, Novi Sad, Serbia, pp. 1052–1058.

Tulini, F.L., Souza, V.B., Echalar-Barrientos, A.M., Thomazini, M., Pallone, E.M.J.A., Favaro-Trindade, C.S.A. 2016. Development of solid lipid microparticles loaded with a proanthocyanidin rich cinnamon extract (*Cinnamomum zeylanicum*): Potential for increasing antioxidant content in functional foods for diabetic population. *Food Research International* 85:10–18.

Turchiuli, C., Fuchs, M., Bohin, M., Cuvelier, M.E., Ordonnaud, C., Peyrat-Maillard, M.N., Dumoulin, E. 2005. Oil encapsulation by spray drying and fluidised bed agglomeration. *Innovative Food Science & Emerging Technologies* 6:29–35.

Tylingo, R., Mania, S., Szwacki, J. 2016. A novel method for drop in drop edible oils encapsulation with chitosan using a coaxial technique. *Reactive and Functional Polymers* 100:64–72.

Urbanska, A.M., Bhathena, J., Prakash, S. 2007. Live encapsulated *Lactobacillus acidophilus* cells in yogurt for therapeutic oral delivery: Preparation and in vitro analysis of alginate-chitosan microcapsules. *Canadian Journal of Physiology and Pharmacology* 85:884–893.

van der Graaf, S., Schroen, C.G.P.H., Boom, R.M. 2005. Preparation of double emulsions by membrane emulsification: A review. *Journal of Membrane Science* 251:7–15.

van der Zwan, E.A., Schroën, C.G.P.H., Boom, R.M. 2008. Premix membrane emulsification by using a packed layer of glass beads. *AIChE Journal* 54:2190–2197.

Verbelen, P.J., Nedović, V.A., Manojlović, V., Delvaux, F.R., Laskošek-Čukalović, I., Bugarski, B., Willaert, R. 2010. Bioprocess intensification of beer fermentation using immobilised cells. In: *Encapsulation Technologies for Active Food Ingredients and Food Processing*, eds. N.J. Zuidam and V.A. Nedović, pp. 303–326. New York: Springer.

Vervaeck, A., Saerens, L., De Geest, B.G. et al. 2013. Prilling of fatty acids as a continuous process for the development of controlled release multiparticulate dosage forms. *European Journal of Pharmaceutics and Biopharmaceutics* 85:587–596.

Wandrey, C., Bartkowiak, A., Harding, S.E. 2009. Materials for encapsulation. In: *Encapsulation Technologies for Food Active Ingredients and Food Processing*, eds. N.J. Zuidam and V.A. Nedovic, pp. 31–100. Dordrecht, the Netherlands: Springer.

Wang, T., Ma, X., Lei, Y., Luo, Y. 2016. Solid lipid nanoparticles coated with cross-linked polymeric double layer for oral delivery of curcumin. *Colloids and Surfaces B: Biointerfaces* 148:1–11.

Watson Inc. 2016. Custom nutrient premixes and ingredients. http://www.watson-inc.com. Accessed October 15, 2016.

Wechtersbach, L., Poklar Ulrih, N., Cigic, B. 2012. Liposomal stabilization of ascorbic acid in model systems and in food matrices. *LWT: Food Science and Technology* 45:43–49.

Whelehan, M., Marison, I.W. September 2011. Microencapsulation by dripping and jet break up. *BRG Newsletter: Bioencapsulation Innovations* 4–10. http://bioencapsulation.net/220_newsletter/BI_09_2011/Bioencap_Innov_sept_2011.pdf. Accessed October 15, 2016.

Yamamoto, C., Neoh, T.L., Honbou, Y., Furuta, T., Kimura, S., Yoshii, H. 2012. Evaluation of flavor release from spray-dried powder by ramping with dynamic vapor sorption–gas chromatography. *Drying Technology* 30:1045–1050.

Yang, S., Liu, W., Liu, C., Liu, W., Tong, G., Zheng, H., Zhou, W. 2012. Characterization and bioavailability of vitamin C nanoliposomes prepared by film evaporation-dynamic high pressure microfluidization. *Journal of Dispersion Science and Technology* 33:1608–1614.

Yang, Y., McClements, D.J. 2013. Encapsulation of vitamin E in edible emulsions fabricated using a natural surfactant. *Food Hydrocolloids* 30:712–720.

Zeeb, B., Saberi, A.H., Weiss, J., McClements, D.J. 2015. Retention and release of oil-in-water emulsions from filled hydrogel beads composed of calcium alginate: Impact of emulsifier type and pH. *Soft Matter* 11:2228–2236.

Zhang, M., Gao, B., Chen, J., Li, Y., Creamer, A.E., Chen, H. 2014. Slow-release fertilizer encapsulated by graphene oxide films. *Chemical Engineering Journal* 255:107–113.

Ziar, H., Gerard, P., Riazi, A. 2012. Calcium alginate-resistant starch mixed gel improved the survival of *Bifidobacterium animalis* subsp. *lactis* Bb12 and *Lactobacillus rhamnosus* LBRE-LSAS in yogurt and simulated gastrointestinal conditions. *International Journal of Food Science and Technology* 47:1421–1429.

Zomorodi, S., Asl, A.K., Rohani, S., Razavi, M., Miraghaei, S. 2011. Survival of *Lactobacillus casei*, *Lactobacillus plantarum* and *Bifidobacterium bifidum* in free and microencapsulated forms on Iranian white cheese produced by ultrafiltration. *International Journal of Dairy Technology* 64:84–91.

Zuidam, J.N., Heinrich, E. 2010. Encapsulation of aroma. In: *Encapsulation Technologies for Food Active Ingredients and Food Processing*, eds. N.J. Zuidam and V.A. Nedovic, pp. 127–160. Dordrecht, the Netherlands: Springer.

Zuidam, N.J., Shimoni, E. 2010. Overview of microencapsulation use in food products or processes and methods to make them. In: *Encapsulation Technologies for Food Active Ingredients and Food Processing*, eds. N.J. Zuidam and V.A. Nedovic, pp. 3–29. Dordrecht, the Netherlands: Springer.

Index

Printed and bound by CPI Group (UK) Ltd, Croydon, CR0 4YY

01/11/2024

01782622-0003